A Neuroarthistory of The Painters of Modern Life

Using a transdisciplinary method combining art history, literary studies, and neuroaesthetics, this book examines the modern urban experience of nineteenth-century Paris through language and images of fragmentation and transformation.

The volume includes new empirical research conducted in collaboration with neuropsychologists, which tracks present-day viewers' physical and psychological responses to nineteenth-century painting and photography, thus providing data to model an experiential aesthetic for Baudelairean modernity. Weingarden reframes our understanding of Haussmannization, the demolition and rebuilding of the city into a modern metropolis, as witnessed by nineteenth-century Parisians, while also shedding new light on writers' responses, particularly those of Charles Baudelaire, and of visual artists like Édouard Manet, who contemplated and theorized this modernity and its impact. Using a unique word-and-image methodology, the author illustrates the development of ironic parody as a pictorial device that represents the rupture, fragmentation, and transmutation experienced by the artists and their viewers, revealing how art historians can utilize nineteenth-century neuropsychological practices and current neuroscience methods to reconstruct the lived, embodied experiences of nineteenth-century Paris.

This book will be of interest to students and scholars working in art history, modern art, urban studies and neuropsychology.

Lauren S. Weingarden is Professor Emerita of Art History at Florida State University.

Routledge Research in Art History

Routledge Research in Art History is our home for the latest scholarship in the field of art history. The series publishes research monographs and edited collections, covering areas including art history, theory, and visual culture. These high-level books focus on art and artists from around the world and from a multitude of time periods. By making these studies available to the worldwide academic community, the series aims to promote quality art history research.

Landscape and Nature in Scandinavian Art
Edited by MaryClaire Pappas and Tonje Haugland Sørensen

Bernini Ephemeral and Preparatory Sculpture
Genevieve Warwick

Conceptual Art After Modernism
Reconceiving Art and Art History
Robert Bailey

The Creation, Diffusion, and Reception of Italian Art in the Early Modern Iberian World
Edited by Corinna Tania Gallori and Benito Navarrete Prieto

Art, Politics, and Palace Eunuchs in Ming China, 1368-1644
Scarlett Jang

The Pictures of Cotton in Eighteenth-Century China
Roslyn Lee Hammers

Contemporary Approaches to Commemorative Public Art
Monumental Developments
Edited by Brenda Schmahmann

A Neuroarthistory of The Painters of Modern Life
Embodying Baudelairean Modernity
Lauren S. Weingarden

For more information about this series, please visit: www.routledge.com/Routledge-Research-in-Art-History/book-seies/RRAH

A Neuroarthistory of The Painters of Modern Life

Embodying Baudelairean Modernity

Lauren S. Weingarden

Routledge
Taylor & Francis Group
NEW YORK AND LONDON

Designed cover image: Bibliothèque nationale de France, Paris. © Archives Charmet / Bridgeman Images.

First published 2026
by Routledge
605 Third Avenue, New York, NY 10158

and by Routledge
4 Park Square, Milton Park, Abingdon, Oxon, OX14 4RN

Routledge is an imprint of the Taylor & Francis Group, an informa business

ISBN: 978-1-138-33750-3 (hbk)
ISBN: 978-1-041-12915-8 (pbk)
ISBN: 978-0-429-44231-5 (ebk)

DOI: 10.4324/9780429442315

Typeset in Sabon
by SPi Technologies India Pvt Ltd (Straive)

In loving memory of my sisters
Dayle Prinstein
Donna Pearlman
My devoted parent Libby Weingarden

Standard English translations of the works of major writers are referenced in this volume. All other translations are by the author.

Contents

Figures

Author Biography

Lauren S. Weingarden is Professor Emerita of Art History at Florida State University in Tallahassee, Florida, USA. Weingarden's career spans four decades of teaching and mentoring, international conference presentations, keynote lectures, academic awards, and service to professional organizations. Her publications focus on the interrelations between written and visual texts and their functions within nineteenth-century cultural contexts, including the monograph *Louis H. Sullivan and a 19th-Century Poetics of Naturalized Architecture* and numerous articles on French modernity, as defined by Charles Baudelaire, as well as essays on contemporary installation art. The present volume combines Weingarden's transdisciplinary approach to word-and-image studies with her current research in neuroaesthetics.

Acknowledgments

This book has come to fruition thanks to my mentor, Richard Shiff, who planted the seed of irony in my still-evolving graduate student mind. That seed took root in 1994 when I was awarded the first research grant dedicated to my word-and-image investigation of Baudelaire's "The Painter of Modern Life" as a progenitor of Edouard Manet's visual ironic parody. Florida State University (FSU) awarded that seed money and continued, for over eighteen years, to support the research and writing of this book. In 2001 The Dedalus Foundation Senior Fellowship Research Grant supported an extended period of in-depth archival research at the Bibliothèque Nationale de France, the Musée Carnavalet, the Musée d'Orsay, and other Paris-based institutions, and beyond. Two additional grants supported the preparation stages of this book: The Dedalus Foundation Fund for Past Fellows and Awardees, awarded in 2023, and the Textbook & Academic Authors Association Publication Grant, awarded in 2024. I am grateful to each of these funding organizations for cultivating my book project and for their continued faith in my work.

The support and encouragement of my colleagues have greatly enriched the academic journey that led to this book. Aimée Boutin, FSU professor of French, has been a constant source of inspiration with her camaraderie, intellectual engagement, and guidance through nineteenth-century French literature. Her advocacy and counsel, whether in one-on-one discussions, collaborative team teaching, or conference planning and presentations, have been instrumental in shaping my research. Matthew Pelowski, Vienna University, has been a guiding light in my exploration of empirical aesthetics and cognitive psychology. His scientific guidance has significantly influenced my work, especially as a project collaborator in three empirical studies, two of which appear in this book. On each occasion, Matthew's graduate students provided additional support, particularly Corinna Kühnapfel, who co-authored our study of present-day viewers of Marville's photographs of Old and New Paris.

As my book title indicates, I am indebted to John Onians for laying the foundational methodology and nomenclature of neuroarthistory. I benefited from his published scholarship and his research material, which he generously shared during my visit to his Norwich home in 2013. Subsequently, John became a frequent endorser of my book project. During a 2013 sabbatical, I delved into the literature on neuropsychology and aesthetics and corresponded with scientists who generously responded to my "layperson" inquiries. Anjan Chatterjee, professor of neurology at the Perelman School of Medicine at the University of Pennsylvania, was especially generous in responding to my numerous emails. Cognitive poetics scholars, (the late) Reuven Tsur, David Miall, and Philip Davis, shared their expertise in bridging literature and neuroscience. In 2013 Philip hosted my visit to the University of Liverpool where he introduced me to his research collaborators and their laboratory. Since 2014, I have collaborated with art historian Christine Vial Kayser (CY Cergy Paris Université) in developing models and methodologies of embodied/experiential aesthetics. These collaborations in conference and colloquium settings provided springboards for my articles included in Christine's edited volumes, notably my co-authored article on Marville's Old and New Paris series, published in *Art as Experience of the Living Body/L'art comme experience du corps vivant* (2024).

I am especially grateful to my colleagues in the Intermedia Studies Program (ISP) at the Federal University of Minas Gerais in Belo Horizonte, Brazil. Since 2010, the ISP has been a crucial sponsor of my research, a sponsorship spearheaded by Thais Diniz. Under Thais' leadership, I was enveloped by the Program's dynamic intellectual community while I held the Chair in Humanities, Letters and Arts at the university's Institute for Advanced Transdisciplinary Studies, and in 2012 when I was a visiting professor in the Literary Studies department as a Fulbright Scholar. These in-residence positions allowed me to develop my transdisciplinary research in the classroom, faculty discussions, and lecture presentations, and, for over a decade, I have been an invited speaker at ISP conferences and published numerous articles in their journals and edited books. I am deeply appreciative of the support and intellectual stimulation provided by my ISP colleagues, who have been influential in the creation of this book.

My longstanding memberships in interdisciplinary organizations have also afforded me opportunities to develop my research at conferences and publish in conference proceedings. The International Association of Word & Image Studies (IAWIS) is foremost among these affiliations. Having served as an IAWIS Board Member between 1994 and 2023, I was privileged to count literary historians and theorists as my close associates, with whom I chaired many conference sessions and shared lively debates. More recently, the Nineteenth-Century French Studies Association and Interdisciplinary Nineteenth-Century Studies provided conference platforms to present my research and dialogue with like-minded scholars.

Countless individuals have supported my work in preparing this book for publication. Please forgive me if I have not named you in this acknowledgment. I am especially grateful for the assistance I received from innumerable librarians, archivists, and curators in the U.S. and in France. Therese Dolan provided generous letters of support for grant applications and encouragement during the writing process. Patricia Mainardi cheered me on while we flâneused in Paris and as phone buddies in the U.S. I also want to recognize Margaret (Peggy) Wright-Cleveland, FSU's Director of Faculty Development. As I developed my book research as an art history professor at FSU, Peggy remained a tenacious reader of my grant proposals and a guide to institutional research and funding opportunities. For the past three years, freelance editor Diane Miliotes has fostered a safe haven to air my conceptual pondering (via Zoom) and has skillfully tailored my densely packed chapters and prose.

At Routledge, I respectfully acknowledge Alexandra McGregor, Senior Editor – Art History and Visual Studies, Loredana Zeddita, Editorial Assistant, and Alanna Donaldson, Senior Production Editor, for their guidance and flexibility in the preparation and submission processes.

I am forever grateful to my family and friends for their respect for the time and space I needed to complete this project. Your love and support are inscribed in the following pages.

Introduction

A Neuroarthistory of The Painters of Modern Life: Embodying Baudelairean Modernity advances a transdisciplinary approach—combining art history, literary studies, and neuroaesthetics—to examine the modern urban experience of nineteenth-century Paris through language and images of fragmentation and transformation. Using this multifocal lens, I aim to reformulate understandings of Haussmannization (after Paris urban designer Georges-Eugène Haussmann [1809–91]), when nineteenth-century Parisians witnessed the demolition of their old city and its rebuilding as a modern metropolis. This volume sheds new light on writers' and artists' visceral and creative responses to these changes. In doing so, I focus on Charles Baudelaire (1821–67), who theorized this modernity, and visual artists such as Édouard Manet (1832–83), whose work reflected its impact. *A Neuroarthistory of The Painters of Modern Life* is not a mere theoretical proposition but a culmination of years of extensive multidisciplinary research and practical application. It is built on a unique word-and-image methodology, refined over two decades of dedicated research, presentations, and publications on Baudelairean modernity. This methodology views verbal and visual exchanges within a unifying field of discourse grounded in Baudelaire's seminal essay, "The Painters of Modern Life" (1863). To substantiate the embodied experiences of rupture and reform this discourse makes manifest, I have adapted neuroaesthetics to word-and-image analyses of verbal and visual texts. Using this transdisciplinary approach, I demonstrate that artists and writers who shared the discursive space of modernity expressed their lived experiences through comparable ironic practices and parodic art forms, devices that ruptured traditional means of representation.[1]

Through novel scholarly engagement with neuroscience and its application in the humanities, *A Neuroarthistory of The Painters of Modern Life* reveals how current methods of neuroscience can be employed to reconstruct past experiences, thus building upon and reframing more traditional historical analyses of this important chapter in European modernity. As such, I engage in neuroarthistory, defined by John Onians in 2005 as pertaining to art historians who use neuroscience to understand all aspects of the making and viewing processes.[2] In doing so, I seek to model the "lived" experience of nineteenth-century modern urban phenomena.

Overview

This book operates along two axes—the historical and the neuroaesthetic—that weave together and shape the chapters' form and content.

DOI: 10.4324/9780429442315-1

The Historical Axis

A Neuroarthistory of The Painters of Modern Life advances along a historical and historiographic trajectory that unveils new dimensions of Baudelaire's influence on his artistic followers; it explores how the ironic structure of his critical and poetic texts offered strategic approaches for representing modernity in visual forms. Modern irony encompasses here both a cerebral state and a literary and artistic strategy that juxtaposes contrasting concepts as irreconcilable. As D. C. Muecke advances in his history of irony and the ironic, during the nineteenth-century Romantic era, irony shifted from a satirical device (current in the seventeenth and eighteenth centuries) to a continual process of philosophical reflection, perpetual self-transformation, and creative and aesthetic introspection. Muecke explains this shift accordingly:

> Where before irony had been thought of as essentially intentional and instrumental, someone realizing a purpose by using language ironically, it now became possible to think of irony as something that could instead be unintentional, something observable and hence representable in art, something that happened or that one became or could be made aware of (...); from now on irony is double-natured sometimes instrumental, ... it now became possible to see the world as an ironic stage and all mankind merely players and where before irony had been thought of as a finite act or at most an adopted manner (...), it could now also be thought of as a permanent and self-conscious commitment: the ideal ironist would be always an ironist, alert even to the irony of being always an ironist; irony, in short, could be seen as obligatory, dynamic and dialectical.[3]

As I will demonstrate, this interpretation of irony, originating from German Romantic irony, extends to Baudelaire's paradoxical literary practices, critical tenets, and parodic devices. As Linda Hutcheon has shown (explained further below), irony is foundational to modern parody and its art forms, entailing a doubling, dialectical function. This doubling effect promotes artistic renewal since parody targets and devalues the old text in the new text. Given that parodic doubling issues from the creator's and then the reader's/viewer's critical reflection upon conventional art forms, I refer to this practice as "ironic parody." As Baudelaire and later scholars explained, he conceived his prose poems (1857–67; published posthumously as *Le Spleen de Paris/Paris Spleen*) as parodies of verse poetry, a transmutation necessitated by the physical and psychical ruptures of the modern city. Baudelaire's creative shift from philosophical introspection to technical and formal distortion was paradigmatic for his artistic followers, the painters of modern life, and other members of his circle referenced below. This transference from Baudelaire's literary models to pictorial forms is what I call "visual ironic parody" or "visual parody." I examine this trajectory in four of the five final chapters of this volume.

Baudelaire's writings on modernity during Paris's urban transformation—which began sporadically in the 1840s, launched fully in 1853, and extended over the next two decades—are paramount to this study. His critical essays, "De L'héroïsme de la Vie Moderne" ("Of the Heroism of Modern Life"), written in the *Salon of 1846*, "Exposition Universelle de 1855" (Universal Exposition of 1855), and "Le Peintre de la Vie Moderne" ("The Painter of Modern Life"), begun in 1859 and published in 1863, mirror these changes. Viewed against the background of Haussmannization, I treat these texts as an intertextual network that comprises the entirety of his aesthetic, creative, and critical practices.

While not commonly considered among his art criticism, I regard Baudelaire's "De L'essence du Rire et Généralement du Comique dans les Arts Plastiques" (1855; "Of the Essence of Laughter and the Comic in the Plastic Arts") integral to this network (treated at length in Chapter 1). While he does treat contemporary artists in this essay, more important for this volume, Baudelaire specifically defines the ironic human condition as "*se dédoubler*." Here and in subsequent chapters, I translate "*se dédoubler*" as "self-doubling," following Paul de Man's translation as "rapid self-doubling" ("la force de se dédoubler rapidement") in his analysis of Baudelaire's concept of irony. I also adapt the term *dédoublement* from de Man, who uses it to explain the "reflective activity" connoted by Baudelaire's use of the infinitive form "*se dédoubler*," that being an activity of "self-duplication or self-multiplication."[4]

Viewing this intertextual network as a dynamic whole, I maintain that Baudelaire defined modernity as an aesthetic of ironic dualities and self-doubling that captured the visceral response to urban rupture and fragmentation in artistic form. Accordingly, Baudelaire defined the "double composition" of modern beauty as a constant tension between "an eternal, invariable element," and "a relative circumstantial element [of] contemporaneity."[5] Baudelaire brings together these tenets in "The Painter of Modern Life," a manifesto-like text celebrating the exemplary artistic persona of Constantin Guys (1802–92) and his acts of self-doubling during the creative process. He is both a man of the crowd and separate from it, a separation that ensures the artist's reflection upon his sensory experience in order to render its vital essence. In Baudelaire's assessment of Guys' modernity, the sketch-like pen-and-ink drawings and the transparency of watercolor washes "reflect [the artist's consciousness] at every moment in energies, more vivid than life itself, always inconstant and fleeting."[6]

In the first three historiographic chapters, I establish the historical context of Baudelaire's art criticism, contingent with his verse poetry (*Les Fleurs du Mal/Flowers of Evil*, 1857, 1861, 1866) regarding the paradoxes of modernity, as they emerged against the backdrop of Haussmannization. Central to these chapters is the ambiguity—or duality—that Baudelaire and his contemporaries fostered in their nostalgia for old Paris and their optimism toward its rebirth. Chapter 1 sheds light on Baudelaire's engagement with German Romantic irony as an antidote to modernity's ruptures, epitomized in "The Painter of Modern Life" and Guys' sketches of Parisian life (1850s–early 1860s). Chapter 2 presents Baudelaire's short-lived collaboration with the engraver Charles Méryon (1821–68), his celebration of the "paradoxes" exposed in Méryon's engravings of old Paris (1852–4), and Baudelaire's own paradoxical verse poetry *Tableaux Parisiens* (Parisian Scenes; 1861; second edition of *Les Fleurs du Mal*) inspired by the engraver's work. Chapter 4 demonstrates that Charles Marville's (1813–79) photographic series of old and new Paris (1865–6; 1876–7) reflected the collective ambiguity that Baudelaire, and a contemporaneous antiquarianist movement more generally, impelled.

Building on this historical continuum, I develop the mirror as a verbal-visual trope that encapsulates the essence of Baudelairean modernity. The mirror serves as both a metaphor and a metonym, symbolizing the ironic self-reflection that characterizes Baudelaire's modernity. Just as mirrors facing one another can elicit endless reflections, ironic self-reflections are an open-ended, regenerative process, compelling both the individual maker and viewer/reader to reflect upon novel encounters that break with familiar identities and customs, encounters from which new selfhoods can emerge. Analogously, as a regenerative process, ironic self-reflection can compel certain makers to contemplate and subvert conventional means of representation. This strategy impels technical and formal inversions that counter the old in the new, a strategy that I call "ironic parody."

Within this framework, I refer to those artists who followed Baudelaire's mandate for "the painters of modern life" as Baudelairean artists. These artists extended ironic reflection to their visual means of representation. Likewise, they incorporated the mirror—whether present or inferred—into their parodic inversions of pictorial conventions. Édouard Manet figures as the primary Baudelairean artist treated here. I show how Manet forged a new visual language of self-doubling, what I call "visual ironic parody," to convey the transformative experiences of the modern city. In addition to Manet, Edgar Degas (1834–1917) and Gustave Caillebotte (1848–94), among other artists, provide examples across this study of the Baudelairean artist. Baudelaire's friendships with these artists, particularly with Manet, and his influence in their circles will be addressed in relevant chapters throughout this volume.

The Neuroaesthetic Axis

Given its transdisciplinary foundations, this volume is both experimental and exploratory, presenting a model for how art history and the humanities in general can be mutually productive with neuroscience to reify embodied and affective responses to material conditions and the art forms that translate those conditions. As I establish at the outset, Baudelaire's references to the nervous system in "The Painter of Modern Life" and elsewhere provide evidence of his basic knowledge of nineteenth-century brain science. Accordingly, he used neurological precepts as a means of both transcribing the visceral effects of modernity and prescribing artistic techniques to mediate these effects. This evidence, in turn, validates my own investigations and integrations of nineteenth-century, and subsequent, contemporary neuroscience.

My art historical adaptation of neuroscience most closely aligns with the recently established subfield of neuroaesthetics, as is Onians's neuroarthistory. Neuroaesthetics itself is an interdisciplinary research field. According to Marcus T. Pearce et al., a team comprised of pioneering researchers, neuroaesthetics is "the cognitive neuroscience of aesthetic experience, drawing on long traditions of research in empirical aesthetics on the one hand and cognitive neuroscience on the other." Traditionally, neuroaesthetics identifies relations "among neuroscientific investigations of aesthetics, beauty, and art."[7] Because my book expands the aesthetic spectrum to include rupture, fragmentation, and even disgust, I benefitted from researchers such as Pearce et al., whose approach "takes as its object of study a wide spectrum of aesthetic experiences resulting from interactions of individuals, sensory stimuli, and context." Specifically, this research team seeks to develop a methodology "mutually complementary to humanities." They thus propose drawing on parent neuroscience fields to establish "a cognitive neuroscience of aesthetics [that] would investigate the complex cognitive processes and functional networks of brain regions involved in those experiences without placing a value on them."[8] These neural processes and regions figure widely in my adaptation of neuroaesthetics to a nineteenth-century experiential embodied aesthetics.

I myself began developing a neuroarthistorical model that extends beyond traditional aesthetics of beauty and reward in 2013.[9] My endeavors were motivated by a quest to give substance to lived experiences of nineteenth-century Parisians during Haussmannization. Coincidently, I sought to align these embodied experiences with the creation and reception of ruptured images, parodic distortions, and ironic subject matter.

At the outset and during the following decade, I encountered very few neuroscientific investigations into "the complex cognitive processes and functional networks of brain regions" addressing the wider spectrum of aesthetic experiences that pertain to the creation and reception of irony and parody as disruptive of artistic conventions. I therefore dug deeper and wider into parent neuroscience fields and their offspring—namely, empirical aesthetics and cognitive poetics. Steven Brown et al.'s and Matthew Pelowski et al.'s studies were foundational to guiding my investigations into empirical aesthetics[10] that extended beyond measuring beauty and reward systems. Their studies provided data for registering viewers' affective and neural responses to non-beautiful and non-traditional artworks. This approach was especially pertinent to tracking viewers' embodied responses to pictorial rupture and fragmentation, ranging from anger to accommodation, that irony and parody compelled, as I explore in Chapters 3, 5, 7, and 10.

Given that empirical aesthetics focuses on viewers' subjective and cognitive responses to artworks, I have joined this methodology with reader reception theory, which issues from literary criticism. This cross-disciplinary union informs each facet of *A Neuroarthistory of The Painters of Modern Life*. Because such a union is unique in art historiography, here I present an explanation of my approach.

Reader reception theory "maintains that textual meaning occurs within the reader in response to a text and recognizes that each reader is situated in a particular manner, including factors such as ability, culture, gender, and overall experiences."[11] In this model, the reader is the receiver and interpreter of the text. Correspondingly, in the visual arts, the viewer is the receiver and interpreter of the artwork. Such an approach, which bridges the maker's technical means of representation with the reader/viewer's aesthetic/embodied reception, is premised on the "ideal reader." The ideal reader is a conduit of the maker/creator's production mode into a reception mode that corresponds with the maker's intentions. In reality, the ideal reader (or viewer) cannot exist. However, in theory, the ideal reader can be constructed by way of reader reception theory in that the maker and receiver share the "same overall experiences."[12] This model underscores Baudelaire's art criticism, particularly as set forth in "The Painter of Modern Life" (1863), where he stated "the viewer becomes the translator of [the artist's] translation[.]"[13] Likewise, the ideal reader also manifests in the Baudelairean artists' dialogic relation with Baudelaire's texts, which I attribute to their shared discursive space shaped by Haussmannization.

Another branch of reader reception theory focuses on the impossibility of an ideal reader, given the unique existence each reader brings to a text.[14] To a limited extent, my dialogic model accommodates this assumption as it applies to the creative interpretation of a text. This reader-creator neither distorts nor critiques a text (as in ironic parody) but translates it according to their own means of artistic production. Within this translation chain, the creator/maker is also a reader/viewer of pre-existing (paradigmatic) texts and images. Such is the case when Baudelairean artists adapted Baudelaire's critical and literary texts to pictorial means of representation. Similarly, as discussed in Chapter 2, Charles Méryon's etchings of pre-Haussmann Paris provided Baudelaire with a memory bank of sensory experiences that shaped the *Tableaux Parisiens* poems. In Chapter 4, I suggest that Charles Marville's photographic techniques (framing, lighting, and viewpoint) developed for his Old Paris series and demolition scenes were motivated by the nostalgic turn of Baudelaire's *Tableaux Parisiens* and the poet's haptic experience of navigating Paris in ruins.

Cognitive poetics also facilitated my joining a word-and-image methodology with neuroaesthetic findings to further model how irony and parody manifest the ways both artist and viewer embody societal and physical ruptures. Notably, cognitive poetics, a branch of literary theory that analyzes the mental processes at work during literary reading,[15] is contingent on reader reception/response theory. Literary historians Reuven Tsur, Philip M. Davis, and David S. Miall, each in collaboration with neuroscientists, demonstrated how the reading/perception of textual ruptures affect cognitive processes that are founded in brain functions (as opposed to interpreting the meaning within the text). These techniques include the grotesque, functional grammatic and semantic shifts, and syntactic dislocations, requiring readers' delayed cognitive processing and accommodation of novel forms. I adapt these cognitive patterns of rupture and delay to the viewers' aesthetic processing of visual parody and its ironic effects, compelling self-reflection upon ruptured artistic norms and the societal and physical ruptures they bear. George Lakoff's pioneering work in cognitive poetics guided me in analyzing how linguistic metaphors are directly produced by embodied sensations and affective associations. I adapt Lakoff's explanations of metaphors and the targets they mask to analyze the arousal and anger shrouded by the critical rhetoric of dirt and obscenity that Manet's paintings *Le Dejeuner sur L'herbe* (*Luncheon on the Grass*) (1863) (Figure 6.1) and *Olympia* (1863) (Figure 6.2) impelled.

By 2017, I had initiated and participated in neuroscientific experiments that are integral to the present volume. This research entailed collaboration and consultation with empirical psychologist Matthew Pelowski. Such collaborations are rare in art history scholarship, and I consider myself fortunate to have forged this opportunity to work with Pelowski, his associates, and graduate student research assistants. Two chapters in *A Neuroarthistory of The Painters of Modern Life* are based on my hypothetical questions, which resulted in experiments Pelowski and his team performed, detailed below. I first worked with Pelowski on a separate project to investigate art viewers' emotive and empathic responses to contemporary installation art (autonomous enclosures within gallery spaces, incorporating various media to create immersive experiences), the results of which were published in 2020.[16] I also independently tested my new interpretive model of "experiential neuroaesthetics."[17] As a researcher at Inhotim (Brumadinho, Minas Gerais, Brazil), a museological complex featuring installation art pavilions in botanical gardens, I hypothesized that Baudelaire's idea of modernity as fleeting and fragmentary has relevance to the twenty-first century. Accordingly, I traced postmodernism's performative turn and immersive aesthetic to its Baudelairean antecedents.[18] Baudelaire and Baudelairean artists sought to put viewers at the center of the pictorial space to engage them in the transitory experience that compelled the artist's creative response to modernity. In *A Neuroarthistory of The Painters of Modern Life*, I turn back to nineteenth-century Paris to extend my modeling of experiential neuroaesthetics to both the creation and reception of pictorial rupture and fragmentation. As a result, this book proposes a mobilized urban spectator and their embodied responses to urban chaos and clamor as both the creator and the viewer of Baudelairean art forms.

The Neuroarthistorical Format and Chapter Preview

The chapter format of *A Neuroarthistory of The Painters of Modern Life* reflects this dialogue between neuroscience and the humanities. Chapters comprising art historical and literary analyses are interwoven with chapters that explore, adapt, and apply neuroscientific research to my modeling of experiential, embodied neuroaesthetics. In the latter category, I examine the creators' and viewers' physiological, affective, and neural responses to modernity and modern art forms. My independent study of neuroscience and neuroaesthetics has facilitated my modeling of an experiential aesthetic to reify the mobile creator-as-flâneur (one who strolls through a city)[19] and the embodied experience they shared with the mobile urban spectator. These interval chapters are experimental and exploratory, comprising both hypothetical models of interpretation and applied scientific models derived from real-time experiments in which I collaborated. In Chapter 3 and Chapter 7, such embodied experiences are hypothetical, based on nineteenth-century psychology and contemporary psychology and neuroscience. In Chapters 5 and 10 (a combined historical and neuroaesthetic chapter), I report on my collaborations with empirical psychologists, in which we conducted research regarding contemporary viewers' responses to nineteenth-century photography and painting, respectively. The results of these laboratory- and computer-based experiments offer tangible data to substantiate the lived experience of nineteenth-century Parisians and its manifestations in art forms.

Within this dialogic format between historical and neuroscience chapters, Chapters 1 and 2, for example, correlate with Chapter 3. In Chapter 1, "Neuroaesthetics and Cognitive Poetics: Mapping Baudelairean Modernity in Neural Processing of Word and Image," I analyze Baudelaire's "Le Poéme du Haschisch" (1858; "The Poem of Hashish") and "The Painter of Modern Life" (1863) within the context of nineteenth-century medical science to expose the writer's embodied reactions to the injurious effect of Haussmannization on the nervous system. Nineteenth-century neurologists and psychologists examined the nervous system as the basis of mind-body synergies. For the first time, I establish that Baudelaire had direct access to these sciences through his friend Dr. Jacques-Joseph Moreau (1804–84) and adapted them to his theory of modernity and his aesthetic of fragmentation and rupture. I also argue that Baudelaire used irony derived from German Romantic ironists as a remedy for these sensory injuries, enabling the poet-critic to translate cerebral ruptures into poetic and prose forms.

This neural injury and recovery theme continues in Chapter 2, "Mirroring Haussmannization in Baudelairean Word and Graphic Image." I focus on Baudelaire's critical praise for etchings by Charles Méryon of pre-Haussmann Paris views in his essay *Salon of 1859*, viewed in tandem with his lyric poems collected in *Tableaux Parisiens* (1861). Baudelaire's nostalgic reading of Méryon's images and his lyric poems reveal ironic contrasts between old and new Paris, relieving his lamentations over Paris in ruins. Viewed alongside Baudelaire's celebration of modernity in "The Painter of Modern Life," his reading of Méryon underscores a collective ambiguity toward modernized Paris. I also adapt recent scholarship in sensory studies, a field that provides empirical evidence to substantiate Parisians' visceral response to Haussmannization, as manifested in sights and sounds of demolition and reconstruction that further suggest this collective ambiguity. I argue that Méryon's etchings register this sensory mapping and resonate in Baudelaire's poetry and his ironic translations of Méryon's macabre visions.

In Chapter 3, "On Baudelairean Modernity: Neuroaesthetic Modeling of Ironic Self-Reflection," I explore how scientific and physiological markers that populate Baudelaire's writings provide the basis for extending contemporary neuroscience to a nineteenth-century aesthetic of rupture and fragmentation. In this chapter, I align ironic self-reflection and its corresponding mirroring metaphor with twenty-first-century neuroaesthetics. Here, neuroaesthetics offers models to correlate the viewer's reception of formal and figural elements of ironic texts and images with the psychic and cognitive processing of non-art environments, such as changed cityscapes. The psychological and neuroscientific models employed in this chapter are ground-breaking for their investigations of the aesthetic experience of unconventional (non-beautiful) artwork with self-reflection and negative valence (as opposed to positive responses to conventional "beauty"). From this perspective, we can hypothesize the nineteenth-century subject's neural activity in response to the urban environment, the maker's mediation of the sensorial experiences, and the viewer's affective and cognitive responses to that mediation in unconventional art forms.

Chapter 4, "Mirroring Haussmannization in Baudelairean Word and Photographic Image," examines the makers' and viewers' affective and haptic responses to the urban environment through the lens of the photographic series of old and new Paris by Charles Marville. This perspective reveals the dynamics between word and photographic image that codified modernized Paris as a national *lieu de mémoire* (memory space), a symbol of technological achievement, cultural continuity, and orderly social reform. However, I assert that the new photographic technology (spawned in the 1850s)[20] also mirrors the cultural ambiguity to which Baudelaire gave voice in his coeval texts, the melancholic *Tableaux Parisiens* (1861) and the celebratory "The Painter of Modern Life" (1863). Photographers captured the disappearing city and its transformation, mediating nostalgia for its past and commending its contemporaneity. Marville is exemplary among these practitioners; the city of Paris commissioned him to record streets to be demolished, demolition sites, and Haussmann's new boulevards. At the same time, Marville's Old Paris series provided foundational collections for newly established public institutions that emerged from a burgeoning antiquarian movement borne from Haussmann's wrecking ball. Here, I establish that Marville's photographic archive illuminates the kaleidoscopic matrix of Paris transformed. His series not only marks the capital city as *un lieu d'expérience* (an experiential space) but also embodies the physical, sensory, and affective vicissitudes that arise from paradoxical encounters with the old and the new and the demolitions that joined the two.

Chapter 5, "A Collaborative Investigation into Affective and Embodied Aesthetic Responses to Charles Marville's Old Paris and New Paris Series," delves into the experiential embodied effect of Marville's photographs on present-day viewers, drawing from responses collected in a real-time psychological experiment. The objective of this investigation was to answer the question, "Can present-day photographic viewing and, by extension, empirical aesthetics be used to corroborate a multisensory past?" This question undergirds this entire book. By using a transdisciplinary method that combines art history and neuroaesthetics, can we identify and understand the mobilized spectator's affective and cognitive responses to the urban and cultural disruptions incurred by Haussmannization? Pursuant to this inquiry, I initiated a collaboration with two empirical psychologists, Pelowski and Corinna Kühnapfel,[21] to survey present-day viewers' affective and embodied responses to Marville's photographs of old Paris and new Paris.[22] This chapter summarizes the empirical data to address an unexpected outcome of our study: how to reconcile the misalignment between our empirical evidence, wherein present-day viewers favored new Paris photographs, and the art historical evidence and historiographic record, which favored old Paris photographs. In this case, reader reception theory offered a method for modelling the "curatorial gaze," evidenced in museum curators' affective, mobilized, and embodied responses to Marville's Old Paris series and their dismissal of his New Paris series.

The chapters that follow deploy a word-and-image approach to understanding irony and parody. In Chapter 6, "Manet's Ironic and Erotic Gaze: Photography, Pornography, and Censorship," I treat Manet's scandalous paintings *Le Déjeuner sur L'herbe* (1863) and *Olympia* (1863) as parodic responses to pornographic photography and its legal censure. Offended critics, however, framed their responses in the pejorative language of social and bodily contamination, rhetoric attached previously to literary and artistic Realism, and prosecutorial language of two 1857 censorship trials: of *Madame Bovary* (1856/57) by Gustave Flaubert (1821–80) and of Baudelaire's *Les Fleurs du Mal* (1857). As I contend earlier, Manet, the primary Baudelairean artist in this art historical study, among other artists, forged a new visual language of self-doubling, namely, visual ironic parody, to convey the transformative experiences of the modern city. *Le Déjeuner sur L'herbe* and *Olympia* are foundational works in Manet's subsequent efforts to extend these literary motifs into visual forms.

While scholarship on literary irony and parody flourished in the 1990s, there is a dearth of comparable analyses in art history. *A Neuroarthistory of the Painters of Modern Life* marks a critical intervention: it is the first art historical study to chronicle parody as a nineteenth-century pictorial device. Linda Hutcheon's *A Theory of Parody: The Teachings of Twentieth-Century Art Forms* (1985) is foundational to my understanding of verbal and visual parody. Like verbal parody, visual parody comprises both references to and inversions of traditional artistic forms—such as perspective, illusionistic modeling, and smooth finish. Like verbal parody, visual parody subverts narrative closure. Such inversions result from the maker's ironic self-reflection upon their medium and, in turn, trigger the reader/viewer's ironic self-reflection during their aesthetic response. While I introduce this strategy in Chapter 1, I go on in later chapters to demonstrate how Manet represented Baudelaire's ironic self-reflections in enigmatic poses, de-centered gazes, and fragmented spatial settings.

The critical discourse surrounding Manet's paintings is the subject of Chapter 7, "Brain Mapping Censorship and Desire in Second Empire Paris." As a result, word-and-image methods converge with neuropsychology and cognitive linguistics to examine the vitriolic language against "obscenity" and the hostile critical reception of Manet's *Le Déjeuner sur L'herbe* and *Olympia*. These paintings function within a broader spectrum of popular outcry against the public display of prostitutes and pornographic photographs along Haussmann's new boulevards. I argue that the critical and legal hostility was not only institutionally motivated but, crucially, was triggered by embodied visceral responses to pornographic texts and images. Here, I integrate neural research on internet pornography addiction that establishes how "neurobiological mechanisms underlie neuropsychological operations."[23] Together with other neurological studies, this research registers the spectrum of arousal and disgust that accompanies pornographic viewing, shedding new light on the critical reception of Manet's portrayals of the model-cum-prostitute and, in the public's eye, the models' proxy for publicly displayed erotic photographs. In turn, cognitive linguistics, adapting George Lakoff's experientialist epistemology, registers the neuropsychological operations manifest in the pejorative language that functioned to mask male sexual arousal.

In Chapter 8, "Baudelaire's Prose Poems: A Paradigm for Ironic Parody," I focus on Baudelaire's prose poems as the corollary of the critic's affirmation of modernity in "The Painter of Modern Life." Building on the continuum between irony and parody and their fusion as ironic parody, this chapter explores Baudelaire's prose poems, published in *Le Spleen de Paris* as the model that generated verbal parody and, more importantly for this study, visual parody. The prefatory *lettre-dédicace* to the prose poems, published in 1862, figures as the foundation for Baudelaire's parodic literary practices. He presented the prose poem as a means of expressing the experiences of modern life and overcoming the limits of verse poetry. To further substantiate the word-and-image foundation of this study, I identify the formal attributes of literary parody that manifest in visual parody. Analogously, I develop the functions of the mirror as the medium and mediator of parody and its ironic effects. This exegesis also serves as a foundation for the next chapter, where I investigate the dialogic relation between literary and critical texts by Émile Zola (1840–1902) and Manet's pictorial representations. To facilitate this transaction/translation, I propose thematic groupings that align with Baudelaire's practice of ironic parody and, subsequently, with the Manet-Zola dialogic exchanges.

Chapter 9, "Verbal and Visual Parody: Zola's and Manet's Ironic Encounters," maps the transformation of ironic parody from a literary device into a pictorial device. I investigate Manet's parodic practices through the lens of Baudelaire's art criticism and then in alignment with Zola's Naturalist novels, particularly *L'Assommoir* (1877) and *Nana* (1880). Both trajectories place Baudelaire's "The Painter of Modern Life" (1863) as a point of departure for arguing that the mirror functions as both metaphor and metonymy of modernity and its dualities—at once fleeting and fragmentary yet eternal and immutable, and as a mandate for artistic self-doubling in response to "multifarious" modern urban conditions. Establishing Baudelaire's model as Manet and Zola's shared contact for extending ironic parody to their respective mediums, I analyze the doubling devices and their mirror functions in Baudelaire's prose poems, the subject of the preceding chapter, but now through the double lens of the prefatory *lettre-dédicace* to the prose poems and "The Painter of Modern Life," published in 1862 and 1863, respectively. In the declarative spirit of "The Painter of Modern Life," the *lettre-dédicace* is a manifesto on the prose poem as well as a parody of the same genre.

In this chapter, I also extend Hutcheon's model of parody as double-coded to a comparative analysis of Manet's paintings and Zola's novels, with the mirror-motif as a parodic device. Modern parody is inherently double-coded since it inscribes both "continuity" with and "critical distance" from the old (targeted) text in the new text, inducing an ironic attitude or situation for the viewer/reader.[24] Thus, I examine Manet's paintings in which mirrors are implied or represented, followed by a discussion of the mirror-motif in Zola's novels. In each case, the mirror engages the reader/viewer in ironic parody as a device for aesthetic disruption and critical self-reflection. I go on to demonstrate how Zola used the mirror motif as a critical standard to champion Manet's art as analogous with Naturalism and how Manet, in turn, devised a critical, pictorial response to that interpretation. In the last analysis, I establish that Manet's *Un Bar aux Folies-Bergère* (*A Bar at the Folies-Bergère*) (1882) (Figure 1.9) best demonstrates Baudelairean parodying devices and their ironic effects and Manet's contestation against Zola's appropriation of his artistic practices to the novelist's Naturalist aesthetic theories. As such, *A Bar at the Folies-Bergère* signifies Manet's manifesto on the independence of his visual language and its aesthetic impact, which does not illustrate literary texts but participates in a shared discourse on modernity and its experiential effects.

Chapter 10, "Reflections on Baudelairean Modernity," combines word-and-image methods with the neural, cognitive, and perceptual responses to Manet's *A Bar at the Folies-Bergère*. This final chapter focuses on modeling an embodied aesthetic to reify the lived experience of nineteenth-century Paris during Haussmannization. In doing so, I adapt empirical psychology (qua empirical aesthetics) models that track the sensory, affective, and cognitive events that comprise aesthetic processing of non-beautiful or non-traditional art forms. These models also facilitate my exploration of the liminal spaces that connect the physiological and neural processes as sites that register rupture, ambiguity, and dissonance, attributes consonant with Baudelairean modernity.

To establish the liminal spaces wherein rupture and ambiguity occur during aesthetic processing, I adapt cognitive poetics models that focus on processing rupture and defamiliarization as modalities unique to literary as opposed to non-literary reading. I have extracted cognitive and neurological data from these studies that apply to experiential neuroaesthetics as manifested by literary irony and visual ironic parody, analogous with grammatic and syntactic modes of delayed categorization, disorientation, and reorientation registered during the reading process.

Concluding the neuroscientific axis of this volume, in Chapter 10, I address current neuroaesthetic research on mirror neurons and the viewer's re-embodiment of the painter's gestural marking in abstract art.[25] From this perspective, I consider whether the viewer's mirror neurons are activated by the ruptured, dissonant formal qualities of Manet's painting. If this is so, to what degree do these qualities dynamize the act of looking? To answer this question, I again collaborated with empirical psychologists at the University of Vienna to track participants' eye movements while viewing digitized images of Manet's *A Bar at the Folies-Bergère*. This experiment measured participants' eye movements while viewing the original image and a modified image in which the mirrored reflection of the man in the painting had been removed. Our data confirmed that eye movements were more rapid, complex, and prolonged for the original image than the modified image, providing a measure of the viewer's effort to resolve the spatial ambiguity produced by the painting.

This eye-tracking data supports my hypothesis that we can simulate the nineteenth-century perceptual processing of mirror reflections and painterly ruptures. These perceptual responses also align with a cognitive poetics modeling of irony and parody as aesthetic modes of delayed categorization, disorientation, and reorientation.

By centering *A Bar at the Folies-Bergère* in this chapter, I acknowledge the painting as a framing device for the entire book. Introduced in Chapter 1 and examined at length in Chapter 9, *A Bar at the Folies-Bergère* coordinates the themes and methods presented in previous chapters. A word-and-image approach provides the overarching methodology with which to integrate Baudelaire and Baudelairean artists, particularly Manet, into a network of shared responses to Haussmannization. This lived experience manifested in writers' and painters' use of irony and ironic parody as literary and pictorial devices. Within this network, the mirror figures as a literal and metaphoric conveyor of rupture, fragmentation, and ironic self-reflection in both the formal structures and thematic representations of Baudelairean modernity. I have also identified contemporaneous scientific and critical texts to substantiate this experiential and embodied aesthetics. As coalesced in *A Bar at the Folies-Bergère*, the reimagined lived experience of Haussmannization, its literary and pictorial representations, and its psychical and formal attributes of irony and ironic parody provide a platform for systematizing a neuroarthistory of nineteenth-century Parisian culture.

As reviewed above, my neuroaesthetic approach to the lived experience of Haussmannization aims to bridge the maker/creators' and the viewer/readers' visceral responses to the physical, psychic, and sensory effects of rupture and fragmentation. An important outcome of bridging maker and receiver is that *A Neuroarthistory of The Painters of Modern Life* establishes how Baudelairean painters assimilated Baudelaire's adaptations of science. These artists represented their sensory responses to the urban environment with visual devices—such as broken brushwork, flattened surfaces, and spatial distortion—that embodied physical and aesthetic rupture. These visual devices, I argue, affected the viewer's own experience of urban rupture and transformation.

By weaving together these varied approaches, *A Neuroarthistory of The Painters of Modern Life* contributes to the interdisciplinary humanities and the field of neuroaesthetics and provides a model for collaborations between humanities and science. As a word-and-image study of Baudelairean visual language, this book aims to enrich cultural and literary studies of nineteenth-century modernity by substantiating the discursive spaces through which artists and writers exchanged technical and formal devices for representing the rupture, fragmentation, and cultural renewal experienced during Haussmannization. In engaging with the field of neuroaesthetics, the book also proposes a historical image bank of modern art forms with which scientists can widen the neural and emotive spectrum of aesthetic responses. Such embodied responses relate both to art and to our own "real" world experiences and demonstrate nineteenth-century antecedents of our postmodern condition.

Notes

1 Throughout this volume I use the terms "interdisciplinary" and "transdisciplinary" to denote distinctive, but related, methodologies. Interdisciplinary refers to the analysis, synthesis, and harmonization of links between related disciplines (for example, within the humanities) into a coordinated and coherent whole. Transdisciplinarity integrates unrelated disciplines, here brain sciences in a humanities context, and transcends their traditional boundaries. These definitions are adapted from Bernard C. K. Choi and Anita W. P. Pak, "Multidisciplinarity, Interdisciplinarity and Transdisciplinarity in Health Research, Services, Education and Policy: 1. Definitions, Objectives, and Evidence of Effectiveness," *Clinical and Investigative Medicine. Medecine Clinique et Experimentale* 29, no. 6 (2006): 351.

2 See John Onians and Eric Fernie, "Neuro Ways of Seeing: Neuroarthistory, 1 May 2008 [interview]," *Tate Etc*, 13: Summer 2008. https://www.tate.org.uk/tate-etc/issue-13-summer-2008/neuro-ways-seeing; and John Onians, *Neuroarthistory: From Aristotle and Pliny to Baxandall and Zeki* (New Haven: Yale University Press, 2007).

3 D. C. Muecke, *Irony and the Ironic* (London: Methuen & Co., 1970): 19.

4 See "The Rhetoric of Temporality," in Paul de Man, *Blindness and Insight: Essays in the Rhetoric of Contemporary Criticism*, 2nd rev. ed. (Minneapolis: University of Minnesota Press, 1983): 212–3.

5 Baudelaire, "Of the Heroism of Modern Life," *Salon of 1846*, in *Selected Writings on Art and Literature*, ed. and trans. P. E. Charvet (London/New York: Penguin Classics, 1992): 106.

6 Baudelaire, "The Painter of Modern Life," in *Selected Writings*, 400.

7 Marcus T. Pearce et al., "Neuroaesthetics: The Cognitive Neuroscience of Aesthetic Experience," *Perspectives on Psychological Science* 11, no. 2 (March 2016): 265. As a pioneering researcher in neuroaesthetics (who labeled the field in 1999), Semir Zeki focused on the neural and brain activity that controls the production and reception of aesthetic beauty, including the appraisal-reward-response system. See T. Ishizu and S. Zeki, "Toward a Brain-based Theory of Beauty," *Plos ONE* 6, no. 7 (2011): e21852. For reviews of early neuroaesthetic research and critics, see Anjan Chatterjee and Oshin Vartanian, "Neuroaesthetics," *Trends in Cognitive Sciences* 18, no. 7 (July 1, 2014): 370–5, https://doi.org/10.1016/j.tics.2014.03.003; and Helmut Leder and Marcos Nadal, "Ten Years of a Model of Aesthetic Appreciation and Aesthetic Judgments: The Aesthetic Episode – Developments and Challenges in Empirical Aesthetics," *British Journal of Psychology* 105, no. 4 (October 3, 2014): 443–64, https://doi.org/10.1111/bjop.12084.

8 Pearce et al., "Neuroaesthetics," 265.

9 While mentioning research authors here, I do so only briefly and refer to their direct impact in the chapter summaries that follow. Readers can find more information on the authors and their works in the Bibliography section.

10 Empirical aesthetics comprises "a research area at the intersection of psychology and neuroscience that aims to understand how people experience, evaluate, and create objects aesthetically." Aenne Brielmann, "Aesthetics, Empirical," *The Internet Encyclopedia of Philosophy*, n.d., https://iep.utm.edu/empirical-aesthetics/.

11 Susan Browne et al., "Reader Response Theory," in *Oxford Bibliographies Online: Literary and Critical Theory*, ed. Eugene O'Brien (Oxford University Press, 2021), n.p.

12 This "shared experience" approach to reader reception theory closely aligns with Stanley Fish's attempts to situate the reading process in a broader, institutional context; see *Is There a Text in This Class?: The Authority of Interpretive Communities* (Cambridge: Harvard University Press, 1980).

13 Baudelaire, "The Painter of Modern Life," in *Selected Writings*, 407.

14 Even adherents to this interpretation, such as Wolfgang Iser, hold that even though "the reader projects meaning onto the text, the reading activities through which we construct that meaning are prestructured by, or built into, the text. In other words, Iser believes that the text itself guides us through the processes involved in interpreting (projecting meaning onto) it," according to Nasrullah Mambrol, "Transactional Reader Response Theory," *Literary Theory and Criticism* (October 28, 2016). In Chapter 5 I differentiate between the "informed" and the "uninformed" viewer, the former of which corresponds to Iser's "Transactional Reader Response Theory," as explained in Iser, "The Reading Process: A Phenomenological Approach," *New Literary History* 3, no. 2 (1972): 279–99.

15 David S. Miall, "Cognitive Poetics: From Interpreting to Experiencing What Is Literary," in *Anglistentag 2007 Münster: Proceedings*, ed. Klaus Stierstorfer (Trier: Wissenschaftlicher Verlag, 2008): 187.

16 See Matthew Pelowski et al., "Do You Feel Like I Do? A Study of Spontaneous and Deliberate Emotion Sharing and Understanding between Artists and Perceivers of Installation Art," *Psychology of Aesthetics, Creativity, and the Arts* 14, no. 3 (August 2020): 276–93.

17 See Lauren S. Weingarden, "Framing an Experiential Aesthetics: The Affective Turn in Contemporary Installation Art," in *Notions Esthétiques (II): La Perception Sensible Organisée*, ed. Véronique Alexandre Journeau and Christine Vial Kayser (Paris: L'Harmattan, 2015): 197–210. I am grateful to The Fulbright Foundation for funding in 2012 that allowed me to develop this material.

18 See Lauren S. Weingarden, "The Performative Turn at Inhotim: Installation Art and Baudelairean Modernity," *Aletria: Revista de Estudos de Literatura* 23, no. 3 (December 31, 2013): 13–30.

19 Throughout this volume I have minimized using the "flâneur" trope as overdetermined in literary theory and art history, referring to the ambivalent male stroller who wanders the boulevard detached from society. Instead, I connect the excursive modes of representation and writing with embodied experiential aesthetics. Notwithstanding social restrictions on single women strollers, men and women shared this immersive urban experience. For a critique of the flâneur trope, see Richard Wrigley, "Unreliable Witness: The Flâneur as Artist and Spectator of Art in Nineteenth-Century Paris," *Oxford Art Journal* 39, no. 2 (July 28, 2016): 267–84.

20 From photography's inception in the 1830s and continuing through the 1850s, its functions transformed from merely aiding scientific experiments or graphic print technologies to an industry of mass production that served news media, tourism, portraiture, and fine arts reproduction. In less than two decades, photographic technologies were transformed from the early-modern, fixed camera obscura to a moveable mechanical device for recording immediate impressions of lived reality. Popular and commercial demands for photography's precise realism drove the rapid development of techniques and materials, transitioning from single images produced and fixed upon metal plates to multiple images printed on paper from glass-plate negatives. Photographers transitioned from lone experimenters to highly trained professionals and entrepreneurs. See Naomi Rosenblum, *A World History of Photography*, 3rd ed. (New York: Abbeville, 1997): 192–9.

21 Doctoral Candidate (2022–23), Department of Psychology, University of Vienna.

22 Lauren S. Weingarden et al., "Embodying the Photographic Image: Charles Marville's Old Paris and New Paris Series," in *Art as Experience of the Living Body/L'art Comme Experience du Corps Vivant*, ed. Christine Vial Kayser (Wilmington, DE: Vernon Press, 2024): 261–92.

23 Todd Love et al., "Neuroscience of Internet Pornography Addiction: A Review and Update," *Behavioral Sciences* 5, no. 3 (September 18, 2015): 388–433.

24 Linda Hutcheon, *A Theory of Parody: The Teachings of Twentieth-Century Art Forms* (New York: Methuen, 1985): 6, 8, 10.

25 See David Freedberg and Vittorio Gallese, "Motion, Emotion and Empathy in Esthetic Experience," *Trends in Cognitive Sciences* 11, no. 5 (May 2007): 197–203, and M. Alessandra Umilta' et al., "Abstract Art and Cortical Motor Activation: An EEG Study," *Frontiers in Human Neuroscience* 6 (2012): 1–9.

1 Neuroaesthetics and Cognitive Poetics

Mapping Baudelairean Modernity in Neural Processing of Word and Image

Introduction

Charles Baudelaire first pronounced his modernist tenets in the *Salon of 1846*, notably in "Of the Heroism of Modern Life" and, subsequently, in his essay "The Painter of Modern Life," written between late 1859 and mid-1860[1] and published in 1863. In these essays, he established an aesthetic of dualities and acts of self-doubling as a visceral response to the transmuting urban experience and its expression in artistic form. As explained in the Introduction to this volume, under Baudelaire's example and subsequent influence, these aesthetic tenets were manifested in the visual and literary arts as irony and parody.

In this chapter, I lay the groundwork for a chronological and conceptual pathway through this multifaceted study. In doing so, I focus on irony as foundational to Baudelaire's modernist tenets, art criticism, and verse poetry (which preceded his parody-informed prose poems). Irony is not only historically primary in Baudelaire's oeuvre, but is likewise so for modern parody as it emerged in French art and literature during the 1860s. While I more fully develop Linda Hutcheon's model of modern parody as "ironic visual parody" in Chapters 6, 8, and 9, here it is instructive to recognize the intrinsicality of irony to the doubling effect of parody as a structural device. For our purposes, parody's doubling effect pertains to both verbal and visual forms. As Hutcheon explains:

> Parody, then, in its ironic "trans-contextualization" and inversion, is repetition with difference. A critical distance is implied the backgrounded text being parodied and the new incorporating work, a distance usually signaled by irony. But this irony can be playful as well as belittling; it can be critically constructive as well as destructive. The pleasure of parody's irony comes not from humor in particular but from the degree of engagement of the reader [or viewer] in the intertextual "bouncing" (to use E. M. Forster's famous term) between complicity and distance.[2]

Notably, as I discuss below (and in Chapters 6, 7, and 10), the degree to which parody elicits pleasure or displeasure depends on the reader's/viewer's knowledge of the backgrounded text/image, or lack thereof, in the new work.

In view of modern parody's contiguity with irony, I continue to premise my discussion of irony on D. C. Muecke's description of Romantic (qua modern) irony as distinguished from seventeenth- and eighteenth-century modes. Briefly stated, irony is presented "as a permanent and self-conscious commitment: the ideal ironist would be always an ironist, alert even to the irony of being always an ironist; irony, in short, could be seen as obligatory, dynamic and dialectical."[3] Given this definition, in this chapter I establish that Baudelaire's ironic

DOI: 10.4324/9780429442315-2

tenets and practices are grounded in German Romantic irony and its instructions for realizing self-doubling, and mirroring, as both an ideology and an aesthetic practice.

These changes began in 1853 when the recently installed Emperor Napoléon III (formerly Charles Louis Napoléon Bonaparte) appointed Georges-Eugène Haussmann as prefect of the Seine to modernize medieval Paris. Haussmann's comprehensive urban plan, subsequently called Haussmannization, demolished old neighborhoods and replaced them with uniformly designed multistoried buildings and geometric thoroughfares. This process continued after 1870, the year marking Haussmann's dismissal and Napoléon III's dethronement, and was completed in 1927 during the Third Republic.[4] While one cannot discount the human suffering that the Haussmannization of Paris engendered, these ruptures and upheavals inspired Baudelaire's and then his followers' celebration of incessant change and cultural renewal. At every moment of every day, Parisians encountered visual and aural cacophonies bursting from demolition and construction sites and the freshly laid boulevards and sidewalks. New building blocks generated a more ephemeral spectacle of transformation along these same conduits. Plate glass windows at street level reflected transient atmospheric and material conditions. Both Edgar Degas' *Un Café, boulevard Montmartre* (1877) and Jean Beraud's *Le boulevard Montmartre, devant le théâtre des Variétés, l'après-midi* (Boulevard Montmartre, in Front of the Théâtre des Variétés, in the Afternoon) (c. 1885) suggest how indoor spaces mingled with outdoor spaces and spaces in between, turning all Paris into a kaleidoscopic spectacle of mirrors (Figures 1.1 and 1.2).[5] Degas' pastel rendering of the boulevard Montmartre at night

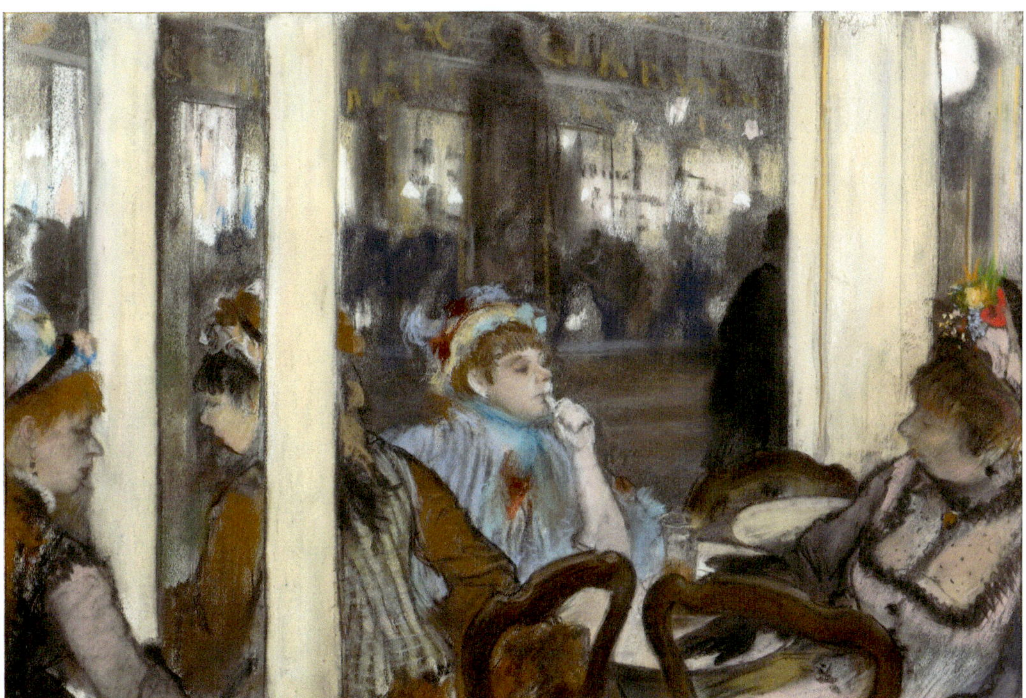

Figure 1.1 Edgar Degas, *Un Café, Boulevard Montmartre*, 1877, pastel, 16 1/8 x 23 5/8 in. Musée d'Orsay, Paris, France. © RMN-Grand Palais/Art Resource, NY. Photo: Hervé Lewandowski.

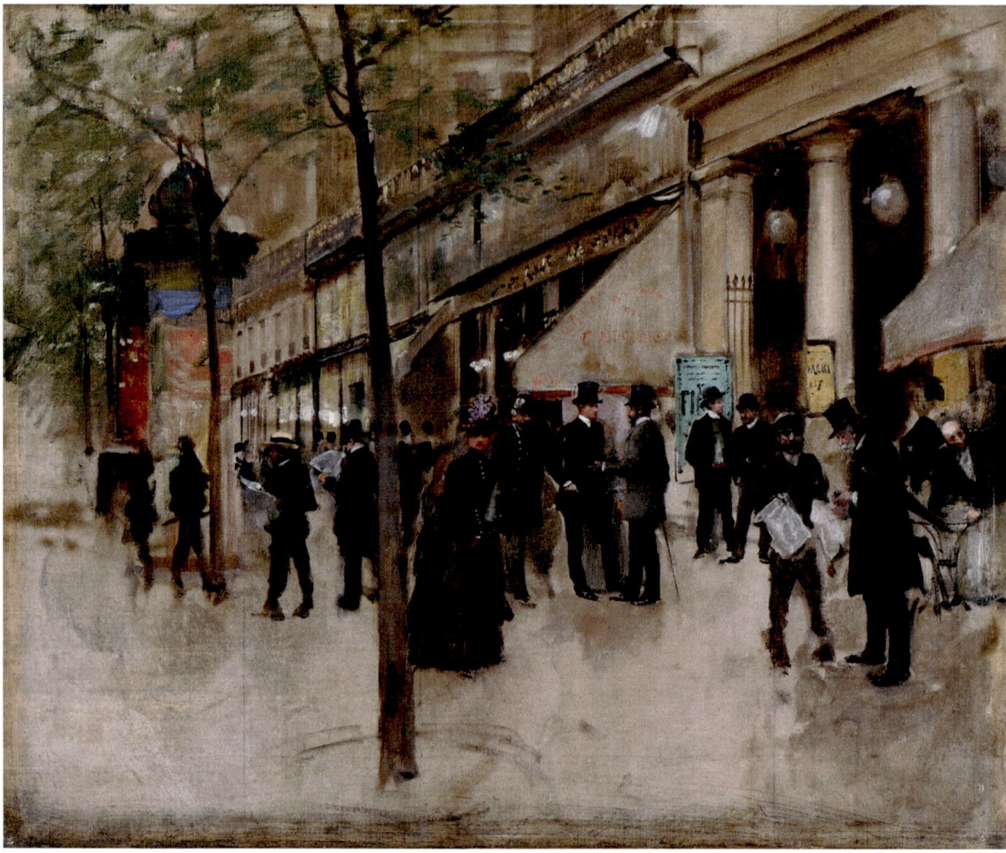

Figure 1.2 Jean Beraud, *Le Boulevard Montmartre, Devant le Théâtre des Variétés, L'après-midi* (Boulevard Montmartre, in Front of the Théâtre des Variétés, in the Afternoon), c. 1885, oil on canvas, 17 7/8 x 21 5/8 in. Paris Musées/Musée Carnavalet, Histoire de Paris.

brings the intermingling of street lamps and interior lighting to the fore, a spectacle of shimmering lights that competes with the display of prostitutes in the viewers' space.[6] Background boulevard lights, featuring street lamps reflected in plate glass shop windows, merge with the foreground terrace lights, made up of globe light fixtures suspended from vertical pillars. Ironically, however, the illuminated shop windows dissolve the figures in transit, while the foreground terrace lights sharply define the women's postures, profiles, and gestures. In Beraud's painting, our gaze follows the receding orthogonal of building facades that culminates with plate glass shop windows. Depicted in a late afternoon overcast atmosphere, the artificial lights add to the dissipation of figures in transit and register our focal point on an animated, yet mundane, urban space. Beraud has captured the fleetingness of modernity, in which new urban technologies, rather than human dramas, play a central role. As I contend in this and the following chapters, if modernist writers and painters subverted traditional narrative and representational devices, they did so as a response to and a means of mirroring the daily reality of fragmentation and transmutation.

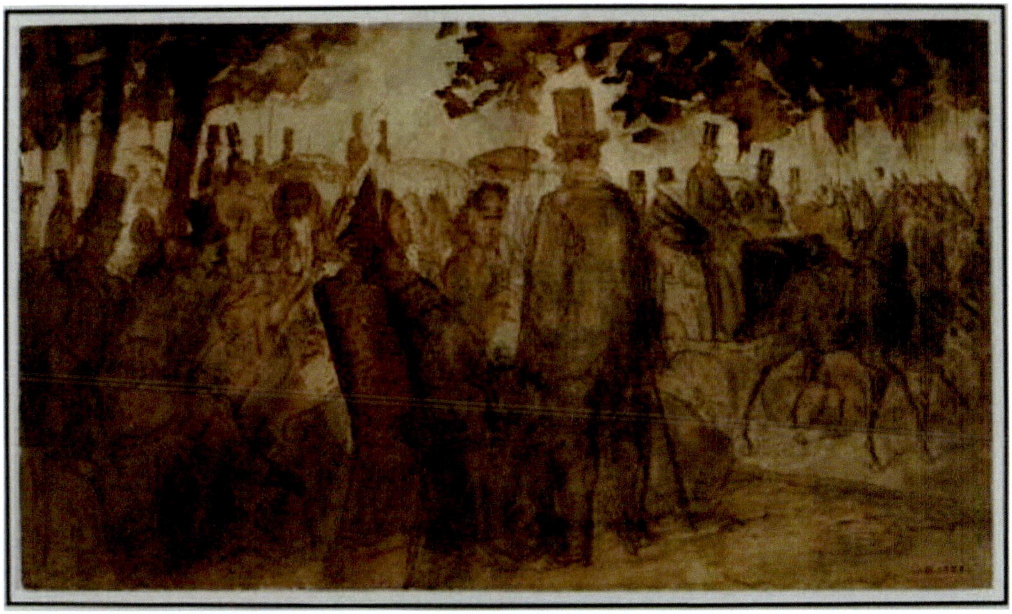

Figure 1.3 Constantin Guys, *L'Avenue des Champs-Elysées, Voitures et Promeneurs* (Avenue Champs-Elysées, Carriages and Strollers), 1852–60, watercolor, wash, and ink on wove paper, 9 x 16 1/4 in. Paris Musées/Musée Carnavalet, Histoire de Paris.

When Baudelaire wrote "Of the Heroism of Modern Life" in 1846, Paris was marked by a densely overpopulated urban fabric and disparate urban renewal sites. In this context, Baudelaire first defined modern beauty as possessing "something eternal and something transitory" and extolled Paris as a site of creative and cultural renovation. As Baudelaire wrote, the artist discovers the "heroism of modern life" in "the scenes of high life and of the thousands of uprooted lives that haunt the underworld of a great city, criminals and prostitutes; ...we have only to open our eyes to see and know the heroism of our day."[7]

Subsequently, in "The Painter of Modern Life" Baudelaire assigned the French painter and illustrator Constantin Guys the dual attributes of the artist: a man of the crowd and a man distant from it (Figures 1.3 and 1.4). In the first instance, the artist is a:

> passionate observer, [for whom] it becomes an immense source of enjoyment to establish his dwelling in the throng, in the ebb and flow, the bustle, the fleeting and the infinite... [He] is an ego athirst for the non-ego, and reflecting it at every moment in energies, more vivid than life itself, always inconstant and fleeting.[8]

Accordingly, in his review of the 1855 Universal Exposition, Baudelaire redefined the "double composition" of beauty as a constant tension between "an eternal and invariable element," and "a relative circumstantial element [of] contemporaneity." He does so by rendering the familiar unfamiliar and by presenting "the ever-new which eternally elud[e]s," but still refers to, "the rules and analyses of the school."[9] But to achieve this end, the artist must first experience a process of "self-doubling," which for Baudelaire characterized the artistic creative process. As noted in the Introduction, I translate

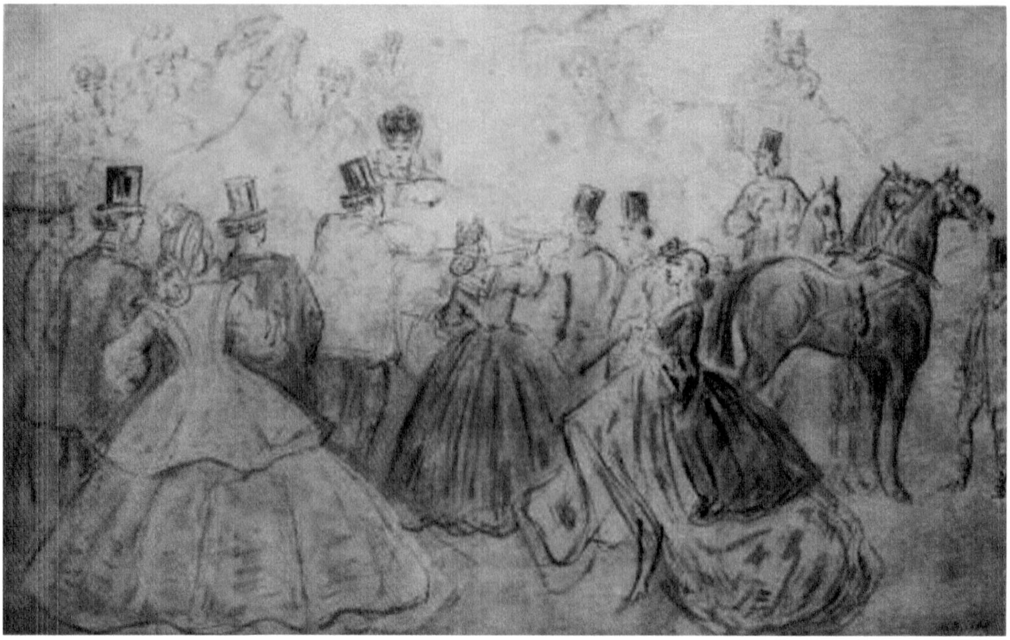

Figure 1.4 Constantin Guys, *Promenade aux Champ de Courses* (Promenade to the Racetrack), 1852, watercolor, wash, and ink on paper, 11 1/2 x 18 3/8 in. Paris Musées/Musée Carnavalet, Histoire de Paris.

"*se dédoubler*" as "self-doubling" per Paul de Man's translation as "rapid self-doubling" ("la force de se dédoubler rapidement") in his analysis of Baudelaire's concept of irony.[10] Baudelaire's description of Guys' activities is a testimony of this self-doubling creative process.

A man of the crowd from dawn to dusk, the artist retreats from the crowd at night. Alone in his studio, he reflects upon his sensory experiences and renders their essence: "All the materials, stored higgledy-piggledy by memory, are classified, ordered, harmonized, and undergo that deliberate idealization, which is the product of a childlike perceptiveness that is acute and magical by its very ingeniousness."[11] In Baudelaire's assessment of Guys' modernity, the sketch-like pen-and-ink drawings and the transparency of watercolor washes convey fleeting cognitive experiences of the dynamic cityscape and physical encounters with the myriad urban populace. As seen in Guys' *L'Avenue des Champs-Elysées, Voitures et Promeneurs* (Avenue Champs-Elysées, Carriages and Strollers) (1852–60), standing and seated fashionably dressed figures and horse-drawn carriages are arranged in parallel layers across the horizontal picture planes, intermingling frontal and background spaces and spaces in between. Like Baudelaire, viewers behold embodied fleetingness conveyed by the contraposed figures and striding horse limbs that combine with the dynamic marks of brushwork and pen-and-ink lines and washes. Similar techniques animate the figures who stride along the racecourse in Guys' *Promenade aux Champ de Courses* (Promenade to the Racetrack) (1852). Here, however, the fashionable attire of fabrics, ruffles, and flairs dominate the picture field and define the fleetingness of modern life.

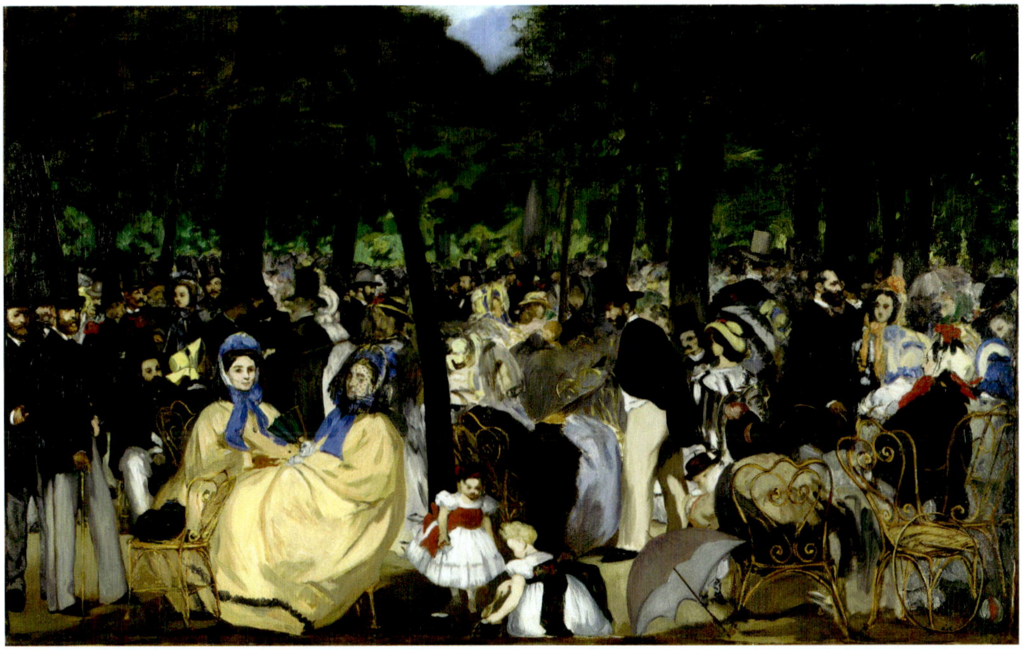

Figure 1.5 Édouard Manet, *Music in the Tuileries*, 1862, oil on canvas, 30 x 46 in. Sir Hugh Lane Bequest, 1917, The National Gallery, London. In partnership with Hugh Lane Gallery, Dublin.

Baudelaire's articulation of the affective experience of modernity is central to his aesthetic mandate and to this study. In "The Painter of Modern Life," Baudelaire further asserts that the artist must submerge himself in the urban crowd and embody its vital essence: "The crowd is his domain, [he] establish[es] his dwelling in the throng, in the ebb and flow, the bustle, the fleeting and the infinite."[12] As the embodiment of the crowd, Baudelaire explains, the artist "may also be compared to [...] a kaleidoscope endowed with consciousness, which with every one of its movements presents a pattern of life, in all its multiplicity."[13] Finally, Baudelaire suggests how the artist can transmit his lived experience to viewers through an artistic technique that is itself fleeting and fragmented: such a technique must "reflect [the artist's consciousness] at every moment in energies, more vivid than life itself, always inconstant and fleeting."[14] Paintings such as Manet's *La Musique aux Tuileries* (*Music in the Tuileries*) (1862), Degas' *Place de la Concorde* (or *Viscount Lepic and his Daughters Crossing the Place de la Concorde*) (1876), and Caillebotte's *Au Café* (*At the Cafe*) (1880) exemplify subsequent artistic responses to Baudelaire's mandate (Figures 1.5, 1.6, and 1.7).

Here we might ask, just what were the relationships between Baudelaire and his followers? On the one hand, it is well known that Manet and Baudelaire were close friends, even though the latter wrote little about the former. Personal letters were exchanged between the two, and they were known to frequently meet in person (I treat their friendship in Chapter 9). Degas was more indirectly influenced by Baudelaire. He and Manet had a sporadic, sometimes hostile, friendship.[15] They circulated among and often met with mutual artist and writer friends, for whom Manet was a leader and advocate of

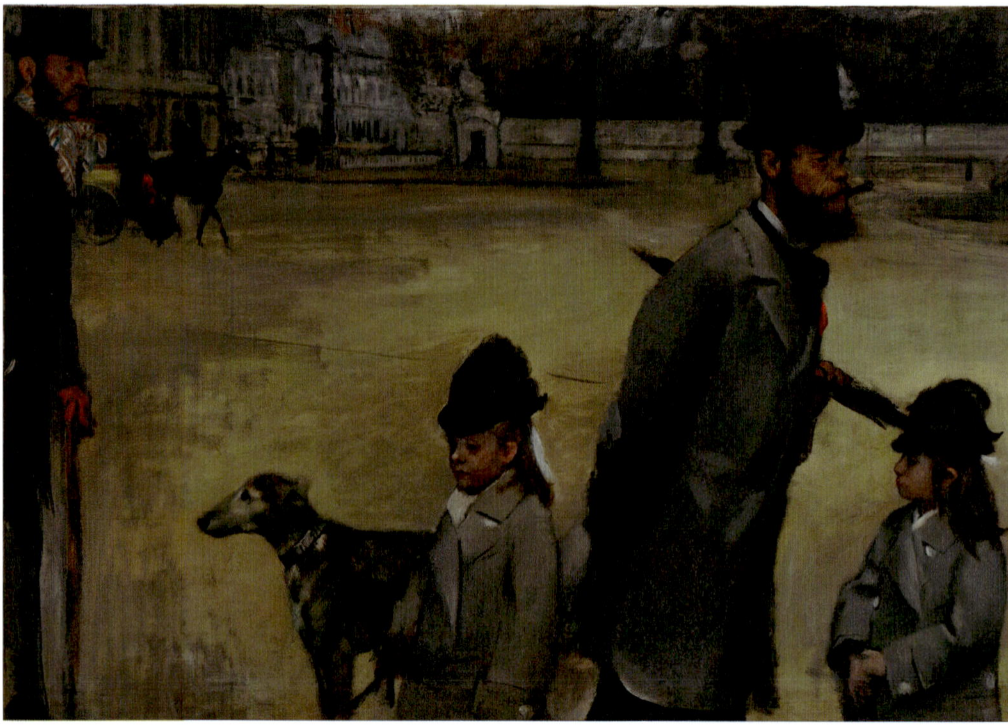

Figure 1.6 Edgar Degas, *Place de la Concorde*, or *Viscount Lepic and his Daughters Crossing the Place de la Concorde*, 1876, oil on canvas, 31 x 46 in. Hermitage Museum, Saint Petersburg. Image: Wikimedia Commons.

Baudelaire's "The Painter of Modern Life."[16] Caillebotte also frequented these meetings, often held in the members' homes and studios.[17]

Manet's *Music in the Tuileries* signals his adherence to Baudelaire's mandate for capturing the transitoriness of modern life and signaled a departure point for his Baudelairean artist-friends. In this painting the viewers' gaze is doubly fleeting. They could embody not only the contemporary figures' conversant and animated poses; the viewers' perspective is also that of a passerby. Manet, moreover, parodied conventional history paintings, such as Thomas Couture's *Romans of the Decadence* (1847; Musée d'Orsay, Paris).[18] He replaced historical figures with contemporary ones and eliminated traditional group paintings' compositional order, spatial perspective, and dramatic focus. We are confronted with these inversions in the loosely articulated mass at the center of the canvas, which foregrounds the overall painterly surface instead of the conventional smooth finish and tonal gradation.

Viewers of Degas' urban painting are equally embodied in the figures' transience and the spectators' fleeting view of the scene. In *Place de la Concorde*, the figures' movements are signified by the opposing directions of their postures and gazes. The relation between figure and field likewise indicates the inversion of conventional narrative and compositional devices. The actual Place de la Concorde appears stripped of its monuments; thus, the spatial void functions as a neutral background against which fragmented, chance encounters occur.

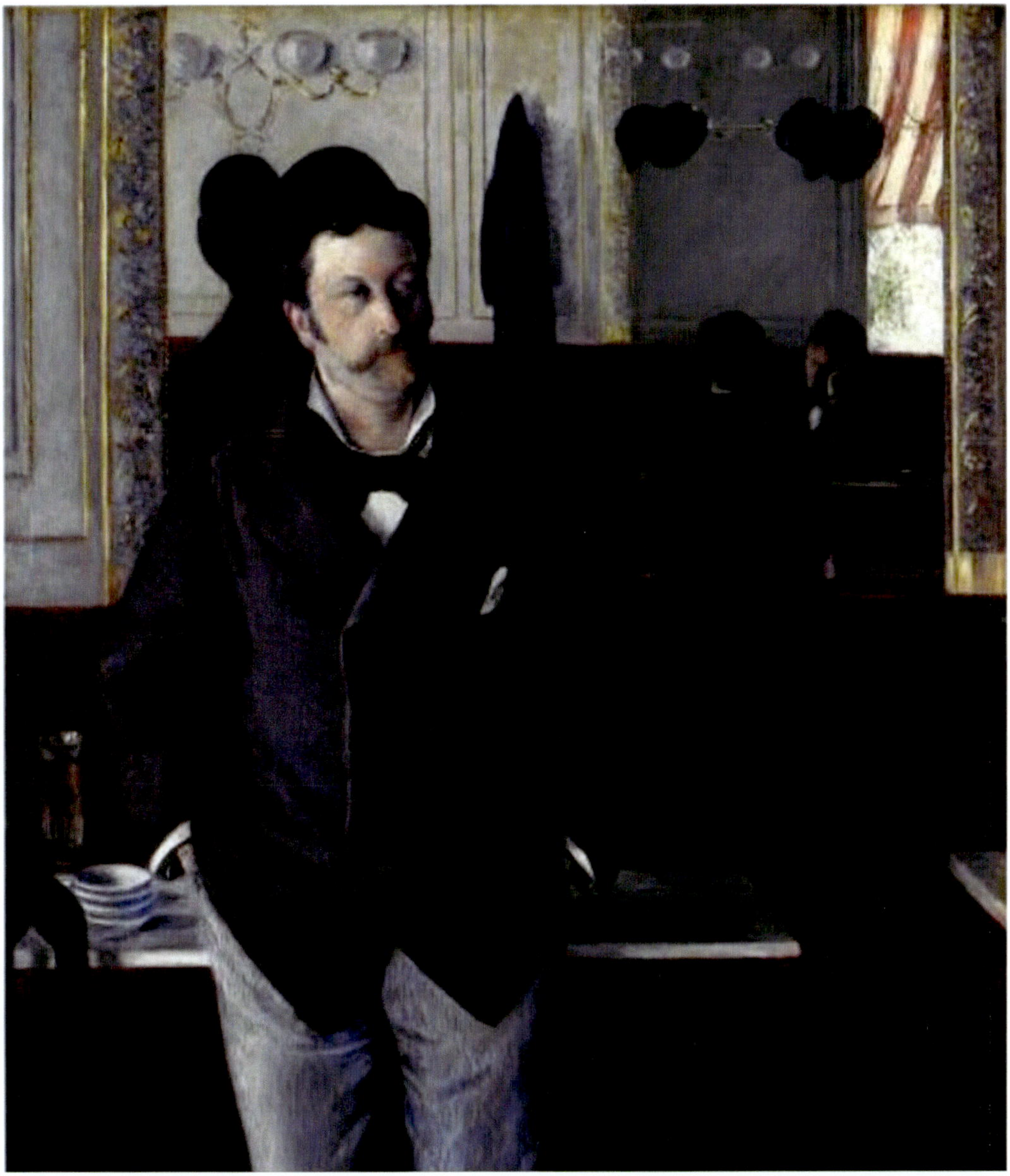

Figure 1.7 Gustave Caillebotte, *At the Café*, 1880, oil on canvas, 61 x 45 1/4 in. Dépôt de l'Etat, 1946, Inv. D.1946. Rouen, Musée des Beaux-Arts © C. Lancien, C. Loisel/Réunion des Musées Métropolitains Rouen Normandie.

In a related fashion, Caillebotte's *At the Cafe* subverts viewers' expectations of illusionistic mirror reflections.[19] In this painting, there is not one mirror, but two: one behind the solitary figure, reflecting the men across the room, and another mirror behind these two men (as indicated by the reflection of their hats hung on brass hat-racks above their heads). This second mirror reflects, in turn, an actual window located on the wall opposite the card players; the actual window is on the wall with the mirror behind the central figure. This play with conventional perspectival elements follows Baudelaire's mandate for "shock and surprise." Here, the interior and exterior collapse, forcing our uneasy confrontation with enigmatic figures and labyrinthine spaces.

Taken together, these kinds of paintings compel ironic encounters. First, viewers engage with the parodic subversion of their aesthetic expectations, necessitating a reconciliation of old conventions with innovative transformations. Such subversions cause rupture and defamiliarization.[20] In each case, these effects are inscribed in the painting's fabric and embodied in viewers' experiences. Consequently, viewers inhabit an affective, ambiguous space of self-reflection. They may feel agitated by the fragmentation of pictorial surface and space and the absence of narrative closure. Alternatively, viewers may positively engage with rupture and defamiliarization to reconcile the opposition and, thereby, become one with cultural renewal. Both responses can also be traced to Baudelaire's alternately positive and negative responses to modernity and their reconciliation in irony's self-reflective gaze.

Historicizing the Ironic Gaze: Mirroring Interiority

Baudelaire's ironic gaze and its persistence in Baudelaire's modernity align his self-doubling tactics with earlier nineteenth-century German Romantic irony.[21] "The Painter of Modern Life" exemplifies how artistic self-doubling correlates with an internal mirror reflection, a topos that is consonant with the German Romantics' ironic gaze.

For the Romantic ironists, irony consists of an active state of critical self-reflection upon the constant contradictions of the finite and the infinite, the self and the non-self, and the past and present. This ironic state likewise entails a self-induced mirroring effect: a subject's recognition of a double self that leads to spiritual growth and, ultimately, cultural progress. As Novalis, the German Romantic polymath, explained, irony is "the power to be *beside himself*, to be conscious and at the same time beyond [one's] senses"[22] (original emphasis). Baudelaire's notion of the artist as a "kaleidoscope with a conscience" embodies this dualism of the self and other. For the philosopher Friedrich Schlegel, irony arises from "the indissoluble antagonism between the absolute and the relative."[23] This dialectic resounds in Baudelaire's definition of modern beauty as joining the eternal and the relative. For the German theorist Karl Solger, irony consists of the co-presence of opposites and contradictions "to be grasped from both sides at the same time."[24] Such simultaneity of opposites is what Baudelaire attributed to viewers' "shock of surprise," as well as to the comic.

The Romantic ironists also transposed the artist's ironic condition to viewers to ensure that subjective transformation translates into the social sphere. They held that ironic oppositions are irresolvable yet reconcilable because irony provides a unifying perspective between the work of art and its reception. For Schlegel, irony "hovers over the whole of the work,"[25] and his analogous discourse on "transcendental poetry" embraces an aesthetic of irony as "the artistic reflection and beautiful self-mirroring" in a work of art.[26]

As Baudelaire characterized Guys' self-doubling during the creative process, the painter of modern life first loses himself in the external subject of representation and then detaches himself from this communion to formulate its visual representation. In prescribing this self-doubling, Baudelaire echoed the precept of Ludwig Tieck, the German Romantic poet, that "[i]rony is the power that constitutes a poet's [qua-artist's] control over his material."[27] Further, Tieck adds, "the poet [artist] should not lose himself in his work but remain above it. Irony protects him from one-sidedness and empty idealizing."[28] Likewise, Baudelaire's idea of this double phase of the creative act (absorption and detachment) is predicated upon the artist's ontological self-doubling achieved before artistic production. In these philosophical terms, Baudelaire defined "*se dédoubler*" as the (ironic) condition inherent in all great artists and their art.

Like the Romantic ironists before him, Baudelaire also defined the artist's strategy for self-detachment or "*dédoublement*" as a method of self-criticism cloaked in the comic.[29] Similar to the earlier model, the comic is presented as an aesthetic means for gaining metaphysical insights. In his essay "De L'essence du Rire et Généralement du Comique Dans les Arts Plastiques" ("Of the Essence of Laughter and Generally of Comedy in the Plastic Arts"), written in 1855, Baudelaire situates laughter as humanity's response to its dualistic condition:

> [S]ince laughter is essentially human it is essentially contradictory, that is to say that it is at one and the same time a sign of infinite greatness and of infinite wretchedness in relation to the absolute being, of whom man has an inkling, infinite greatness in relation to the beasts. It is from the constant clash of these two infinites that laughter flows.[30]

For Baudelaire, comic art is the form through which laughter is released because the comic mirrors humanity's dualism and activates its contradiction:

> [I]n order for the comic, in other words, an emanation, an explosion, an emergence of the comic, to exist, there must be two in the presence of each other: that it is particularly in the laugher, the spectator, that the sense of the comic resides.[31]

To this matrix of human contradictions, Baudelaire adds the distinction between "the significative comic" or "ordinary comic" and its "antithesis," "the absolute comic." The "laugher," as defined above, partakes in the former manifestation of the comic. As Baudelaire explains, "[T]he significative comic speaks a language that is clearer, easier for the common man to understand, and above all easier to analyse, its element being visibly double: art and the moral idea." Conversely, the absolute comic is of a higher order in form and effect: "[T]he absolute comic, coming much closer to nature, appears as a unity that must be grasped intuitively."[32] Not only does the absolute comic activate the spectator/reader's higher faculty of intuition, but it also reflects the artist's subjective intelligence:

> Artists create the comic; having studied and brought together the elements of comic, they know that such and such a creature is comic, and that he is comic only on condition that he is unaware of his own nature, just as, by an inverse law the artist is an artist only on condition that he is dual and that he is ignorant of none of the phenomenon of his dual nature.[33]

Thus, in "Of the Essence of Laughter," the comic attains its noble status because it issues from philosophical irony.[34] As Baudelaire states, the "comic sensibility" belongs to "all artistic phenomena which indicate the existence of a permanent dualism in the human being; that is, the power of being oneself and someone else at one and the same time." In this statement, "comic sensibility" can easily be replaced with "ironic self-reflection." However, Baudelaire's distinction between the ordinary comic and the absolute comic prevails—while the former ranks as banal, the absolute comic approaches irony's supremacy. This distinction between the banal and the sage resonates with Schlegel's similar distinctions when discussing Socratic irony. In fragment 108 of *Lyceum [Critical] Fragments*, Schlegel catalogs the multiple dualities between irrational and rational processes and parodic outcomes that Socratic irony entails:

> It contains and incites a feeling of the insoluble conflict of the absolute and the relative, of the impossibility and necessity of total communication. It is the freest of all liberties, for it enables us to rise above our own self; and still the most legitimate, for it is absolutely necessary. It is a good sign if the harmonious dullards fail to understand this constant self-parody, if over and over again they believe and disbelieve until they become giddy and consider jest to be seriousness, and seriousness to be jest.[35]

In Baudelaire's scheme, the absolute comic is denied irony's status. This is so because, whereas the (outward) comic event induces laughter in the spectator/reader, ironic self-reflection (an internal event) precludes outward laughter. As Baudelaire suggests, only the philosopher can laugh at himself, thanks to his ironic self-reflection:

> The comic, the power of laughter, is in the laugher, not at all in the object of laughter. It is not the man who falls down that laughs at his own fall, unless he is a philosopher, a man who has acquired, by force of habit, a power of getting outside himself quickly and watching [*se dédoubler*], as a disinterested spectator, the phenomena of his ego. But such cases of that sort are rare.[36]

Baudelaire's self-reflective philosopher performs according to Schlegel's distinction of the ironic as a higher form of cultural practice, specifically in philosophical poetry. Poetry becomes philosophical and transcendent when irony hovers over the entire work to exceed the specificity of rhetorical devices. Schlegel posits these distinctions in fragment 42 of *Lyceum [Critical] Fragments*:

> Philosophy is the real homeland of irony, which one would like to define as logical beauty: for wherever philosophy appears in oral or written dialogues—and is not simply confined into rigid systems—there irony should be asked for and provided.

For Schlegel it follows that ironical (transcendent) poetry must be distinguished from ironical (buffoonish) performativity:

> Only poetry can also reach the heights of philosophy [...], and only poetry does not restrict itself to isolated ironical passages, as rhetoric does. There are ancient and modern poems that are pervaded by the divine breath of irony throughout and informed by a truly transcendental buffoonery. Internally: the mood that surveys everything and rises infinitely above all limitations...; externally, in its execution: the mimic style of an averagely gifted Italian *buffo*.[37]

Given the rarity of philosophical self-doubling and its ironic effects, Baudelaire attributes this "power" to those he regards as "true artists." In "Of the Essence of Laughter," Baudelaire identifies E. T. A. Hoffmann as the philosopher-artist *par excellence*.[38] In this context, Baudelaire ascribes to Hoffmann the invention of the absolute comic. Evoking Schlegel's esteem for transcendent poetry and ironic poets, Baudelaire lauds Hoffmann's transmission of his metaphysical insights through the comic characters of his tales.

> The exalted nature of the absolute comic makes of it the special preserve of superior artists who have in them the necessary receptivity for absolute ideas. Thus, the man who until now has apprehended these ideas better than anyone else, and who has exploited some of them in his purely aesthetic, as well as his creative work, is Theodore Hoffmann. He was always careful to distinguish between ordinary comics and the comic he called 'innocent comics.' He often sought to express in works of art the learned theories he propounded didactically or given expression to in the form of inspired conversations or of critical dialogues.[39]

For Baudelaire, Hoffmann's invention of the absolute comic issued from the writer's self-aware doubleness that permeates the entire work (or, in Schlegel's words, "the mood that surveys everything and rises infinitely above all limitations"):

> I submit that when Hoffmann engenders absolute comic he is surely aware of the fact; but, equally, he also knows that the essence of this type of the comic is to appear to be unaware of oneself and to instill in the spectator, or rather the reader, a feeling of joy at his own superiority and the joy of man's superiority over nature.[40]

Baudelaire also suggests how irony's doubleness is mirrored "internally" in the formal structure of Hoffmann's tales. He observes, "What particularly distinguishes Hoffmann is his involuntary, and sometimes very voluntary mixture of a certain element of the significative comic with the most absolute comic."[41] Baudelaire also recognizes how Hoffmann's blending of the ordinary and absolute produces a drunken giddiness, akin to what Schlegel ascribed to Socratic irony:[42]

> [Hoffmann's] most supernatural fleeting comic conceptions, which often resemble the visions of a drunken man, have the very evident moral implications; it is as though we were dealing with the profoundest type of physiologists or psychiatrists, who was amusing himself by clothing his abstruse science in poetic forms, like a scholar speaking in fables and parables.[43]

Given the ironic lens through which Baudelaire views Hoffmann (and subsequently Guys), it is safe to assume that Hoffmann was Baudelaire's point of entry to the German Romantic Ironists.[44] Hoffmann's own assimilation of Schlegel and German Ironists has been widely documented. As Maria Tatar observes, "Of all the German Romantics, E. T. A. Hoffmann perhaps most successfully translated the theoretical pronouncements of his colleagues into literary practice. He persistently asserted his right to exercise the power to confuse[.]"[45] Based on this evidence, it is likely that once Baudelaire discovered Hoffmann's tales, he was eager to read, in his words, "[the] learned theories which [Hoffmann] had put forth didactically, or thrown out in the form of inspired conversations or critical dialogues[.]"[46] If this was the case, when Baudelaire wrote "Of the Essence of Laughter," he had already assimilated the essential tenets of German Romantic irony. Herein, irony provided the poet-critic with an ontological schema, a critical methodology, and a creative strategy.

Four years later, Baudelaire redefined the comic-ironic as the ironic—pure and simple—constituting the visceral and psychological essence of the painter of modern life. In his essay of the same name, Constantin Guys merits the status of a "philosopher-artist." Here, Baudelaire asserts that he would designate Guys as a philosopher, but for the artist's said resistance to metaphysical thinking. Given this resistance, Baudelaire struggles to find an appellation that fits his artist-genius. First, however, he rejects "the dandy" because of Guys' double nature:

> The dandy is blasé, or affects to be, as a matter of policy and class attitude. M. G. [Monsieur Guys] hates blasé people. Sophisticated minds will understand me when I say that he possesses that difficult art of being sincere without being ridiculous. I would confer on him the title of philosopher, to which he has a right for more than one reason; but his excessive love of visible, tangible things, in their most plastic form, inspires him with a certain dislike of those things that go to make up the intangible kingdom of the metaphysical. Let us therefore reduce him to the status of the pure pictorial moralist, like La Bruyère.[47]

Having thus characterized both Hoffmann and Guys as "moralists," Baudelaire made the ironic gaze equivalent to the artist's moral ethos for portraying the two-fold "heroism of modern life" in "the scenes of high life and of the thousands of uprooted lives that haunt the underworld of a great city, criminals and prostitutes[.]"[48] Herein lies Baudelaire's self-reflection on modernity's double nature: both the cause for celebration and cultural renewal and the cause of fear, anxiety, and a turning inward.

Modernity and the Threat of Madness

While Baudelaire's "The Painter of Modern Life" and the ironic behavior it entails can be viewed as a celebratory mandate for cultural renewal, a more troubling picture of modernity unfolds when it is read alongside a concurrent text, "Le Poéme du Haschisch" ("The Poem of Hashish" in *Artificial Paradises* [1860]).[49] Baudelaire's disdain for hashish as the progenitor of madness can be correlated with his fear of modernity. In this context, for Baudelaire the modern city was a source of sensory and psychic assaults, constituting an equally dangerous agent of madness (in texts such as "The Swan" and "Lost Halo," which I will discuss in later chapters). To understand this fear and its impact on his art criticism, it is helpful to examine Baudelaire's references to the nervous system in both "The Painter of Modern Life" and "The Poem of Hashish" in the context of nineteenth-century medical science and the writer's engagement with that discourse. Baudelaire's references, in my view, serve as signals of his attempt to transcribe the visceral effects of modernity, to prescribe artistic techniques to mediate these effects, and to ascribe to painting a retreat from the madness that modernity incurs.

Baudelaire's familiarity with medical knowledge about the brain and nervous system issues from two medical scientists whose theories are echoed in his writing: Johannes Müller (1801–58) and Jacques-Joseph Moreau (1804–84).[50] Müller's textbook on neurophysiology, *Handbuch der Physiologie des Menschen*, published between 1833 and 1840, was immensely influential.[51] Among its numerous translations are five French editions published between 1840 and 1851 (as *Physiologie du Système Nerveux*; *Elements of Physiology*). Müller's importance during both Baudelaire's time and after is that "his research led to the first comprehensive scientific conception of the nervous system as a unit."[52] It was likely through Baudelaire's physician friend, Dr. Moreau, that Müller's ideas were conveyed to the writer, and Moreau is important in his own right for his treatise on the experimental use of hashish, *Du Haschisch et de L'aliénation Mentale: Études Psychologiques* (1845; *Hashish and Mental Illness*).[53]

In charting Baudelaire's references to brain science, I have identified three topoi that he adapts from these sources: the dynamic physiological duality of external and internal stimuli, the force of nervous energy that mediates sensory stimuli and neuronal activity, and their harmful effects on the nervous system, such as shock and drug-induced hallucinations.

In "The Painter of Modern Life," Baudelaire explicitly references the nervous system; from this point, implicit references unfold. Specifically, he compares Guys' visceral impressions of modern life to the convalescent's sensual enjoyment in "the sight of the passing crowd" and delight in "breath[ing] in... all the spores and odors of life."[54] In this light, Baudelaire declares "that inspiration has some connection with congestion, that every sublime thought is accompanied by a more or less vigorous nervous *impulse that reverberates in the cerebral cortex*"[55] (emphasis added). Indeed, we now understand the cerebral cortex to be the brain's "outermost layer of nerve cell tissue.... [It] is responsible for the higher-level processes of the human brain, including language, memory, reasoning, thought, learning, decision-making, emotion, intelligence and personality."[56] Also noteworthy is Baudelaire's reference to "congestion" concerning the convalescent who has "returned from the shades of death."[57]

Baudelaire's understanding of the dualistic physiology of sensory stimuli and brain activity and their vibratory neural patterns resonates with Müller's theories. Equally important for the writer's appreciation of the neural system was Müller's theory of "the phenomenon of reflection" and its corollary of involuntary and voluntary neural activity. Accordingly, the phenomenon of reflection involved "the involuntary transition—occurring in the spinal cord or brain—of excitation from the centripetally conducting nerves to the centrifugally conducting ones."[58] In Müller's words: "Certain muscles [...], which are quite independent of the influence of the will, are nevertheless influenced by particular states of the mind; and [...] the nerves have as great an influence over the involuntary as over the voluntary movements[.]"[59] While external conditions induce involuntary activity in the nervous system, voluntary neural activity occurs once the brain registers the external input and activates both psychic and motor neural responses. Comparing the voluntary nerves to a piano keyboard, "Müller argued that once the thought strikes the brain 'key,' it gives rise to 'currents or vibrations' in outgoing nerves, which then cause muscular motions."[60] As Müller put it:

> The primitive fibres of all the voluntary nerves being at their central extremity all spread out in the brain to receive the influence of the will, we may compare them, as they lie side by side in the organ of the mind, to the keys of a pianoforte, on which our thoughts play or strike, and thus give rise to currents or vibrations of the nervous principle in a certain number of primitive nervous fibres, and consequently to motions.[61]

Baudelaire also seems receptive to Müller's idea that the changing conditions of external phenomena render the human body a site of perpetually changing "energies." According to Müller:

> *The immediate objects of the perception of our senses are merely particular states induced in the nerves and felt as sensations.... [B]ut inasmuch as the nerves... are material bodies... they make known to the sensorium, by virtue of changes produced*

in them by external causes, not merely their own condition, but also properties and changes of condition of external bodies. The information thus obtained by the senses concerning external nature, varies in each sense, having a relation to the qualities or energies of the nerve.[62]

(original emphasis)

As we have seen, the concept of neuronal vibrations figures in Baudelaire's description of the artist's visceral response to the city. However, for Baudelaire, the involuntary transitions that occur in the spinal cord or brain and "give rise to symptoms of irritation or excitation" were a source of psychic anxiety, confusion, and rupture. Baudelaire articulates this physiological event in the prose poem "Le Confiteor de L'artiste" ("The Confiteor of the Artist")[63] published in *Le Spleen de Paris* (1857–67).[64] Here, Baudelaire identifies how external stimuli, if left unchecked, may cause painful neural "vibrations." Accordingly, the poet describes his embodied response to "the end of an autumn day," wherein "the immensity of sky and sea," "the chasteness of the azure heavens," and "the monotonous melody of the ocean sea" cause "the ego" to lose itself "in the vastness of revery." While the reader is lured into thinking that the artist celebrates nature's external stimuli that produce sublime thoughts, Baudelaire rails against their harmful effect on the nervous system:

Yet, these thoughts, whether they come out of me or surge forth from things, quickly become too intense. When energy combines with sensual delight, it causes discomfort and a distinct suffering. My too strained nerves no longer produce anything but clamorous and painful vibrations.[65]

While nature is the source of the poet's suffering in "The Confiteor of the Artist," the clarity with which Baudelaire describes the sensory assaults on the "ego" provides a template for identifying the involuntary excitation of the nervous system during the artist's struggle against sensory assaults in the modern city. In "The Painter of Modern Life," as we will see, this template becomes more nuanced as the city becomes a source of psychic rupture and defamiliarization. To do this, Baudelaire invokes Moreau's analogy between hashish-induced hallucinations and madness.

Baudelaire befriended Moreau as a member of the Hashish Club (active from 1844 to 1849). At the time, the doctor continued experimenting with hashish as a means of understanding madness based on his own hallucinatory experiences and those of the Club members. Thus, Baudelaire knew the physiological effects of hashish and Moreau's theories of madness firsthand.[66] Equally significant, Moreau likely transmitted his understanding of Müller's models to Baudelaire, especially humankind's dualistic physiology, the brain's connection to the central nervous system, and the vital energies that manifest in neurological anatomy. Moreau embedded these tenets in his 1845 treatise *Hashish and Mental Illness*[67] summarizing them in the second chapter titled "Physiology":

So it seems that two modes of mental life are known to man. The first results from our relationship with the external world, with that great whole known as the universe; we share it with the creatures that resemble us. The second is only the reflection of the first, and feeds only, in a sense, on materials supplied it by the first, while remaining completely separate.[68]

In the same chapter, Moreau recast Müller's phenomenon of reflection within his thesis on the correlation between hashish and mental illness:

> Among our many mental faculties there is only one with which we can study the impact of these very faculties on our own physiological condition: it is reflection, the power of the mind to observe itself. It is a mirror of sorts in which we can contemplate ourselves and detect our most intimate movements.[69]

Moreau added, "This power fails us when our faculties are troubled," whether affected by madness or hashish. Significantly, Moreau's physiological tenet regarding the mind's mirror reflection would uphold Baudelaire's ironic ethos.

Specifically, the importance of *Hashish and Mental Illness* for Baudelaire resides in Moreau's narrative descriptions of the effects of hashish in both his case studies and in staging the progressive effects of hashish in the imbiber. Moreau aptly summarizes his hypothesis regarding hashish and madness in the book's first chapter:

> There is not a single, elementary manifestation of [madness] that cannot be found in the mental changes caused by hashish, from simple manic excitement to frenzied delirium, from the feeblest impulse, the simplest fixation, to the merest injury to the senses, to the most irresistible drive, the wildest delirium, the most varied disorders of feelings.[70]

As in madness, so hashish-induced hallucinations: "*It is manic excitement*," Moreau asserts, that joins the two states of mind-body synchronicity (original emphasis). In each case, manic subjects have lost control over their rational faculties:

> How can we accurately describe that simple yet complex state of uncertainty, oscillation, and flight of ideas often accompanied by a profound incoherence? *It is a disintegration, a veritable dissolution of that mental structure known as the mental faculties.*[71]
>
> (emphasis added)

In the third chapter, "Psychological Phenomena," Moreau enumerates the eight stages or "phenomena" that manifest as the increasingly debilitating effects of hashish on the nervous system. The stages begin with ingestion and emotive elation and end with the subjection of all faculties to the involuntary neural activity of hallucinations. In this progression, hashish acts most injuriously on the imbiber's will: "The action of hashish weakens the will—the mental power that rules ideas and associates and connects them together. Memory and imagination become dominant."[72] Moreau later notes:

> Before it acts to isolate completely external impressions from the outside world, the action of hashish, working on all the faculties at once, is characterized… by, for example, *a surplus of mental energy*, a brilliance of memories, and a rapid thought process. Imperceptibly, this produces in the will, in all the instincts, such a relaxation that we become subject to a variety of impressions. The sequence of our thoughts can be broken by the slightest interference; we submit to the most contradictory influences. *We go in every direction.*[73]
>
> (emphasis added)

In the final stage, the imbiber becomes the "hallucinator" who "hears his own thoughts, as he sees and hears the creations of his imagination and as he is moved by impressions he finds in his memory." Accordingly, hallucinations are but a manifestation of unending mental reflections—an embodiment of mirrors:

> The primary origin, the basic cause of hallucinations is invariable and identical with itself. It lies in a special state of the intellectual organ. When this state develops, a multitude of objective and subjective physical and mental factors can become the purely fortuitous and indirect cause of hallucinations. Without this special state of the intellectual organ, these agents are powerless.
>
> Do we need to name this essential cause? It is indeed the cause of all the morbid phenomena of the mind, "excitement."[74]

Given that drug-induced hallucinations mimic madness, Baudelaire wrote "Du Vin et du Haschisch" ("On Wine and Hashish") in 1851 and then "The Poem of Hashish" in 1858 to caution the artist against using hashish as a stimulant for creative production. In these essays, Baudelaire maintains Moreau's analogy between hashish intoxication and madness.[75] Although he does not name Moreau, Baudelaire also adapts the doctor's thesis on the eight-stage progression from ingestion to hallucinations and imitates the doctor's detailed accounts of his own and others' intoxications with hashish. Note that "The Poem of Hashish" is a much more elaborate version of the section on hashish in "On Wine and Hashish." This quantitative difference notwithstanding, each essay is a testament to the imbibers' loss of control over their senses, mobility, and mental reflection and a warning against the resultant waste of "nervous energy." Baudelaire's *Autoportrait Sous L'influence du Haschisch* (*Self Portrait Under the Influence of Hashish*) (ca. 1844) records the writer's discordant sensations (Figure 1.8).

In this ink pen and wash rendering, Baudelaire depicts himself immobilized by the clamorous Parisian cityscape awash in daybreak's twilight. Baudelaire's physique embodies the duality of this transitional moment; his body is both a conduit of radiance and a vessel of assault. Behind his figure, the city lights cast bands of light and dark auras, while this nebulosity is interrupted by a lightning bolt striking from above and the spiked sunrays rising from below. Before the figure, brushstrokes and pen marks threaten to slash through him if he advances. This persistent dualism mirrors the exaltation and madness induced by both modernity and hashish.

In "The Poem of Hashish," Baudelaire summarizes Moreau's account of hashish's progressively debilitating effects in three stages.[76] In Baudelaire's telling, emotional elation and what Moreau called "excitement, dissociation of ideas" (as the second phenomenon)[77] constitute the first stage of hashish effects:

> At first it is a certain hilarity, absurdly irresistible, which possesses you. These excesses of gaiety, without due cause, of which you are almost ashamed, frequently occur and divide the intervals of stupor, during which you seek in vain to pull yourself together. [...] Incongruous likenesses and correspondences, impossible to foresee, interminable puns, comic sketches, spout eternally from your brain.

This combination of gaiety and verbal confusion mimic madness:

> This gaiety, turn by turn idle or acute, this uneasiness in joy, this insecurity, this indecision, last, as a rule, but a very short time. Soon the meanings of ideas become so

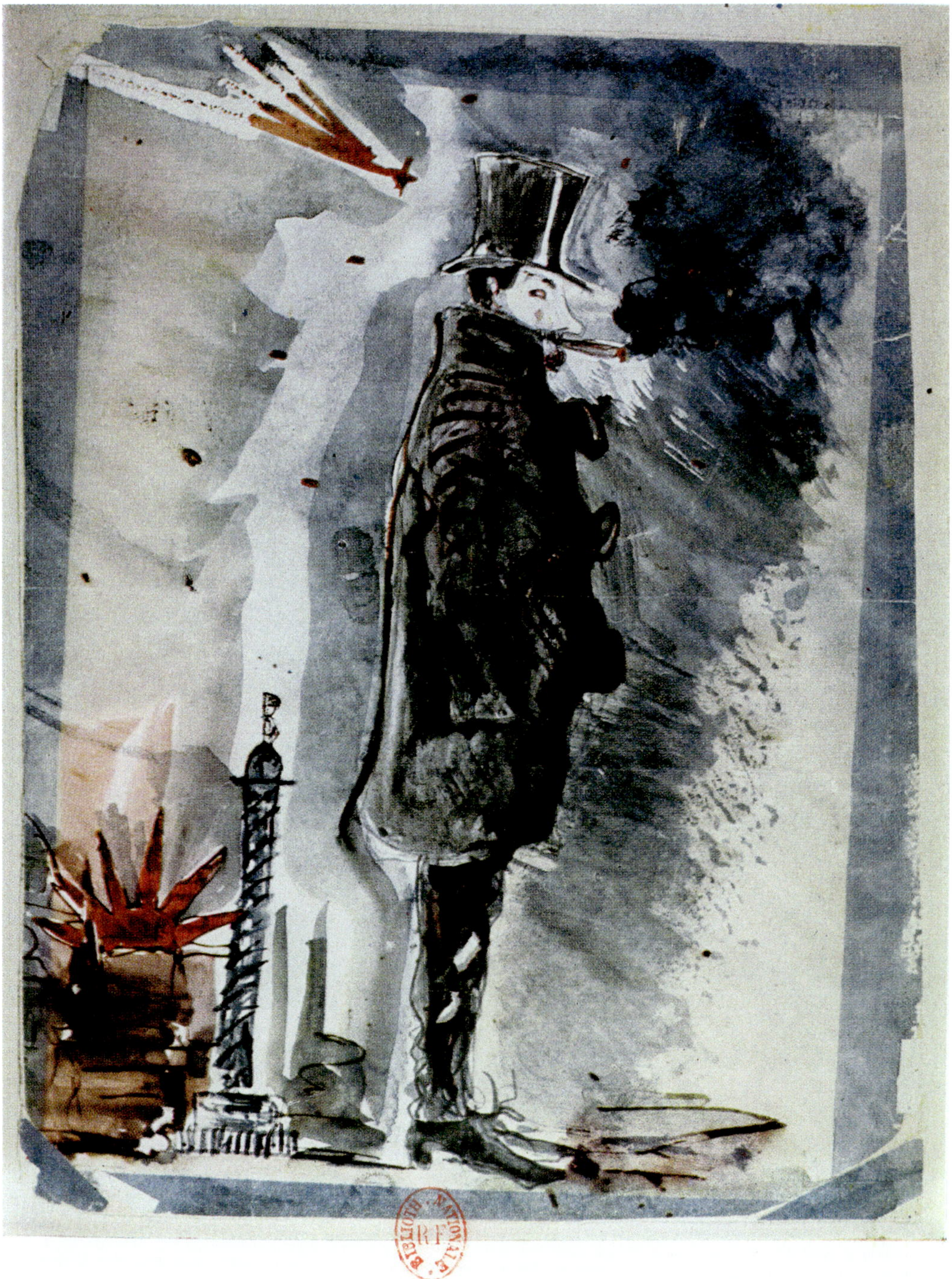

Figure 1.8 Charles Baudelaire, *Autoportrait Sous L'influence du Haschisch* (*Self-portrait Under the Influence of Hashish*) c. 1844, ink pen and wash, 8 1/2 x 6 3/4 in. Bibliothèque nationale de France, Paris. © Archives Charmet/Bridgeman Images.

> vague, the conducting thread which binds your conceptions together becomes so tenuous, that none but your accomplices can understand you. [...] This frivolity, these bursts of laughter, like explosions, seem like a true mania, or at least like the delusion of a maniac, to every man who is not in the same state as yourself.[78]

Baudelaire's second stage concerns the body's sensory system. Here, Baudelaire built on Moreau's description of physical effects, discussed in his first chapter and as the third phenomenon, "development of the sense of hearing; the influence of music."[79] As Baudelaire explains, the second phase announces itself by "a sensation of coolth at the extremities… and a great weakness in all the limbs.… [Y]ou feel an embarrassing stupor and stupefaction."[80] During this stage, the senses become extremely sensitive and randomly associative: "The senses become extraordinarily delicate and sharp. The eyes pierce the infinite. The ear hears the tiniest sounds in the middle of the loudest noises."

Baudelaire's third stage follows Moreau's "third phenomenon: errors of time and space"[81] and the full-blown hallucinations of the most advanced stages of hashish intoxication:

> It is then that the hallucinations begin; external objects take on wholly and successively most strange appearances; they are deformed and transformed. Then—the ambiguities, the misunderstandings, and the transpositions of ideas! Sounds cloak themselves with colour; colours blossom into music. [...] Every poetic brain in its healthy, normal state, readily conceives these analogies.

These synesthetic analogies exceed the "poetic brain" as they penetrate and overcome mindful control.

> But I have already warned the reader that there is nothing of the positively supernatural in hashish intoxication; only those analogies possess an unaccustomed liveliness; they penetrate and they envelop; they overwhelm the mind with their masterfulness.[82]

Later in this section, he notes the loss of mental control accompanies the loss of pain:

> Since long you have not been your own master; but you trouble yourself no longer about that. Pain, and the sense of time, have disappeared; or if sometimes they dare to show their heads, it is only as transfigured by the master feeling, and they are then, as compared with their ordinary form, what poetic melancholy is to prosaic grief.[83]

For Baudelaire, this euphoric condition is a dangerous one, as witnessed upon awaking in a deleterious state the next morning.

Baudelaire concludes his testimony by asserting that the aftermath of hashish intoxication is as disabling as the hallucinations, which make the imbiber powerless to even pick up a pen or pencil to record his own observations while under intoxication. First, the formidable fatigue that ensues the following day:

> [which] does not show itself at once, and when you are obliged to acknowledge its existence you do so not without surprise: for at first, when you are really assured that a new day has arisen on the horizon of your life, you experience an extraordinary sense of well-being; you seem to enjoy a marvelous lightness of spirit. But you are

> scarcely on your feet when a forgotten fragment of intoxication follows you and pulls you back; it is the badge of your recent slavery. Your enfeebled legs only conduct you with caution, and you fear at every moment to break yourself, as if you were made of porcelain. A wondrous languor... possesses itself of your spirit[.]

He thus admonishes the hashish user (and, by extension, himself) for wasting his vital energies.

> There, then, you are, for some hours yet, incapable of work, of action, and of energy. It is the punishment of an impious prodigality in which you have squandered your nervous force. You have dispersed your personality to the four winds of heaven -- and now, what trouble to gather it up again and concentrate it!

Echoing Moreau, Baudelaire further laments that in hashish's aftermath, "the will, the most precious of all faculties, is above all attacked."[84]

When Baudelaire began writing "The Painter of Modern Life" in 1859, he had imbued his aesthetic tenets with his scientific knowledge of hallucinations and madness. From this perspective, modernity's rupture and defamiliarization effect negative valence. In this context, the myriad urban sights and sounds assault the artist's nervous system, threatening to shatter the cerebral control of his sensations. Baudelaire describes such overloading of the cerebral cortex as "manic excitement."

Just as Baudelaire had previously portrayed the poet's struggle against nature's assault on the senses, in "The Painter of Modern Life" he describes a similar duel between the artist and the urban landscape. As he put it, when confronted by "the multiplicity of detail," it is as though "their main faculty is confused and paralyzed." Next, a progressive mental confusion unfolds that corresponds with the hashish-induced hallucinations.

> Then begins a struggle between the determination to see everything, to forget nothing, and the faculty of memory, which has acquired the habit of registering in a flash the general tones and shape, the outline pattern. An artist with a perfect sense of form but particularly accustomed to the exercise of his memory and his imagination, then finds himself assailed, as it were, by a riot of details, all of them demanding justice, with the fury of a mob in love with absolute equality. Any form of justice is inevitably infringed; any harmony is destroyed, sacrificed; a multitude of trivialities are magnified; a multitude of little things become usurpers of attention. The more the artist pays impartial attention to detail, the greater does anarchy become.[85]

Baudelaire's narrative of the artist's struggle against madness induced by external stimuli of the city echoes Moreau's description of how the physiological effects of hashish and madness obstruct the mind's mirroring, self-reflective power: "This power fails us when our faculties are troubled, when there is chaos in their midst, in a word, when there is madness."[86]

Baudelaire invoked a similar analogy between the artist's sensory/psychical assault and madness when he declared that Guys' intoxication with external stimuli "has some connection with congestion, that every sublime thought is accompanied by a more or less vigorous nervous impulse that reverberates in the cerebral cortex."[87] Here, the reference to congestion as a neurological condition conforms to Moreau's definition of the unhealthy, "troubled" brain, whether caused by madness or hashish. According to

Moreau, physical sensations of "hot flashes," lightheadedness, and fainting are symptoms of "cerebral congestions." These congestive symptoms result from "the primordial psychic change that we have identified as the basic, necessary source of all mental aberrations, particularly hallucinations, or the hallucinatory state…. [Such] cerebral congestions constitute, in a great many cases, the preliminary state of madness[.]"[88]

Having glimpsed the precipice of madness—in both his writing and his psyche—Baudelaire prescribes a remedy for the artist's neural recovery. In so doing, he assimilates his scientific knowledge into his aesthetic strategy of self-doubling. As witnessed in Guys' practice, noted early in this chapter, while the artist is a man of the crowd during the day, he takes refuge from the fray at night. He processes his experiences artistically while alone in his studio:

> And things seen [in the streets] are borne again on the paper, natural and more than natural, beautiful and more than beautiful, strange and endowed with an enthusiastic life, like the soul of their creator.… All the materials, stored higgledy-piggledy by memory, are classified, ordered, harmonized.[89]

Scientifically speaking, the artist's receptive nerves are harmonized with the brain's self-reflective power and outgoing motor, albeit artistic, responses. This neurological balance accords with Moreau's description of the healthy brain's recovery from external stimuli through sleep. As noted earlier, Moreau posits "two modes of mental life… known to man," one regarding the "external world" and the other a separate internal existence. Sleep mediates and keeps the balance between these two modes: "Sleep is like a barrier raised between these two lives: the physiological point where external life ceases and internal life begins. / So long as this situation exists, mental health prevails[.]"[90]

Significantly, Moreau's description of the healthy brain also involves the recovery of mental reflection, a dynamic condition he compares to the kaleidoscope:

> In a normal state—that is, in a state of reflective self-control and perfect independence—those ideas play in our mind, as if they were in some way alien to us. The slightest impulse of our will makes them vary endlessly like images in a kaleidoscope that we can stop any time.[91]

Thus, Baudelaire's comparison of the painter of modern life with "a kaleidoscope endowed with consciousness" signifies both the artist's recovered psyche and his heroic conquest of modernity's assault on his person.

In Baudelaire's assessment of Guys' modernity, he poses the artist's repaired psyche as the mirror through which the spectator recovers from the multitude of urban sensory, physical, and mental assaults.

> M. G., faithful interpreter of his own impressions, notes with instinctive vigour the culminating features or highlights of an object (…) or its main characteristics, sometimes even with a degree of exaggeration useful to human memory; and the imagination of the viewer, undergoing in its turn the influence of this imperious code, conjures up in clear outline the impression produced by the objects on the mind of M. G. In this case, the viewer becomes the translator of a translation, which is always clear and always intoxicating.[92]

In this exchange, Guys' sketch-like techniques and ephemeral transparency convey fleeting cognitive experiences and physical engagement with the myriad urban spectacles. With this portrayal of the artist and his art, Baudelaire evoked a celebratory tone, affirming modernity and its ever-changing fleetingness.

Aestheticizing the Ironic Gaze as a Mode of Healing and a Mode of Escape

Did other artists and, more generally, the urban populace share Baudelaire's fear of madness? In answering this question, we return to Hoffmann's ironic practices to expose how Baudelaire and Baudelairean artists used irony to heal psyches made "unhealthy" by modern urban phenomena. Baudelaire's recommendation of Hoffmann's acts of self-reflection warrants this approach. As he stated in "The Poem of Hashish," "Those who know how to observe themselves, and who preserve the memory of their impressions, those who, like Hoffmann, have known how to construct their spiritual barometer[.]"[93]

Alexander Schlutz's study of self-reflection and healing in Hoffmann's *Princess Brambilla* provides a model for posing Baudelaire as Hoffmann's 'ideal reader.' As such, Baudelaire recognizes his psychic self-reflection in the tale's ironic unfolding. Schlutz's summary of the story captures the convoluted, hallucinatory complexity and, notably, the embedded German Romantic irony, to which Baudelaire responded:

> Set during the irreverent days of the Roman carnival, *Brambilla* delights in a disorienting chaos of masks, costumes, dream images and hallucinations as Hoffmann pulls all the stops in a fantastical narrative of extraordinary complexity. The text unfolds a carnivalesque universe that—in the spirit of Early German Romantic irony—plays with the inescapable mediatory "distortions" brought about by the structures of human consciousness itself.[94]

Schlutz elaborates explicitly on the congruence between Schlegel's philosophical tenet of ironic self-reflection and Hoffmann's comic devices. Hoffmann's dizzying conflation of "narrative levels, variations and repetitions thus double and mediate each other in a fantastical *mise en abîme*," or hall of mirrors, through which the reader "can see her own inverted image and the tomfooleries of her own self and thought."[95] Herein lies the threat of madness from which the reader must retreat and recover:

> Just as the carnival is a ritual of life-affirming renewal, Hoffmann's self-reflexive narrative, too, is a story of change and healing, in which the self comes to understand that it has no access to a stable and unmediated self-knowledge. If this moment of insight initially appears as a near-psychotic threat to the very basis of identity, the laughter it triggers is ultimately a laughter of liberation. The self, now reconstituted in the mirror of laughter, the textual movement of meta-narrative irony, comes to embrace itself as a process of continuous change, a recognition that then allows it to be made whole in the mediatory gaze of another.[96]

Schlutz further explains how Hoffmann's ideal reader must be able to transcend the story's contradiction, tragedy, and jest that never resolve:

> It would be a mistake to attempt to resolve [these] contradiction[s].... The envisioned ideal reader of *Brambilla* would thus be one who could follow both reading strategies

> simultaneously, who could accept that the text is to be taken as both jest and seriousness at the same time, someone, in other words, who would be able to read the text...ironically, as an inverted, laughter-inducing mirror that reflects reflection and reveals all such binary distinctions as mere illusions. Such an ironic perspective would require one not simply to tolerate and sustain but to relish the dizziness and the intellectual vertigo that result from being unable to tell apart jest and seriousness.[97]

Given this "dizzying play of ironic doubling," and the "mirror [it] produces—in whomever the text induces a 'Selbstthätigkeit,' a self-activity, as Fichte and the Early German Romantics would have called it," the ideal reader "may find themselves again in a moment of joyous recognition and spontaneous, healing laughter."[98]

As the ideal reader that Schlutz describes, Baudelaire lauded *Princess Brambilla* as a prime example of absolute comic; in his words, *Princess Brambilla* "is a veritable catechism of high aesthetics."[99] As previously argued, Baudelaire extended Hoffmann's absolute comic into an aesthetic of *dédoublement*. Moreover, in conceiving *dédoublement*, Baudelaire conflated a triad of dizzying effects—hashish intoxication, comic laughter, and urban assaults on the sensorium, each triggering ironic self-reflection. In this case, when *dédoublement* manifests in the making and viewing of modern painting, the interior gaze replaces comic-induced laughter. Thus, Baudelaire restored the healing power of ironic laughter to philosophical/meditative irony, thereby modeling an escape from the maddening effects of modernity.

While Baudelaire's Parisian contemporaries may not have had hashish and madness as their frames of reference, we can assume that they experienced similar sensory and psychic assaults on their beings as they navigated the streets of Haussmannized Paris. Let us now survey Manet's *A Bar at the Folies-Bergère* (1882), as well as several Degas paintings, to examine how the aesthetic of the ironic gaze provides viewers with both a mode of escape and recovery from the embodied ruptures of Haussmannization (Figure 1.9). What follows maps viewers' potential aesthetic experience of the simultaneous affliction and affirmation of modernity.

Assuming the viewers' gaze in Manet's *A Bar at the Folies-Bergère*, we identify with the barmaid's ironic condition: she and we are both of the crowd and separate from it. In this position, we share the barmaid's internal self-reflection as a mode of escape. Her psychological isolation is compounded by the mirrored spectacle that unfolds before her. At the same time, the viewers' ironic self-doubling—being "*beside [one]self*"—is re-doubled in relation to the mirrored space wherein they occupy the client's position. As such, viewers are both outside the painting and within, a duality further signified by the client's absence/presence in front of/within the mirror reflection.

Likewise, the mirror image is a fragment of an overall ambiance and in turn fragments the objects and figures reflected therein (note, for example, the cropped legs of the trapeze artist). At the same time, these painterly and figurative fragments cohere as the mirror-image of the kaleidoscopic cognitive experience of modernity, "reflect[ed] at every moment in energies, more vivid than life itself, always inconstant and fleeting."[100] Irony "hovers" over this technical coherence, immersing viewers in endless contradictions and transformations.

Degas' paintings—*Place de la Concorde* (1875) and *Mme. Jeantaud au Miroir* (*Mme. Jeantaud at the Mirror*) (1875), alongside *Mary Cassatt at the Louvre* (1880)—further buttress readings that map viewers' interior self-transformations from physical-psychic

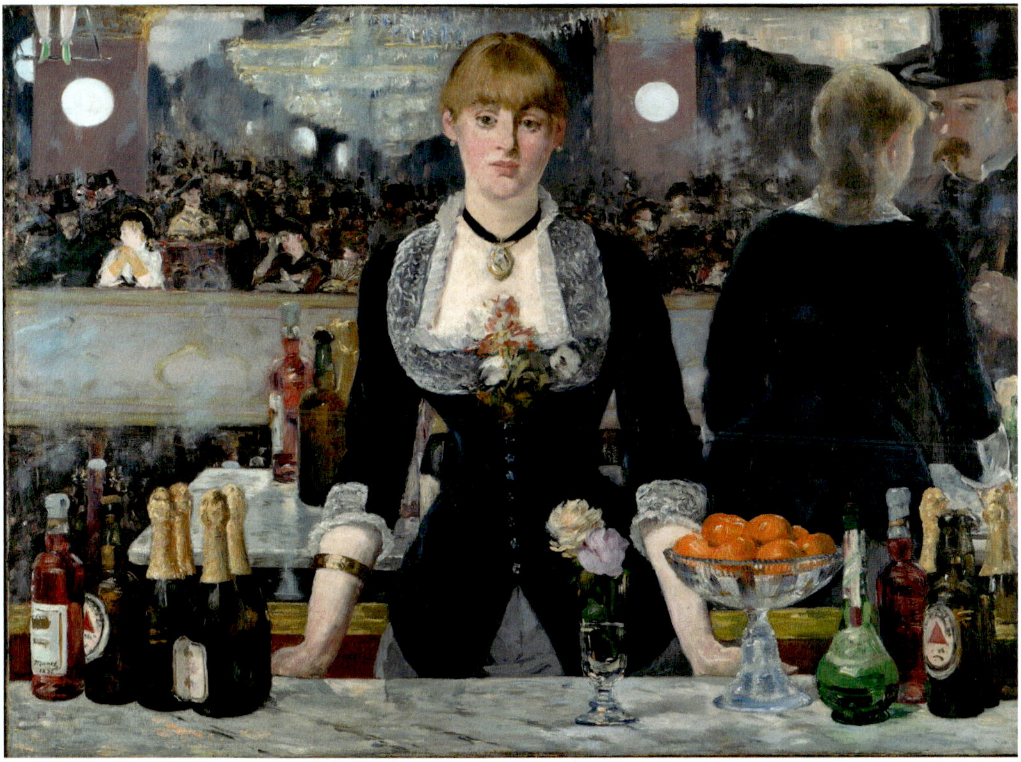

Figure 1.9 Édouard Manet, *A Bar at the Folies-Bergère*, 1882, oil on canvas, 3 1/8 x 4 1/6 in. The Courtauld, London (Samuel Courtauld Trust © The Courtauld). Photo © The Courtauld/Bridgeman Images.

affliction to aesthetic reconciliation (Figures 1.10 and 1.11). As noted earlier, viewers of Degas' urban paintings embody both the figures' transience and the standing spectators' fleeting view of the scene. Yet, standing before these paintings, the viewer's gaze repeatedly returns to the center, where the void reflects our ironic condition: we are physically absent yet psychically present. In *Mme. Jeantaud at the Mirror*, for example, a woman in street clothes stops to reflect upon the impending or recent self-rupture inflicted by the urban crowd: she is both within and beside herself as she gazes at her reflection as an "other," at once fleeting and transformed. Likewise, viewers caught between Mme. Jeantaud's actual gaze and the reflected one are thereby immersed in the interiority of her ironic gaze.

Similarly, *Mary Cassatt at the Louvre* offers the museum interior and the interior aesthetic experience as the loci of escape and cultural renewal. Although we stand outside and behind the painted viewer, we share her contemplative gaze. Cassatt's picture gazing, like ours, is ironic. She is a woman of urban manners and fashion, musing on the past from the perspective of the present. As such, her aesthetic gaze mirrors the internal dynamic of "an eternal, invariable element" and "a relative circumstantial element [of] contemporaneity,"[101] an ironic opposition through which Baudelairean modernity endures.

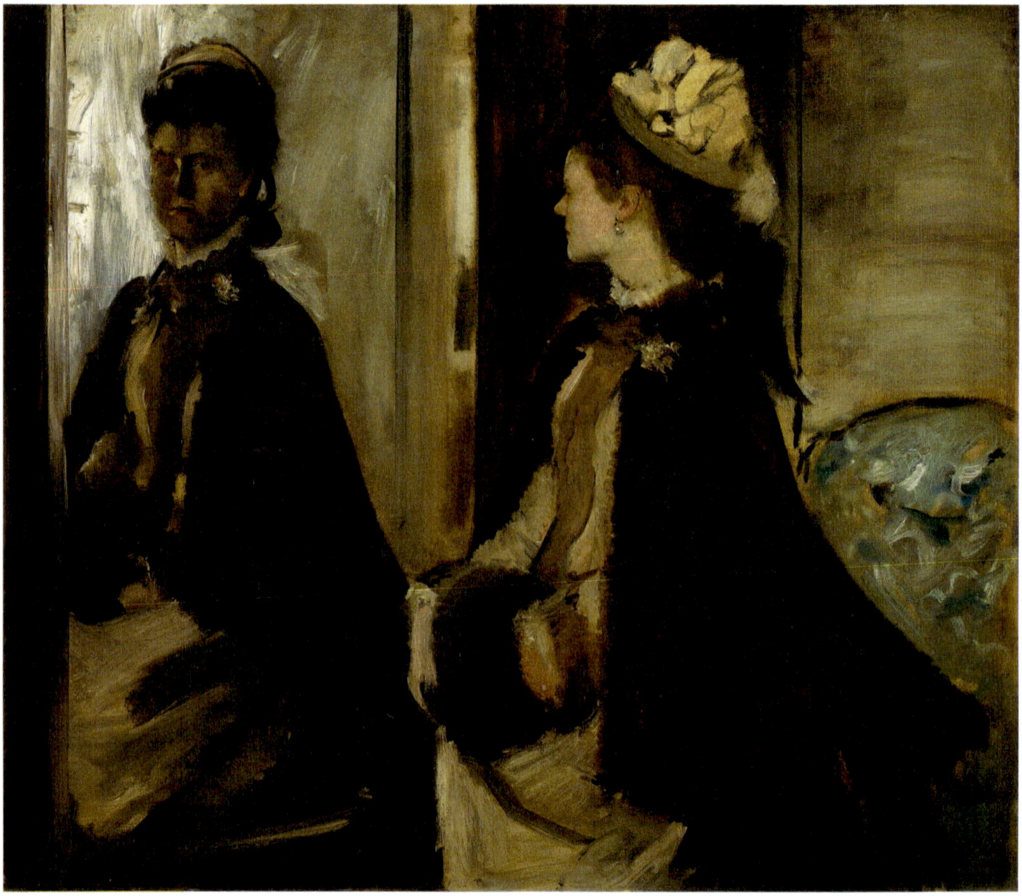

Figure 1.10 Edgar Degas, *Madame Jeantaud at the Mirror*, c. 1875, oil on canvas, 30 x 33 5/8 in. Musée d'Orsay, Paris, France. © RMN-Grand Palais/Art Resource, NY. Photo: Adrien Didierjean.

In Chapter 2 and Chapter 4, I track a collective ambiguity toward Haussmannization in Charles Meryon's engravings of old Paris and Charles Marville's photographs of old and new Paris. Viewed through the double lens of Baudelaire's fear and celebration of modernity, captured in his art criticism and verse poetry, Meryon's and Marville's works serve as precursors to the ironic parody exemplified in Baudelaire's later prose poems and Manet's paintings of Baudelairean modernity. By examining these precursory works as mediators of the shared phenomena of Haussmannization, I aim to locate aesthetic irony in the visceral, lived experiences of writers, artists, and viewers. Thereby, here and throughout this volume, I follow Baudelaire's lead in identifying the transference of the artist's experience to viewers. To legitimize Baudelaire's tenets further, in Chapter 3 I draw upon current psychological and neuroscientific models of aesthetic processing of artworks that comprise rupture and defamiliarization. This empirical chapter also lays the groundwork for Chapters 6 through 10, identifying the embodied experience of Haussmannization in artworks that defy conventional techniques of surface finish, compositional harmony, and narrative clarity.

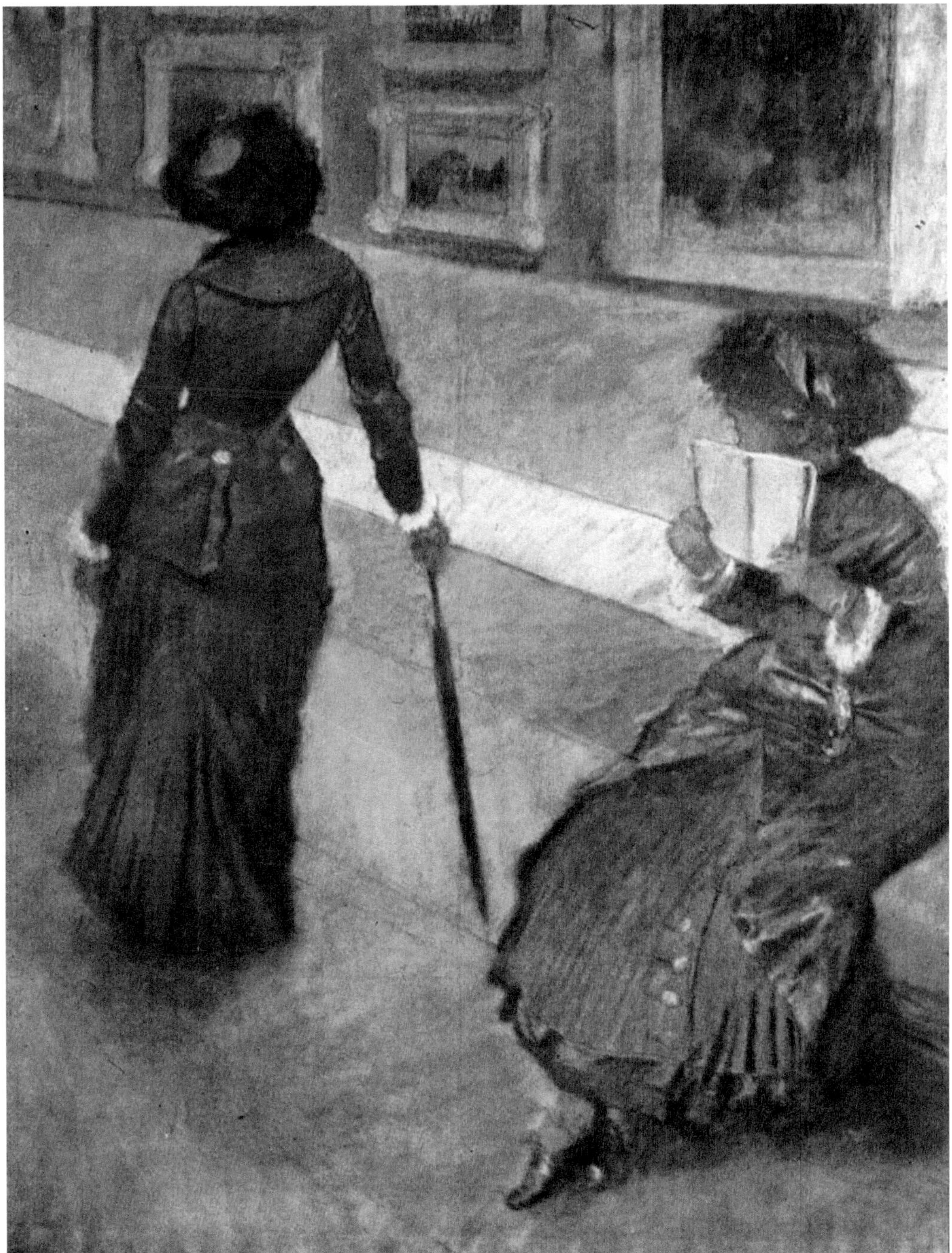

Figure 1.11 Edgar Degas, *Mary Cassatt at the Louvre*, c. 1880, pastel, unknown dimensions. Private Collection. Image: WikiArt.org.

Notes

1 According to P. E. Charvet in Charles Baudelaire, *Selected Writings on Art and Literature*, ed. and trans. P. E. Charvet (London/New York: Penguin Classics, 1992): 458, n. 1, "The Painter of Modern Life" "probably dates from November 1859 to February 1860." According to Claude Pichois, the first draft of "The Painter of Modern Life" dates from the summer of 1860. See Claude Pichois, with Jean Ziegler, *Baudelaire*, trans. Graham Robb (London: H. Hamilton, 1989): 261.
2 Linda Hutcheon, *A Theory of Parody: The Teachings of Twentieth-Century Art Forms*, 32.
3 Muecke, *Irony and the Ironic*, 19.
4 In 1927 the opening of Boulevard Haussmann marked the completion of Haussmann's *grands travaux*; see Patrice de Moncan and Claude Heurteux, *Le Paris d'Haussmann* (Paris: Editions du Mécène, 2002): 64.
5 Although I only treat Eugène Atget briefly in this study's final chapter, his photographs of window reflections best capture this transience and transparency, as I have discussed in "The Mirror as a Metaphor of Baudelairean Modernity," in *Orientations: Space/Time/Image/Word, Word & Image Interactions 5*, ed. Claus Clüver et al. (Amsterdam/New York: Editions Rodopi, 2005): 16–36. I have summarized these findings in Chapter 7 of this book; see pp. 241–2 and Figures 7.4 and 7.5.
6 On the prostitutes' identity and gestures of solicitation, see Hollis Clayson *Painted Love: Prostitution in French Art of the Impressionist Era* [orig. 1991] (Los Angeles: Getty Research Institute, 2003): 105–8.
7 Baudelaire, "Of the Heroism of Modern Life," in *Selected Writings*, 106.
8 Baudelaire, "The Painter of Modern Life," in *Selected Writings*, 399–400.
9 Baudelaire, "The Exposition Universelle (1855)," in *Art in Paris, 1845–1862: Salons and Other Exhibitions*, ed. and trans. Jonathan Mayne (Ithaca: Cornell University Press/Phaidon, 1965): 124–5.
10 As explained in the Introduction, Baudelaire used the term "*se dédoubler*" in "De L'essence du Rire et Généralement du Comique dans les Arts Plastiques" (1855), in *Oeuvres Complètes* (Paris: Bibliothèque de la Pleiade-Gallimard, 1961): 694; "Of the Essence of Laughter, and the Comic in the Plastic Arts," trans. in *Selected Writings*, 154. The dictionary translation of *se dédoubler* is "to split into two" (https://www.deepl.com/se dédoubler), which Charvet uses in his translation of "Of the Essence of Laughter." In addition to following de Man's translation of "*se dédoubler*," I am also adapting the term *dédoublement* from de Man, who uses it to explain the "reflective activity" connoted by Baudelaire's use of the infinitive form *se dédoubler*, that being an activity of "self-duplication or self-multiplication." See "The Rhetoric of Temporality," in Paul de Man, *Blindness and Insight: Essays in the Rhetoric of Contemporary Criticism*, 2nd rev. ed. (Minneapolis: University of Minnesota Press, 1983): 212–13.
11 Baudelaire, "The Painter of Modern Life," 402.
12 Baudelaire, "The Painter of Modern Life," 399.
13 Baudelaire, "The Painter of Modern Life," 400.
14 Baudelaire, "The Painter of Modern Life," 400.
15 See Jean Sutherland Boggs, et al., *Degas* (New York/Ottawa: The Metropolitan Museum of Art/National Gallery of Canada, 1988): 140.
16 Henri Fantin-Latour's *Un Atelier aux Batignolles* (A Studio at Les Batignolles; 1870; Musée d'Orsay, Paris; https://www.musee-orsay.fr/fr/oeuvres/un-atelier-aux-batignolles-210) is a testament to Manet's primacy among the avant-garde artists who called themselves the Batignolles Group and later became Impressionists. Shown in the studio of Manet in the Batignolles Quarter, the group gathers around Manet at his easel. Six years earlier, Fantin-Latour painted *Hommage à Delacroix* (Homage to Delacroix; 1864; Musée d'Orsay, Paris; https://www.musee-orsay.fr/en/artworks/hommage-delacroix-208), a visual manifesto of Delacroix's primacy among the Realist artists, and which featured a standing Manet and a seated Baudelaire side-by-side.
17 See Michael Marrinan, *Caillebotte: Painting the Paris of Naturalism, 1872-1887* (Los Angeles: Getty Research Institute, 2016): 64–5.
18 Since the beginning of his artistic training Manet was in constant dialogue with masterworks of the past and present. While studying with Thomas Couture from 1850 to 1856, Manet

copied the old masters at the Musée du Louvre. He also copied works by "living artists," then housed at the Musée du Luxembourg. See Françoise Cachin et al., *Manet, 1832–1883* (New York: Metropolitan Museum of Art/Abrams, 1983): 47; and Gary Tinterow et al., *Manet/ Velazquez: The French Taste for Spanish Painting* (New York: Metropolitan Museum of Art; New Haven, 2003): 204–6. In Chapter 9, I discuss Manet's *Music in the Tuileries* in dialogue with Jacques-Louis David's (1748–1825) *The Intervention of the Sabine Women* (1799). I discuss other instances of Manet's parodying of historical masterpieces and history paintings in Chapter 6 and elsewhere in Chapter 9.

19 As Kirk Varnedoe observed, this painting presents a "'picture within a picture' of the mirror view" in *Gustave Caillebotte* (New Haven: Yale University Press, 1987): 139.

20 Here I borrow the term "defamiliarization" from literary theory as established by Russian Formalists, referring to "the literary device whereby [verbal or visual] language is used in such a way that ordinary and familiar objects are made to look different," Nasrullah Mambrol, "Defamiliarization," *Literary Theory and Criticism* (March 17, 2016), https://literariness.org/2016/03/17/defamiliarization/.

21 Baudelaire's familiarity with German Romantics is well established. See, for example, Francis S. Heck, "Baudelaire's *La Fanfarlo*: An Example of Romantic Irony," *The French Review* 49, no. 3 (Feb. 1976): 328–36. Heck provides a concise overview, which includes the transmission of irony from the Germans to Baudelaire: "In regard to Baudelaire's contact with German Romanticism, well known is his fondness for the work of E.T.A. Hoffmann (cf. 'Les Correspondances')" (329).

22 Novalis [Friedrich von Hardenberg], "Miscellaneous Writings" ("Vermischte Bemerkungen" [1797]), trans. in *German Aesthetic and Literary Criticism. The Romantic Ironists and Goethe*, ed. Kathleen M. Wheeler (Cambridge: Cambridge University Press, 1984): 86 [fragment no. 23].

23 Friedrich von Schlegel, "Critical Fragments" ("Kritische Fragmente" [1797]), in *German Aesthetic and Literary Criticism*, 43.

24 Karl Solger, *Erwin*, or *Four Dialogues on Beauty and Art* (*Erwin. Vier Gespräche über das Schöne und die Kunst*, [1816])," in *German Aesthetic and Literary Criticism*, 144.

25 Schlegel, "Über Goethes *Meister*" [1798] ("On Goethe's *Meister*"); rpt: Friedrich von Schlegel and Hans Eichner, *Über Goethes Meister: Gespräch über die Poesie* (Paderborn: F. Schöningh, 1985): 108; trans. Wheeler, *German Aesthetic and Literary Criticism*, 67; also Wheeler, "Introduction," *German Aesthetic and Literary Criticism*, 22. Ludwig Tieck also described irony's hovering effects, a concept he attributed to Solger and Schlegel in the preface to *Wilhelm Lovell* (Vorwort zu *Wilhelm Lovell* [1793–96]), trans. in Wheeler, *German Aesthetic and Literary Criticism*, 122.

26 Schlegel, *Athenaeum Fragments* (no. 238) (*Athenäums-Fragmente* [1798]), in Friedrich von Schlegel, *Friedrich Schlegel's "Lucinde" and the Fragments*, intro. and trans. Peter Firchow (Minneapolis: University of Minnesota Press, 1971): 195.

27 Ludwig Tieck, *Nachgelassene Schriften*, ed. R. Köpke (Leipzig: F. A. Brockhaus, 1855): II, 239; qtd. and trans. in Wheeler, "Introduction," *German Aesthetic and Literary Criticism*, 17. As Wheeler argues, what finally determines an appropriate technique is the capacity for the artist's individual "manner" to draw the reader into a universal striving and achieve "the goal of self-discovery and self-cultivation common to all art" and consonant with the artist's own inner knowledge (12). This tenet will also be the basis of Baudelaire's critical standards for an artist's personal technique.

28 Tieck, quoted from *Nachgelassene Schriften*, II, 239; cited and trans. in Wheeler, "Introduction," *German Aesthetic and Literary Criticism*, 17.

29 Baudelaire often echoes Tieck who believed that "detached contemplation of ourselves" is "the ground of all true elevation." Tieck often distinguished between crude or common and high irony, a difference which Baudelaire assigned to comedy. By "high irony," Tieck meant "not derision, mockery, persiflage or what in a similar vein is usually understood by the term. Rather, irony is the most profound seriousness, yet bound up with play and genuine joviality. Irony is not merely negative, it is rather through and through positive" (*Nachgelassene Schriften*, II, 238–9; qtd. and trans. in Wheeler, "Introduction," *German Aesthetic and Literary Criticism*, 19). While these passages are extracted from Tieck's posthumous writings, published in 1855, they nevertheless relate to his extensive literary criticism, especially on Shakespeare, and issue

from his close alliance with other German Romantic writers, including the Schlegel brothers, whom he frequently cites.

30 Baudelaire, "Of the Essence of Laughter," in *Selected Writings*, 148.

31 Baudelaire, "Of the Essence of Laughter," 160.

32 Baudelaire, "Of the Essence of Laughter," 152.

33 Baudelaire, "Of the Essence of Laughter," 161. Each of the attributes that Baudelaire defines as "comic" coincides with the 'basic features' of irony as defined by literary critic D. C. Muecke: a confident unawareness, feigned in the ironist but genuine in the victim of irony, a (resultant) contrast of appearance and reality; a comic element in the victim's unawareness; an element of detachment; and an aesthetic element. See Muecke, *Irony*, 24–48. Baudelaire constructed his artistic duality in *La Fanfarlo*, first published in January 1847. As Francis Heck has shown, *La Fanfarlo* is an example of Baudelaire's ironic writing and his ironic self-reflection as an artist. *La Fanfarlo* thus figures as "a portrait of the artist at the crossroads. There is, however, a certain detachment on the part of the author, which permits him to view his hero and his work in an ironic manner, as well as to parody himself." ("Baudelaire's La Fanfarlo: An Example of Romantic Irony," 328–36.)

34 Baudelaire's analogy between the comic and philosophic irony can be compared with the Romantic ironists' philosophic, self-reflective notion of comedy and humor; see Wheeler, "Introduction," *German Aesthetic and Literary Criticism*, 17–9. Wheeler's discussion includes references to Friedrich Schlegel's attribution of this ironic attitude to Goethe and to Novalis's equating Jean-Paul Richter's humor with Friedrich Schlegel's irony (see note 38).

35 Schlegel, *Lyceum [Critical] Fragments*, in Wheeler, *German Aesthetic and Literary Criticism*, 36.

36 Baudelaire, "Of the Essence of Laughter," 148. Baudelaire's original French text speaks more directly to the force of self-doubling: "la force de se dédoubler rapidement et d'assister comme spectateur désintéressé aux phénomènes de son *moi*." (original emphasis); "De l'essence du rire et généralement du comique dans les arts plastiques," in Baudelaire, *Oeuvres complètes de Charles Baudelaire*, II, *Curiosités esthétiques* (Paris: Michel Lévy Frères, 1868): 370. As per note 3 above, I prefer Charvet's original translation of "*se dédoubler*" as "one who had acquired by habit a power of rapid self-division" ("Of the Essence of Laughter," 154).

37 Schlegel, *Lyceum [Critical] Fragments*, in *German Aesthetic and Literary Criticism*, 41–2.

38 For an in-depth comparative study of the similarities between Baudelaire's and Hoffmann's aesthetics in their literary works, see Rosemary Lloyd, *Baudelaire et Hoffmann: Affinités et influences* (Cambridge: Cambridge University Press, 1979). Most importantly, Lloyd documents Hoffmann's special role in shaping Baudelaire's thought and art regarding his dual vision of existence and his theory of conscious reflection as the keystone of artistic creation. We can also assume that certain of these similarities have more diffuse origins in the French literati's reception of German Romantic literature and philosophy during the mid-nineteenth century, including the writings of Heinrich Heine and Stendhal. Heine's acquaintances overlapped with Baudelaire's, including Gautier, Balzac, George Sand, Gérard de Nerval, Dumas, Victor Hugo, Musset, and Grandville. See Jeffrey L. Sammons, *Heinrich Heine: A Modern Biography* (Princeton: Princeton University Press, 1979): 169. Baudelaire defended Heine in response to Jules Janin's 1865 pejorative review of Heine (see Pichois with Ziegler, *Baudelaire*, 340). That Baudelaire here attributes a duality of opposites—melancholy and mocking—to Heine and related poets' work is revealing. According to Jocelyn Kolb, Heine must be counted among the German Romantic ironists based on his use of the Doppelgänger motif in his poems; see Jocelyn Kolb, "Romantic Irony," in *A Companion to European Romanticism*, ed. Michael Ferber (Malden, MA: Blackwell, 2005): 383–8.

39 Baudelaire, "Of the Essence of Laughter," 153.

40 Baudelaire, "Of the Essence of Laughter," 161.

41 Baudelaire, "Of the Essence of Laughter," 159.

42 Baudelaire's tenet that the comic subject not only must be unaware, but the author may sometimes be unaware of the comic, joins him with Hoffmann. Hoffmann and Baudelaire also align with Socratic irony, as explained by Maria M. Tatar, "E.T.A. Hoffmann's 'Der Sandmann': Reflection and Romantic Irony," *MLN*, 95, no. 3 (1980): 595.

43 Baudelaire, "Of the Essence of Laughter," 159.

44 Rosemary Lloyd dates Baudelaire's literary assimilation of Hoffmann to 1845: Baudelaire names Hoffmann in his 1845 poem "A une Jeune Saltimbanque" ("To a Young Acrobat"), and

his *Salon of 1845*, Baudelaire borrows the phrase "chronic dualism ("dualism chronique") from Hoffmann's tale *Princess Brambilla*. See Lloyd, *Baudelaire et Hoffmann*, 10. Baudelaire's ironic self-construction in *La Fanfarlo*, first published in January 1847, also reflects this early assimilation of Hoffmann.

45 See Tatar, "E.T.A. Hoffmann's 'Der Sandmann'," 585. For Hoffman's ironic practices and affiliation with German Romantic ironists, see Birgit Röder, *A Study of the Major Novellas of E.T.A. Hoffmann*. (Rochester, NY: Camden House, 2003) and Val Scullion, "Kinaesthetic, Spastic and Spatial Motifs as Expressions of Romantic Irony in E. T. A. Hoffmann's *The Sandman* and Other Writings," *The Journal of Literature and Science* 2, no. 1 (2009): 1–22.

46 Jean-Paul Richter, a German Romantic writer and ironist, was one of the German writers Hoffmann named in *Princess Brambilla*; in E. T. A. Hoffmann, *The Golden Pot, and Other Tales*, trans. Ritchie Robertson (Oxford: Oxford University Press, 1992): 223. Notably, in a review of Baudelaire's *Salon of 1846*, his criticism was compared with that of Diderot, Hoffmann, Stendahl, and Heine. However, as Claude Pichois notes, even though the article was published with Henry Murger as the author, the review was likely written under the guidance of Baudelaire himself. See Pichois with Ziegler, *Baudelaire*, 138.

47 Baudelaire, "The Painter of Modern Life," 399. Jean de La Bruyère was a French satiric moralist who is best known for one work, *Les Caractères de Théophraste Traduits du Grec avec Les Caractères ou les Moeurs de ce Siècle* (1688; *The Characters, or the Manners of the Age, with The Characters of Theophrastus*), which is considered to be one of the masterpieces of French literature," "Jean de La Bruyère," *Encyclopedia Britannica*. https://www.britannica.com/biography/Jean-de-La-Bruyere.

48 Baudelaire, *Salon of 1846*, 106.

49 *Artificial Paradises* (*Les Paradis Artificiels*, published on June 2, 1860) analyzes the state of being under the influence of opium and hashish. It comprises three previously published essays: "On Wine and Hashish" ("Du Vin et du Haschisch," in *Le Messager de l'Assemblée*, September 30, and March 11 and 12, 1851); "The Poem of Hashish" (*Le Poème du Haschich* in *Revue Contemporaine*, September 30, 1858, under the title "De l'Idéal Artificiel.—Le Haschisch…"), and "An Opium Eater" (*Un Mangeur D'opium*, in *Revue Contemporaine*, January 15 and 31, 1860, under the title "Enchantement et Tortures d'un Mangeur D'opium"). This publishing chronology is drawn from Charles Baudelaire, *Oeuvres Complètes*, ed. Michel Jamet, pref. Claude Roy (Paris: Bouquins/Robert Laffont, 1999): 966–7, n. 1.

50 Also known as Jacques (-Joseph) Moreau de Tours.

51 Müller supported the idea of neural forces, identifying some with will and others with sensory powers. He spoke of "specific nerve energies" to refer to the different kinds of mental forces associated with varying tracts of nerve. See E. S. Reed, *From Soul to Mind: The Emergence of Psychology, from Erasmus Darwin to William James* (New Haven: Yale University Press, 1997): 84.

52 "Müller, Johannes Peter," *Complete Dictionary of Scientific Biography. Encyclopedia.com* (June 27, 2018. https://www.encyclopedia.com/people/medicine/medicine-biographies/johannes-peter-muller. Müller's research on the human nervous system is at the center of nineteenth-century developments in psycho-physiology. According to Robert M. Young, "[T]he functional division of nerves and physiological phenomena in sensation and motion was obviously not a novelty in the nineteenth century. … [However,] [t]he significance of the nineteenth-century analysis lay first in its experimental demonstration in the central nervous system and second in the progressive extension of the concept as the fundamental explanatory principle in both physiology and psychology;" *Mind, Brain, and Adaptation in the Nineteenth Century: Cerebral Localization and Its Biological Context from Gall to Ferrier* (New York: Oxford University Press, 1990): 93.

53 See Jacques-Joseph Moreau, *Du Haschisch et de L'aliénation Mentale: Études Psychologiques* (Paris: Fortin, Masson, 1845) and in English *Hashish and Mental Illness*, eds. Hélène Peters and Gabriel G. Nahas, trans. Gordon J. Barnett (New York: Raven Press, 1973). Moreau and other French scientists of his generation would have been receptive to Müller's work on the nervous system because of contemporary French scientists' research on the brain and nervous system anatomy. For more extensive discussion, see Young, *Mind, Brain, and Adaptation in the Nineteenth Century*: 55–74 and 74–88.

54 Baudelaire, "The Painter of Modern Life," 397.

55 Baudelaire, "The Painter of Modern Life," 398.
56 "Cerebral Cortex: What It Is, Function & Location," Cleveland Clinic, 2022. https://my.clevelandclinic.org/health/articles/23073-cerebral-cortex.
57 Baudelaire, "The Painter of Modern Life," 398.
58 "Müller, Johannes Peter," in *Complete Dictionary of Scientific Biography*. https://www.encyclopedia.com/science/dictionaries-thesauruses-pictures-and-press-releases/muller-johannes-peter.
59 Johannes Müller, *Elements of Physiology*, vol. II, trans. William Baly (London: Taylor & Walton, 1840–2): 907.
60 Reed, *From Soul to Mind*, 93.
61 Müller, *Elements of Physiology*, vol. I, 737.
62 Müller, *Elements of Physiology*, vol. II, 1073.
63 Confiteor is "a form of Roman Catholic prayer in which public confession of sins is made." "Confiteor," http://en.wiktionary.org/wiki/confiteor; the Latin etymology of *confiteor* is "I confess, initial word of the formula," "confiteor, n.", *OED Online* (Oxford University Press, December 2022).
64 The collection of fifty prose poems written between 1857 and 1867, known as *Le Spleen de Paris*, was published in 1869, two years after Baudelaire's death. The first twenty poems were first published in *La Presse* in August and September 1862. The prose poems are the subject of Chapter 8 of this book.
65 Charles Baudelaire, "The Confiteor of the Artist," in *The Poems and Prose Poems of Charles Baudelaire*, ed. James Huneker (New York: Brentano's Publishers, 1919), n.p. "The Confiteor of the Artist" is among the twenty poems published in *La Presse* in 1862 (see note 64). According to Raymond N. MacKenzie, "'The Artist's Confiteor' is a confession of the sin of pride: the artist's 'rival' is Nature itself, a rival who, when raising him up to the level of the infinite, sometimes seems an accomplice, but who then turns and presents itself as an impenetrable wall between the artist and the infinite;" "Introduction" in Charles Baudelaire, *Paris Spleen and La Fanfarlo*, trans. Raymond N. MacKenzie (Indianapolis: Hackett Pub. Co, 2008): vii-xxv.
66 Théophile Gautier, writing an essay on the poet, noted: "That [Baudelaire] once or twice tried hashish, as a psychological experience, is possible and even probable; but he did not make continuous use of it. This happiness, bought at the chemist's and carried in the pocket, was repugnant to him[.] ... He came but rarely, and then only as a spectator, to the séances at the Hôtel Pimodan, where our circle met to take the 'dawamesk'[.]" Théophile Gautier, *Charles Baudelaire: His Life*, trans. Guy Thorne (London: Greening & Co., 1915): 70.
67 "[Moreau] regarded as his most important work a publication from 1859 entitled *Psychologie Morbide Dans Ses Rapports Avec la Philosophie de L'histoire*. The principal idea of this book is evident on the very first page: 'The dispositions of the mind which make a man different from others by the originality of his thoughts and conceptions and by superior intelligence stem from the same organic conditions that may give rise to insanity.'" Bo Holmstedt, "Introduction," in Moreau, *Hashish and Mental Illness*, xix. *La Psychologie Morbide Dans Ses Rapports Avec la Philosophie de L'histoire, ou, De L'influence des Névropathies sur le Dynamisme Intellectuel* (Paris: V. Masson, 1859).
68 Moreau, *Hashish and Mental Illness*, 21.
69 Moreau, *Hashish and Mental Illness*, 17.
70 Moreau, *Hashish and Mental Illness*, 18.
71 Moreau, *Hashish and Mental Illness*, 18.
72 Moreau, *Hashish and Mental Illness*, 33.
73 Moreau, *Hashish and Mental Illness*, 36.
74 Moreau, *Hashish and Mental Illness*, 86.
75 Moreau's more scientific precaution warns against strictly equating drug-induced hallucinations and those induced by mental illness: "When I speak of hallucinations the word must not be taken in its strictest sense: a very important shade of difference distinguishes pure hallucination, such as doctors often have occasion to study, from the hallucination, or rather of the misinterpretation of the senses, which arises in the mental state caused by the hashish.... The energy, the almost speaking liveliness of hallucination in this form of intoxication in no way invalidates this original difference: the one has root in the situation, and, at the present time,

the other has not." This and the following quotations are from Charles Baudelaire, *The Poem of Hashish*, trans. Aleister Crowley (1895; repr., n.p.: Erowid.org, 2014), hereafter: *Poem of Hashish*.

76 Baudelaire follows Moreau's account of the three phases of visual phenomena that occur in their "barest simplicity" during hashish intoxication. See Moreau, *Hashish and Mental Illness*, 81.

77 These terms—excitement, dissociation of ideas—appear as the heading for section III of Chapter 3: "Psychological Phenomena" of "the physical changes that precede or accompany the mental problems caused by hashish," in Moreau, *Hashish and Mental Illness*, 31.

78 Baudelaire, *Poem of Hashish*.

79 These terms—development of the sense of hearing; the influence of music—appear as the heading for section V of Chapter 3 in Moreau, *Hashish and Mental Illness*, 38.

80 Baudelaire, *Poem of Hashish*.

81 The term—third phenomenon: errors of time and space—appear as the heading for section IV of Chapter 3 in Moreau, *Hashish and Mental Illness*, 36.

82 Baudelaire, *Poem of Hashish*.

83 Baudelaire, *Poem of Hashish*.

84 Baudelaire, *Poem of Hashish*. Although Baudelaire claimed moral neutrality in writing "The Poem of Hashish," he concludes his essay with moral reprehension, comparing hashish intoxication with Satanic powers: "It is easy to grasp the analogy which exists between the Satanic creations of poets and those living beings who have devoted themselves to stimulants. Man has wished to become God, and soon? – there he is, in virtue of an inexorable moral law, fallen lower than his natural state! It is a soul which sells itself bit by bit."

85 Baudelaire, "The Painter of Modern Life," 407.

86 Moreau, *Hashish and Mental Illness*, 16–7.

87 Baudelaire, "The Painter of Modern Life," 398.

88 Moreau, *Hashish and Mental Illness*, 148–9.

89 Baudelaire, "The Painter of Modern Life," 402.

90 Moreau, *Hashish and Mental Illness*, 21.

91 Moreau, *Hashish and Mental Illness*, 50.

92 Baudelaire, "The Painter of Modern Life," 406–7.

93 Baudelaire, *Poem of Hashish*.

94 Alexander Schlutz, "The Mirror of Laughter: Mediation, Self-Reflection, and Healing in E. T. A. Hoffmann's *Princess Brambilla*," *European Romantic Review* 22, no. 3 (2011): 415.

95 Schlutz, "The Mirror of Laughter," 420.

96 Schlutz, "The Mirror of Laughter," 415.

97 Schlutz, "The Mirror of Laughter," 416.

98 Schlutz, "The Mirror of Laughter," 420. Johann Gottlieb Fichte was a German philosopher and a founder of German idealism.

99 Baudelaire, "Of the Essence of Laughter," 159.

100 Baudelaire, "The Painter of Modern Life," 400.

101 Baudelaire, "The Painter of Modern Life," 392.

2 Mirroring Haussmannization in Baudelairean Word and Graphic Image

This chapter continues on from the previous one in its aim to reify an affective context for Baudelaire's ironic strategies as both poet and critic. Here and in Chapter 4, I expand the visual evidence to include graphic art and photography as documents of and responses to the changing Parisian urban landscape. Examining this word-and-image matrix exposes a psychological ambiguity attributed to both Baudelaire and the populace regarding Haussmannization and modernity. In this context, ambiguity manifests, on the one hand, as nostalgia and anxiety about the present and, on the other, as a celebration of societal change and cultural renewal.[1] Here, however, ambiguity takes center stage, as it applies to Baudelaire's double-sided attitude toward modernity as manifested during Haussmannization. No less important, I treat ambiguity as an analogue of irony and *dédoublement*. I do so based on the definition of ambiguity as not only an expression of uncertainty, but as a language structure imbuing at least two differing concepts to verbal (or visual) meanings.[2]

In my argument, I track the creative responses to daily encounters with urban clamor and chaos. This follows two contiguous trajectories on the artistic and sensory map of Haussmannized Paris. The first presents Baudelaire's use of ambiguity as analogous to irony and the sensory experiences from which irony ensues. This line of inquiry focuses on Baudelaire's critical praise for Charles Méryon's etchings of Paris views in his *Salon of 1859*, viewed in tandem with his lyric poems collected in the second edition of *Les Fleurs du Mal*, titled *Tableaux Parisiens* (1861; Paris Scenes). The second trajectory, addressed in Chapter 4, recovers a collective ambiguity by aligning Baudelaire's art criticism and *Le Spleen de Paris*, his prose poems, with the photographic record of old and new Paris.

To gain a deeper understanding of these paths, I delve into the critical aspects of Haussmannization and its impact on Baudelaire and Baudelairean modernity. Additionally, I introduce recent advances in sensory studies in social sciences and humanities. When applied to the nineteenth-century French context, these sensory studies provide tangible proof to substantiate bodily reactions to Haussmannization as depicted in my word-and-image matrix of cultural ambiguity.

Haussmannization

As previously noted, Baudelaire's "The Painter of Modern Life," drafted in 1859 and published in 1863, anticipated and mirrored the actual events of social and physical disruptions that the writer and artist experienced during the Second Empire's modernization of Paris. In 1853, Napoléon III appointed Georges-Eugène Haussmann as the prefecture of Paris, leading to these widespread changes. Haussmann designed a comprehensive

DOI: 10.4324/9780429442315-3

plan that exceeded all earlier, albeit sporadic, efforts to modernize Paris. Haussmannization continued throughout the 1880s and beyond despite Haussmann's resignation six months before the Franco-Prussian War began in July 1870 and Napoléon III's fall. This latter decade is crucial in considering how Baudelairean artists continued to display traits and practices aligned with the concept of "the painter of modern life."

Haussmannization entailed a series of *grands travaux*, the enormous public works for modernizing sanitation, safety, transportation, leisure, and commerce.[3] These projects included widening and straightening streets and boulevards, expanding the underground sewage system, and installing underground networks of gas pipes and above-ground grids of streetlights and *pissoirs* (public urinals). Under Haussmann's plan, two new railway stations were built to complete the national rail system that radiated from Paris to all parts of France. These monuments to modernity featured wide-open networks of glass-and-iron construction, a system repeated for ten pavilions that comprised the new central markets, Les Halles. Four large parks were placed at the city's cardinal points to monumentalize public leisure, while smaller parks and squares were created for residential and civic spaces. Haussmannization also fueled a boom in commercial real estate and private building projects, such as multistoried retail-office-apartment buildings and department stores. A standard system of height, proportions, and classical details unified these structures, as did the treatment of their facades, which were dressed in the creamy white Lutetian limestone quarried below the city. Mansard roofs and uniformly spaced trees planted along the boulevards completed the image of a modern, efficient, and harmonious Paris reborn.

The human cost of such grandiose schemes was significant. The capital city saw a mass migration of laborers and workers in related service industries from the provinces. Builders, pickaxe wreckers, bricklayers, stone-pavers, haulers, metal welders, machinists, stone breakers and carvers, and planters and gardeners were just a few of the laborers employed. The city they had come to build was stretched to its limits by the new labor force. Unable to afford rising rents in newly constructed housing, these migrant laborers and their urban counterparts were relocated to the city's outer rings. This largely rural terrain formed the eight new *arrondissements* (administrative districts) added during Haussmannization to the twelve outwardly spiraling districts that had defined the city since 1795. Herein lies the actual human cost of Napoléon III and Haussmann's urban renewal scheme: the disruption of human lives and family legacies. The Parisian working class and the bourgeoisie were equally disrupted by Haussmannization: old neighborhoods were bulldozed to make way for the newly straightened streets and axial boulevards with uniformly aligned building facades. For two decades, houses, shops, workshops, churches, stables, and taverns disappeared in a single day or night, causing occupants to become nomads in their native city. Indeed, Honoré Daumier's caricatures attest to the rapidly changing cityscape, where widespread demolitions defamiliarized the familiar (Figure 2.1).[4]

In "Un Effet des Démolitions dans Paris." (Result of the Demolishing of Buildings in Paris), a Parisian man returns home after a brief (business?) excursion and finds his domicile obliterated. In the moment, he realizes that his wife has also disappeared because of the demolition process and ensuing displacement. Yet this spousal loss represents much more—the loss of domestic life, a shared past, and a predictable future. Amidst the ruins of Haussmann's wrecking ball, the former household head is destitute of physical and psychical shelter.

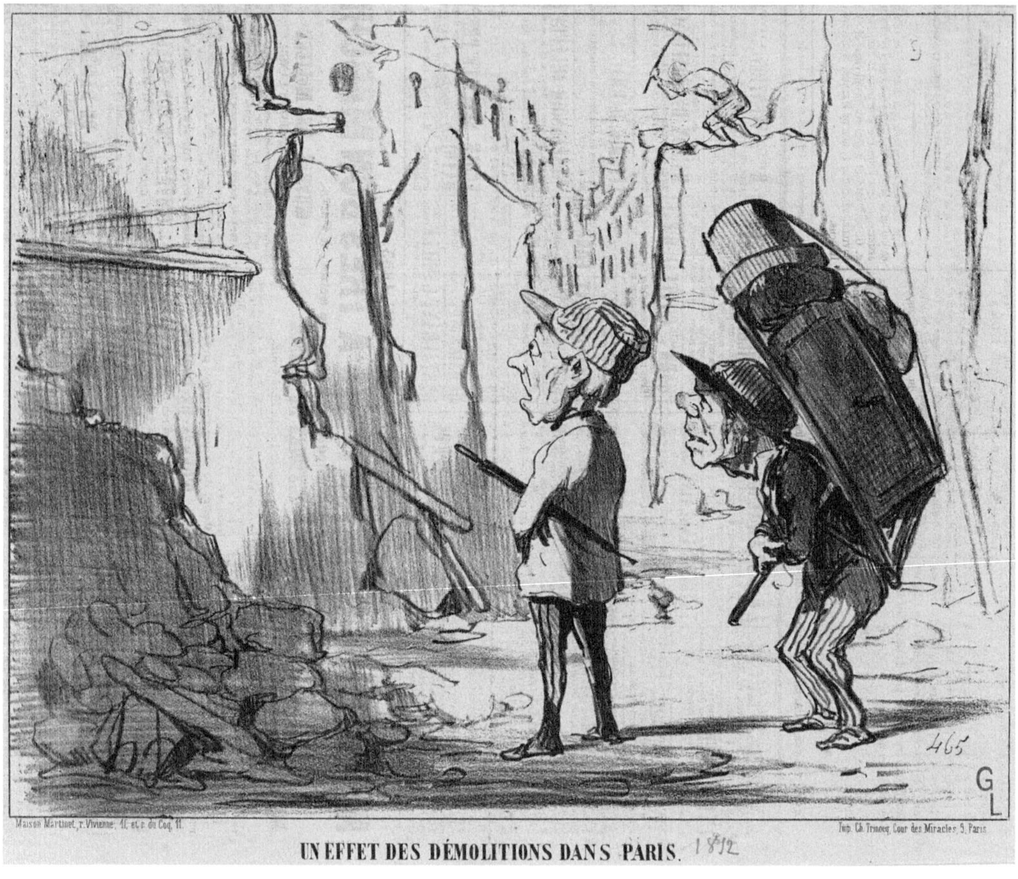

Figure 2.1 Honoré Daumier, "Un Effet des Démolitions dans Paris." (Result of the Demolishing of Buildings in Paris). In *Le Charivari* (December 12, 1852). National Gallery of Art, Washington, DC. Corcoran Collection. Gift of Dr. Armand Hammer. Full caption: "C'est pourtant bien ici que je demeure... et je ne retrouve pas même ma femme!" (After all, this is where I used to live... and I can't even find my wife anymore!)

Sensory Studies and Nineteenth-Century Paris

To consider Haussmannization as a historical event, we must also consider the sensory environment contributing to new artistic practices and aesthetic tenets. Sensory studies can bring back into focus the visceral aspects of historical places that their inhabitants experienced daily. In general, its practitioners "inventory... the sensations that were present at a given moment in history."[5] By extension, literary scholars who engage with sensory studies identify textual tropes that signify sounds, smells, tactile sensations, and perceptions. Nineteenth-century literary studies have significantly benefitted from the "sensual turn," given the acceleration of sensory experiences in urban centers at this historical moment. According to anthropologist David Howes, during the first half of the nineteenth century, "the senses of proximity, of touch, taste, and smell, which govern in depth the affective mechanisms, experienced an increase in their relative power[.]"[6] In this light, the stage was set for an exponential increase in the power of sensory agency during Haussmannization.

The thesis of urban historians Jill Steward and Alexander Cowan is especially apt for recovering the "empire of the senses" in Second Empire Paris. They contend "that sensuous encounters between individuals and environments are produced and structured, not just by their material features, but also by the particular social and cultural contexts in which encounters take place."[7] Accordingly, Haussmannization not only drastically changed and expanded the urban topography but also changed and expanded the sensory terrain. Steward and Cowan further propose that correlative with urban modernization was the modernization of "manufacturing, distribution, and retailing systems" and "the state of technology and forms of commercial and leisure activity," all of which were "factors influencing the nature of the sensory environment."[8]

In obtaining corroborative evidence from sensory studies, literary sensory studies that focus on nineteenth-century Paris and the flâneur undergird my arguments here. In a special issue of *Dix-Neuf*, *Rethinking the Flâneur: Flânerie and the Senses* (2012),[9] guest editor Aimée Boutin and her contributors reverse the hierarchy of vision and the flâneur's "disengaged spectatorship," modalities that had dominated twentieth-century scholarship.[10] Instead, they reintegrate vision with the other senses and re-immerse the flâneur and the flâneuse (feminine form of flâneur) within their sensory urban environment. In the issue's introductory essay, Boutin establishes that while each contributor "tackles a particular sensory system, [taken] together the contributions suggest how the flâneur renders the city a sensuous place, how urban sensations anchor the flâneur onto the streets where he is bathed in the multitude of sights, sounds, smells, tastes, and touches."[11] Boutin's other essay in the volume, "Aural Flânerie," for example, exemplifies how contemporaneous writings on the flâneur, such as "Honoré de Balzac's *Physiologie du Mariage* [1829] and especially Victor Fournel's *Ce Qu'on Voit dans Les Rues de Paris* [1858] represent the city in aural terms as musical score, and as harmonious or cacophonous concert."[12] These concerts often include street singers, musicians, criers, and vendors. While other writer-flâneurs may have been repulsed by these same sounds as aural dissonance, each writer translated urban sounds into "aesthetic creations" that would aurally resonate for their readers.[13]

In many ways, these studies substantiate what I demonstrate in the preceding chapter—Baudelaire's artist-flâneur, Constantin Guys,[14] not only observes the urban crowd; he embodies all the fleeting, transient sensations of the crowd. Equally important, Baudelaire consecrated the new, albeit noisy Parisian boulevards as the artist-flâneur's new home: "The crowd is his domain.... His passion and his profession is to merge with the crowd. For the perfect idler [*sic* flâneur],[15] for the passionate observer it becomes an immense source of enjoyment to establish his dwelling in the throng, in the ebb and flow, the bustle, the fleeting and the infinite."[16]

As "a kaleidoscope with a conscience," Guys savors not only the "ebb and flow" of the boulevards but absorbs its tumults and shocks. As "the lover of universal life," the artist-flâneur "moves into the crowd as though into an enormous reservoir of electricity."[17]

Baudelaire likewise incorporated boulevard sounds and tactility into his poetry.[18] For example, he centered the clamor of the street in the first line of "À une Passante" ("To a Passerby"): "The street about me roared with a deafening sound[.]" In the prose poem "Perte D'auréole" (1869; "The Loss of a Halo"), Baudelaire foregrounded the sensation of touch at its inception. Here, the poet recounts that as he "was crossing the boulevard, in great haste, and as [he] was hopping about in the mud, amid that moving chaos where death arrives at a gallop from all sides at once, [his] halo, in a sudden start, slipped from [his] head into the mire of the macadam."[19] Baudelaire's parodying of the poet's exalted status aside,[20] this scenario captures the tactile sensations of carriage-borne airstreams, slippery, splashing mud puddles, and sticky macadam pavement. Tactile sensations are compounded by human touch. The prose poem "Le Joueur Généreux" (1869; "The Generous Gambler") opens with the narrator recounting the physical sensation of being grazed by a passerby:

> Yesterday, amidst the crowds on the boulevard, I felt myself brushed against by a mysterious Being whom I had always longed to know and whom I recognized immediately, although I had never seen him. He no doubt nurtured a similar desire where I was concerned for as he passed by he gave me a meaningful wink which I hastened to obey.[21]

Ironically, this shared transient touching triggers the narrator's encounter with the devil, from which Baudelaire contemplates the being's deception and the evil it entails.

In his poems about the flâneur's urban sensations, Baudelaire abandons the artist-flâneur's amorous "enjoyment" of crowds and bustling boulevards. Instead, as discussed in Chapter 1, we are reminded of Baudelaire's fear of the maddening assault on his senses caused by the city's clamorous sights and sounds (see Figure 1.8). This deviation from the celebratory tone of "The Painter of Modern Life" is consonant with the overarching ambiguity that Baudelaire fostered in his poems and his art criticism. Here, too, sensory studies pertain, evidencing that ambiguity also registers sensory experiences. In Howes' words, "Sensory studies approaches themselves emphasize the dynamic, relational (intersensory—or multimodal, multimedia) and often conflicted nature of our everyday engagement with the sensuous world."[22]

Baudelaire's Ambiguity

Despite the technological order, Haussmann's transformation of the city of Paris as France's national capital into the capital of international modernity signaled a mutation of Paris as a fixed *lieu de mémoire*—a cohesive monument of centralized rule and national heritage—to a *lieu d'expérience*—a visceral embodiment/environment of fragmentation, dissonance, and change.[23] As noted in the previous chapter, when Baudelaire wrote "Of the Heroism of Modern Life," Paris was already marked by sporadic urban renewal schemes. Baudelaire first defined modern beauty in this context as "something eternal and something transitory." He extolled Paris as *un lieu d'expérience*, where the artist discovers "the heroism of our day" in "the scenes of high life and of the thousands of uprooted lives that haunt the underworld of a great city."[24] As we have seen, later, in "The Painter of Modern Life," the artist Constantin Guys becomes the hero who overcomes the sensory assaults of urban crowds, sights, and sounds.

What needs addressing here, however, is the discrepancy between Baudelaire's verbal renderings of modernity in his art criticism and the visual responses to his texts by Baudelairean artists. Thus, while Manet, Degas, and Caillebotte figured Baudelaire's urban types in the subject matter of their paintings, they each developed techniques of representation that deviated from or conflicted with Baudelaire's own, albeit conventional, aesthetic expectations. Baudelaire not only defined modernity in reference to Guys' ink-and-watercolor sketches of Parisian society and fashions (rendered for the *Illustrated London News*), but he also extolled Charles Méryon's somber (even Rembrandtesque) etchings of Paris *ancien* (which would inspire *Tableaux Parisiens* and

Le Spleen de Paris).[25] Both artists lacked popular and professional notoriety in their own day. We have already reconciled, in Chapter 1, Baudelaire's praise for Guys against the background of his familiarity with brain science, his fear of madness, and the din of the Parisian crowd. In this case, Baudelaire created an artistic persona who enacted irony as an aesthetic response to the visceral experience of psychic and physical rupture. In the following section, we will see that Baudelaire's commentary on Méryon also presents the poet's creation of an artistic persona configured to embody his nostalgia for Paris *ancien*, a yearning that irony also remedies.

Reassessing Walter Benjamin's Baudelaire and Méryon's "Elective Affinity"

In Baudelaire's assessment of Guys' modernity, sketch-like techniques and ephemeral transparency convey fleeting cognitive experiences and physical engagement with the myriad Parisian populace. To be sure, this description could equally apply to urban venues rendered by Manet and Degas, as seen in *Music in the Tuileries* (1862) and *Place de la Concorde* (1876) (Figures 1.5 and 1.6). However, it does not pertain to the etchings of Paris by Méryon, whom Baudelaire championed in his *Salon of 1859*.

In the *Salon of 1859* section on "landscape," Baudelaire problematized the absence of "a genre [of] the landscape of great cities" at the present Salon but did so to recall "a series of etched studies of the most picturesque views in Paris" by "a strange and stalwart man," M. Méryon. Even though Méryon was not an exhibitor at the Salon, Baudelaire praised the former sea-captain-turned-engraver as "a mighty artist, ... who bade farewell to the ocean's solemn adventures to paint the gloomy majesty of this most disquieting of capitals." Not only does Baudelaire extol Méryon's macabre interpretation of the Parisian landscape, but also the artist's technical precision—"the sharpness... of his drawing"—in capturing the bizarre intermingling of old and new.[26]

> Seldom have we seen the natural solemnity of a great city depicted with more *poetic power*: the majesty of the piles of stone; those spires pointing their fingers to the sky; the obelisks of industry vomiting a legion of smoke against the heavens; the enormous scaffolds of the monuments under repair, pressing the *spider-web-like and paradoxical beauty* of their structures against the monuments' solid bodies; the steamy sky, pregnant with rage and heavy with rancor; and the *wide vistas whose poetry resides* in the *dramas* with which one endows them in *one's imagination*–none of the complex elements that compose the *painful and glorious décor* of civilization has been forgotten.[27]
>
> (emphasis added)

In July 1859, when Baudelaire's *Salon* was published, Méryon had been undergoing treatment for insanity at the psychiatric hospital Charenton.[28] Baudelaire's acknowledgment of Méryon's illness is marked by both awe and fear: "But a cruel demon touched M. Méryon's brain; a mysterious delirium has blurred those faculties that seemed as solid as brilliant."[29] From Baudelaire's critical perspective, Méryon's madness displayed his genius for creating shocking and surprising artworks. Personally, Méryon's madness roused his own mental vulnerability. Following a face-to-face meeting with Méryon, Baudelaire wrote to his publisher Auguste Poulet-Malassis on January 8, 1860, wondering how he had escaped a similar fate: "After [Méryon] left me, I wondered how it was that I, who has always been, in the mind and nerves, everything necessary to become mad, I should not become so. Seriously, I sent thanks to the heaven of the Pharisee."[30]

Walter Benjamin was among the first, and subsequently most influential, theorists to recognize Baudelaire's tribute to Méryon and their shared views of Paris transformed. Based on his analysis of Baudelaire's *Salon of 1859*, Benjamin revealed Baudelaire's "elective affinity" with Méryon in his essay "The Paris of the Second Empire in Baudelaire." Benjamin used this phrase to describe the poet's and the engraver's parallel life experiences:

> The two men had an *elective affinity* to each other. They were born in the same year, and their deaths were only months apart. Both died lonely and deeply disturbed—Méryon as a demented person at Charenton, Baudelaire speechless in a private clinic. Both were late in achieving fame.[31]
>
> (emphasis added)

This description merits attention since it emphasizes the negative side of Baudelaire's ambiguity toward Paris's transformation. Here and elsewhere, Benjamin seeks to define Baudelaire from his Marxist critique as the "lyric poet in the era of high capitalism." Accordingly, Baudelaire's poetry mirrors the "decrepitude" and "the horrible social realities" of the Second Empire and its capital city.[32] Given this critical perspective, Benjamin suppresses Baudelaire's celebration of modernity and cultural renewal, further revealed by aligning the poet with the etcher.

In his essay, Benjamin's observations of Baudelaire and Méryon's elective affinities follow a brief discussion of the Haussmannization of Paris. Benjamin suggests that this natural alliance extends to the social forces that bind the two, defining their modernity and shared ambition of modernizing antiquity. Indeed, Benjamin considered Baudelaire's "short piece on Méryon" among the best of his prose work and "an homage to modernism." He continues, "For in Méryon, there is an interpenetration of classical antiquity and modernism; ... in [Méryon], the form of this superimposition, the allegory, appears unmistakably."[33] Conversely, the German critic faults the poet for failing to achieve a comparable synthesis in his poetry writing. He asserts as well that Baudelaire's aim to fuse "the transitory" and "the eternal" constitutes the "weakest" aspect of the poet's "theory of modern art."[34] Notably, Benjamin pays little attention to Guys and the celebratory tenor of "The Painter of Modern Life," the essay in which Guys figures. Because Benjamin aimed to signal to his readers the incipient social ills that issued from Second Empire Paris and persisted in his own time, Baudelaire's verse poem "Le Cygne" ("The Swan" in *Tableaux Parisiens/Fleurs du Mal* [1861]) provided an occasion to interpret an allegory of modern Paris and its "decrepitude." In this poem, Baudelaire bemoans the "Old Paris [which] is no more" where the landscape is strewn with "piles of rough-hewn capitals and shafts," and where, among "new palaces, scaffolding, blocks, /... everything becomes an allegory for me, / And my dear memories are heavier than rocks." In this allegorical respect, Baudelaire's achievement is comparable to Méryon's, an elective affinity suggested by Benjamin's pictorial interpretation of the poem:

> It is not an accident that it is an allegory. ... The stature of Paris is fragile; it is surrounded by symbols of fragility—living creatures and historical figures [whose] common feature is *sadness about what was and lack of hope* for what is to come.... Whenever Paris occurs in the *Fleurs du mal*, it bears signs of this *decrepitude*.
>
> (emphasis added)

Benjamin attributes the enduring value of Baudelaire's Paris verse poems to this essential "decrepitude" and the poet's "reservation against the big city."[35]

We can use Benjamin's interpretation of Baudelaire's *Salon* excerpt on Méryon and his poem "The Swan"—signifying decrepitude, etc.—to delve more deeply into the poet's and engraver's elective affinity and thereby go beyond their shared personal history of madness, muteness, and deprivation. To do so, we will focus on their common nostalgia for Paris *ancien* as manifested in their (failed) efforts to publish, in a single volume, their works portraying Paris *vécu*, the "lived" city they inhabited. By reframing this alliance, I seek to reverse the negative impression left by Benjamin's reading of Baudelaire's elective affinity with Méryon. This reversal entails reviewing Baudelaire's letters regarding the publishing project, registering the impact of Méryon's etchings of Paris on Baudelaire's poetry, and recognizing Baudelaire's ironic mediation of Méryon's macabre visions.

Baudelaire's (Virtual) Collaboration with Méryon

In the 2001 exhibition catalogue *Paris, 1860: Eaux-Fortes sur Paris & "Les Tableaux Parisiens,"* Jacques Damade chronicled Baudelaire's collaboration with Méryon in letters dating from 1859 to 1860.[36] This chronology provides the basis for highlighting Baudelaire's involvement with and support of Méryon and their proposed publishing project. The published album[37] would have assembled the twenty-two etchings Méryon created between 1850 and 1854, the *Eaux-fortes sur Paris* series. Méryon published twenty-one etchings in three series between 1852 and 1854, collectively titled *"Eaux-fortes sur Paris" par C. Méryon, MDCCCLII* (as per the title page Méryon designed).[38] However, the plates were never published as a single set, nor in the album Baudelaire and the artist envisioned.[39]

According to Damade, Baudelaire and Méryon first met at the end of December 1859 or the beginning of January 1860, most likely at a café in the Gare Saint-Lazare neighborhood where they both resided. In the January 8, 1860 letter reporting this meeting to Poulet-Malassis, Baudelaire reveals his attraction to the enigmatic, even the bizarre, content of Méryon's compositions

> In one of his large engravings [*Le Pont-au-Change*],[40] he substituted for a small balloon a swarm of birds of prey, and when I told him that it was unlikely to put so many eagles in a Parisian sky, he replied that this was not without foundation, since *these people* (the emperor's government) had often unleashed eagles to study the omens following ceremonies/rituals–and that it was printed in the newspapers, even in *Le Moniteur*.[41] (Figure 2.2)

To this, Baudelaire added observations that underscored Méryon's fascination with the macabre and the fantastic:

> I must say that he is in no way hiding his respect for superstitious roads, but he says he sees evil and the cabal everywhere. He noted, in another of his plates [*Le Pont-Neuf*], that the shadow carried by the stonework of the Pont Neuf on the side wall of the dock exactly represented the profile of the sphinx - it was, for his part, completely unintentional, and that he had noticed this oddity later in remembering that this

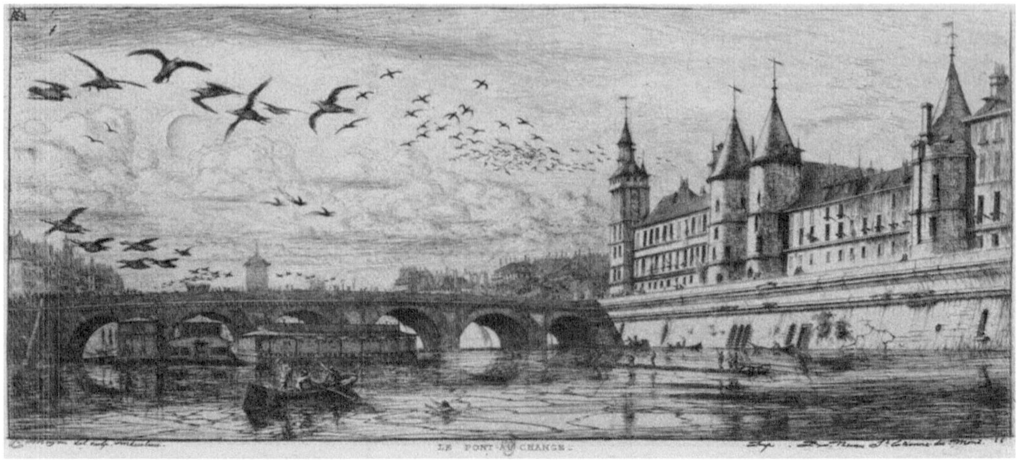

Figure 2.2 Charles Méryon, *Le Pont-au-Change*, from *Eaux-fortes sur Paris* (Etchings of Paris), 1854, etching on laid paper, ninth state, 6 x 13 1/4 in. (plate). Bibliothèque nationale de France, Paris. Image: gallica.bnf.fr/Bnf.

> drawing was done shortly before the coup. But the Prince [Napoléon III] is the actual being who, by his actions and his face, most resembles a sphinx.[42] (Figure 2.3)

At the same meeting, Méryon asked Baudelaire if he had read Edgar Allan Poe's novels and if he believed in Poe's reality. While Baudelaire answered the first question in the affirmative (having translated Poe's works in 1857), Méryon replied to the second query according to his visual translation of Poe's "La rue MORGUE."[43] Baudelaire quoted Méryon:

> I made a drawing of the Morgue. An orangutan. I often compared myself to a monkey. The monkey assassinated two women, the mother and her daughter. I also morally assassinated two women, the mother and her daughter. I have always taken the novel as an allusion to my misfortunes.[44]

This first meeting between the poet and etcher thus established their mutual affinity for the bizarre. Baudelaire had already professed this kinship in his review *Exposition Universelle 1855*:

> *Beauty is always bizarre*. I do not mean to say that it is voluntarily, coldly bizarre, because in that case it would be a monster.... I mean that it always contains a bit of strangeness, naïve strangeness, not intentional but unconscious, and it is this strangeness that causes it to be particularly Beautiful.[45]
>
> (original emphasis)

Less obvious is what Baudelaire would have seen and felt in viewing the two etchings that Méryon showed him at this meeting. Both *Le Pont-au-Change*[46] and *Le Pont-Neuf* portray ancient monuments that, at the time of Méryon's renderings (1854), had yet to be affected by Haussmann's *grands travaux*. By 1859/60, however, Haussmannization

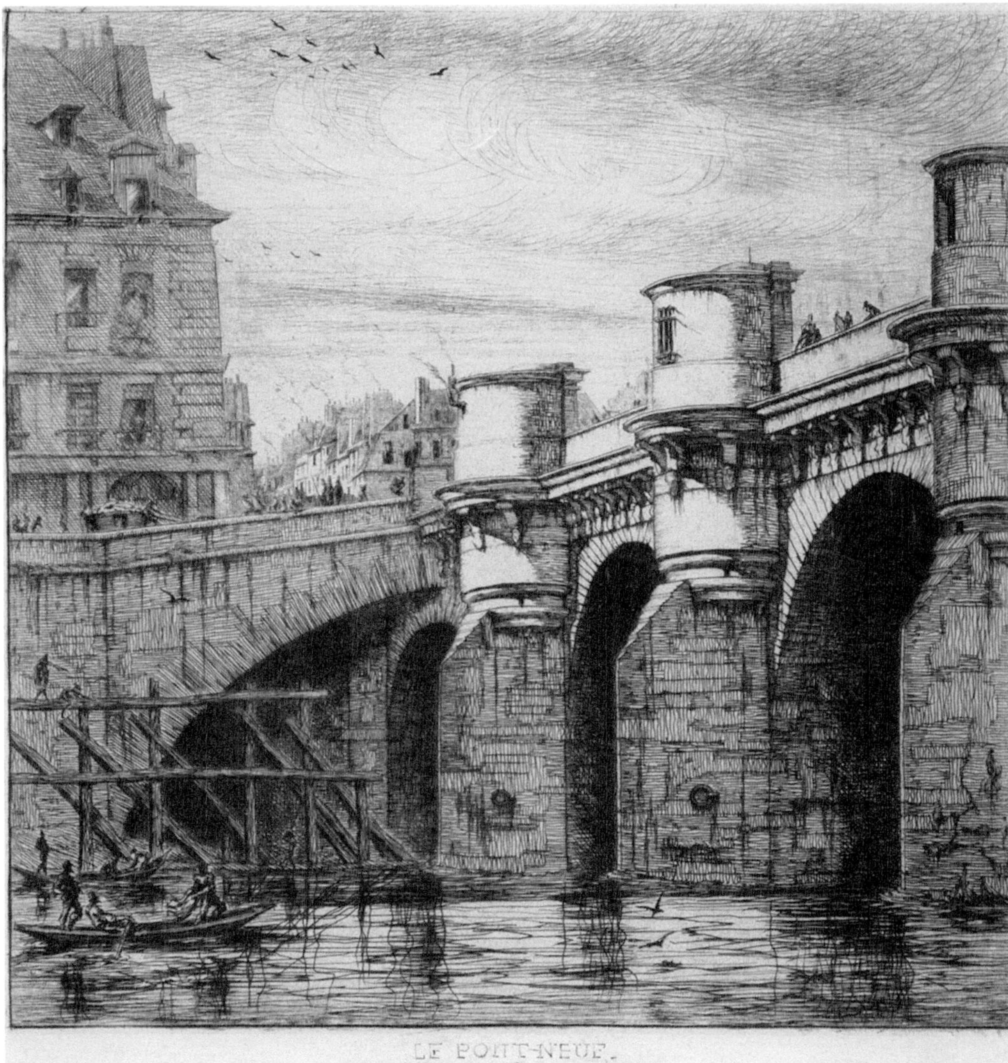

Figure 2.3 Charles Méryon, *Le Pont-Neuf*, from *Eaux-fortes sur Paris* (Etchings of Paris), 1853, etching and drypoint, tenth state of eleven, 7 1/4 x 13 1/4 in. (plate). Metropolitan Museum of Art, New York. Bequest of Julia H. Manges, in memory of her husband, Dr. Morris Manges, 1960.

would have changed both the artist's and poet's perceptions and recollections of these *lieux de mémoire*.

Both bridges linked the Île de la Cité,[47] the Roman birthplace of Paris, to the Right Bank and for centuries served as centers of commerce and entertainment. First built in the ninth century and rebuilt in the seventeenth century, the Pont-au-Change (Exchange Bridge) was dismantled, beginning in 1858, and replaced by Napoléon III's new bridge, completed in 1860. The Pont-Neuf, initiated by Henri III in 1578 and completed by Henri IV in 1606, consists of fortress-like crenellations, massive towers, and half-circle bastions. In Méryon's depiction, the bridge's enormous stonework dwarfs human figures, and its gigantic arched vaults shroud the underpasses in absolute darkness. Although Haussmann's wrecking ball spared the Pont-Neuf, Méryon's etching shows men with pickaxes demolishing the original one-story bastions (built for pedestrian shelter) to make way for new gaslight lampposts. As displayed in Baudelaire's letters, at stake here is the ongoing demolition and construction at the city's historical center. This activity not only rendered Méryon's prints audible but also rendered them sites of emotional conflict.

Given the poet's and artist's mutual interests—an admiration for Poe, melancholia for old Paris, and a penchant for the bizarre—Baudelaire responded enthusiastically to the proposed joint publication; by mid-February 1860 he had started planning the project, after first discussing it with Méryon's publisher, Auguste Delâtre. However, the poet may have already conceived of a similar project weeks before his initial meeting with Méryon. In a letter to Victor Hugo, dated December 13, 1859, Baudelaire had preemptively promoted his creative contribution. Here, Baudelaire speaks of the delivery of Méryon's etchings to Hugo but only to direct the writer to the letter's attachment—the excerpt on Méryon from his *Salon of 1859*:

> Sir, I learned with great pleasure that M. Méryon and his publisher and friend, M. Delâtre, propose to send a copy of the beautiful compositions of some views of Paris, drawn and etched and carefully printed by the artist himself. I take this opportunity to attach an excerpt of a work on the arts where your name again came to offer itself to my pen.[48]

This gesture implies that his review of Méryon will enhance the meaning of the images, not only for Hugo but for viewers in perpetuity.

In a letter to Poulet-Malassis dated February 16, 1860, Baudelaire refers to the plan as initiated by Delâtre. He recounts that Delâtre had asked him to write text for the album, which Baudelaire describes as "some reveries of ten, twenty or thirty lines [written] on the beautiful prints, philosophical reveries of a Parisian flâneur."[49] But Baudelaire already voices frustration with Méryon, "And Méryon! Oh! it's intolerable." Despite Delâtre's initiative, Méryon delays the publication due to his capriciousness and inattentiveness:

> But M. Méryon, who doesn't see things in the same way, intervenes. One should say, 'On the right, you see this, on the left, you see that.' One should look for notes in old books. One should say, 'Here there were originally twelve windows, which the artist reduced to ten,' and finally one is supposed to go to City Hall to inquire about the exact time of the demolitions. M. Méryon talks, gazing at the ceiling, ignoring all remarks.

One week later, on February 23, Méryon wrote to Baudelaire informing him that he had sent him a portfolio of his "*Vues de Paris*" in acknowledgment of the poet's devotion rendered to him. Here, too, Méryon attests to their mutual imagining of "curious things of the past," hoping that his etchings will have the same "dreaming" effect on his friend as the etcher had experienced while making them.[50] Méryon ends by assuring Baudelaire that he will send the notes about his etchings to help him with his writing for the album.

Baudelaire continues to have faith in the project. Writing to his mother, Mme. Aupick, on February 28, he is confident that the long-awaited album of Méryon's "*vues de Paris*" (i.e. *Eaux-fortes sur Paris*) will please her, especially the views taken before the "*grandes demolitions*" (major demolitions).[51] Writing to his mother several days later, Baudelaire accounts for purchasing three copies of Méryon's album, one of which he will send to her and another he will keep for himself. In this letter, dated March 4, Baudelaire justifies the high price he paid for the albums, given his admiration for Méryon and his prolonged search for the images. More importantly, Baudelaire invokes the affective resonance of Méryon's images that straddle the past and present, an evocation that constitutes the etchings' *poetic* resonance:

> Relative to the etchings [...], I told you that I wanted them and that I searched for them for several years. The first time that I saw them, I judged that this man had genius. He had just left Charenton and had not recovered. I promised to write a text for his etchings.... You are mistaken in calling these the *old* Paris. They are poetic views of Paris, such as they were before the immense demolitions and road repairs ordered by the Emperor. In some scenes, you will even see buildings enveloped in scaffolding (for example the clock tower of the *Palace of Justice*).[52]
>
> (original emphasis)

Baudelaire's distinction between mere mimetic representations of old Paris and poetic views is essential. It correlates poetry's transformative function with its power to evoke memories of pre-Haussmann Paris. As we shall see, this distinction underscores Baudelaire's efforts to extend Méryon's visual poetics into textual verse.

One week after writing to his mother, Baudelaire's optimism wanes. On March 11, 1860, he informs Poulet-Malassis that Méryon has reneged on the project, fearing the intrusion of the poet's text on his images.[53] Admitting his embarrassment regarding "*l'affaire*" Méryon, Baudelaire explains: "M. Méryon is repelled with a sort of horror by the idea of text comprised of twelve small poems or sonnets; he has refused the idea of poetic meditations in prose." Finally, Baudelaire agreed to provide Méryon with a "text in the style of a guide or manual, unsigned," an offer that the etcher later rejected.[54]

Although Baudelaire's creative role in the album's production had ended, as had his communication with Méryon, the poet remained active in realizing the publication of a bound volume of the album and in promoting the etcher's genius.[55] In a letter from Méryon to Baudelaire, dated December 21, 1860, the artist expresses his resentment that Baudelaire had not responded to his recent letters and feared an end to their relationship. Méryon also reminds Baudelaire of his promise to write an article about the artist.[56] Although no further communication exists between the two,[57] Baudelaire made good on his promise in 1862 when he re-published his *Salon of 1859* excerpt on Méryon in *Peintres et Aquafortistes*. Still more revealing of Baudelaire's enduring attention to Méryon's etchings are his creative works from their short-lived collaboration.

Tableaux Parisiens: Melancholy of the Senses

When Baudelaire published the second edition of *Les Fleurs du Mal*, released in February 1861, he re-arranged the first 1857 edition to include a new subdivision titled *Tableaux Parisiens*. Eight of the eighteen poems collected under this heading were retained from the first edition, and ten were composed between the second half of 1857 and the end of 1860. A chronology of the new poems, which had been published separately in diverse literary and art journals, reveals that Baudelaire wrote all but one of these verses ("Paysage" ["Landscape"]) after he had seen Méryon's *Vues de Paris*.[58] The five poems that depict Paris and its inhabitants and their proximity to the *Salon of 1859* excerpt on Méryon are of interest and provide a starting point for tracking Baudelaire's engagement with the etchings. Published in the June/July issue of *La Revue Française*, the *Salon of 1859* was followed by the publication of "Les Sept Vieillards" ("The Seven Old Men") and "Les Petites Vieilles" ("The Little Old Women") on September 15, 1859. Notably, "The Swan" appeared on January 22, 1860, shortly after Baudelaire and Méryon's first meeting. "Rêve Parisien" ("Parisian Dream") was published on May 15, and "Les Aveugles" ("The Blind") and "To a Passerby" were published in October 1860, both appearing just months after Baudelaire had purchased the three sets of Méryon's *Vues de Paris* (March 1860).

That Baudelaire physically handled and scrutinized Méryon's etchings alongside the artist, with his friends, and by himself is crucial for establishing how the poet translated his embodied experiences into poetic forms. This ocular flânerie furnished Baudelaire with a memory bank and an imaginary that compelled his creative work and his actual traversals through the city. *Tableaux Parisiens* manifests as "the philosophical reveries of a Parisian flâneur" that Baudelaire had planned for Méryon's *Vues de Paris*. Indeed, the pictorial title for this verse collection evokes the poet's debt to Méryon's "tableaux." As displayed in Baudelaire's letters, Méryon's etchings also substantiated the poet's cognitive ambiguity toward modern Paris. Populated by enigmatic human figures, these urban scenes evoke the "mysteries" of Paris that lingered in the medieval remnants surrounding *les lieux de mémoire* (such as the Conciergerie) and lay hidden behind Haussmann's renovations.

To extend this inquiry into Baudelaire's affinity for Méryon's etchings, let us closely examine *Le Pont-au-Change* and *Le Pont-Neuf*, the two plates that Baudelaire viewed alongside Méryon at their January meeting. Perhaps they were aided by a magnifying glass or loupe,[59] necessary apparatuses (then and now) for discerning in these images human life in action and buildings in decay, demolition, or repair. Here, as in the other plates, Baudelaire would have seen inhabitants of Paris *ancien*. These figures are minuscule in scale, dwarfed against the monumental structures or the vast Seine River. Their physiques are rendered ephemeral by the loose hatchings that mark their presence. Often, figures appear in shadows cast by scaffolded and fissured stone walls. Others appear as shadow figures silhouetted against sunlit pavement, quays, and buildings. Others catch points of light in deeply shadowed boats, barges, archways, and streets.

These figures, their activities, and modes of transport imbue the scenes with clamorous urban noise, resounding with wheels, hoofs, and boots on cobblestones, pickaxes on stone, river currents, and marine work. In *Le Pont-Neuf*, three figures animate and resonate with the scene. On the left side, the urban populace is obscured by shadows cast by a Henri IV-era house, a pavilion that connects the bridge to the Place Dauphine. On the right side, figures on the bridge are silhouetted against distant buildings lit by raking sunlight. Although the figures seem incidental to the built environment, closer inspection reveals laborers on quay-side scaffolding, boat-rowers, horse-pulled carts and cargo drivers, and pedestrian strollers and spectators. Mystery imbues the commonplace, however. Are the spectators on the bridge witnessing a criminal act amidst the boaters? In the boat closest to the picture plane, a standing man twists and turns as he throws something—or someone—overboard.

Mystery is also present in *Le Pont-au-Change*. Here, the sunlit Seine River reflects the apparent calm of fishermen, bargemen, and boatmen at work. But a drowning man breaks the river's mirror surface. Barely visible at the bottom center of the plate, the man frantically flails and shouts for help while boatmen, who focus on the hot air balloon above, ignore him. A crowd on the bridge also hails the balloon with raised arms and joyful shouts. Another harrowing note upsets the celebration. Although a unified crowd seemingly populates the bridge, a hearse and a line of mourners move across the overpass in the opposite direction of the celebrants.[60]

Méryon's *La Morgue* (*The Morgue*) provides an actual instance of Baudelaire's close looking (Figure 2.4). In the letter to his mother regarding the delivery of the Méryon

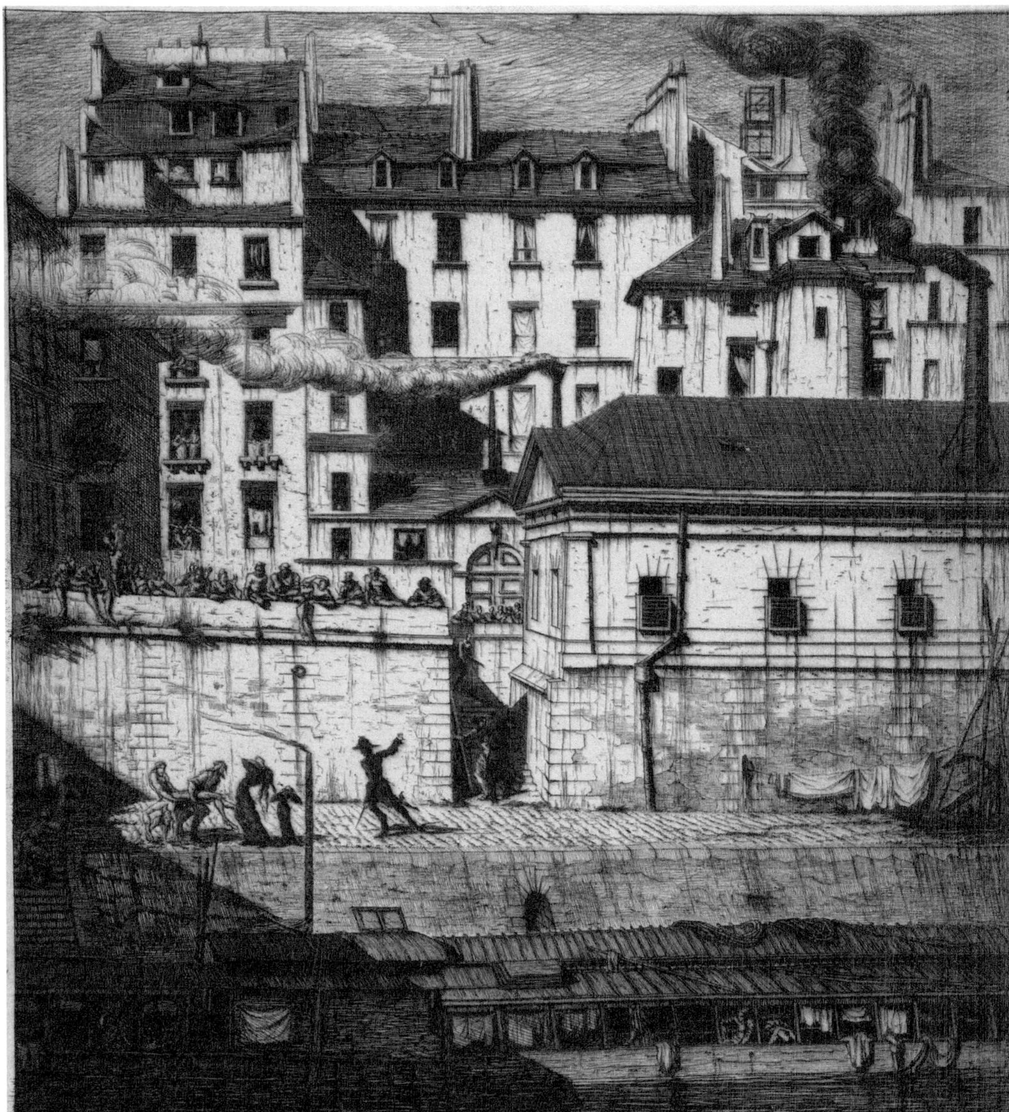

Figure 2.4 Charles Méryon, *La Morgue* (*The Morgue*), from *Eaux-fortes sur Paris* (Etchings of Paris), 1854, etching and drypoint on laid paper, third state of seven, 8 3/4 x 7 3/4 in. (sheet). Metropolitan Museum of Art, New York. Bequest of Susan Dwight Bliss, 1966.

album, Baudelaire described this plate by noting the figures and the drama that unfolds, silhouetted against the Morgue's massive stone wall:[61] "*La Morgue*. A dead body has just been pulled from the river. Women grieve. A policeman. Crowd."[62] In "Baudelaire and Méryon, Painters of the Urban Landscape," Lois B. Hyslop and Francis E. Hyslop expand on the drama that unfolds in the etching. Their analysis, in effect, simulates Baudelaire's close looking and his ensuing "dreaming" before Méryon's scene:

> In none of Méryon's etchings is the drama more somber than in that of *La Morgue*. On the street in front of the massive buildings, tiny figures act out a tragedy that the artist may have witnessed during his lonely walks along the Seine. Sailors carrying a corpse that has been dragged from the river, are being directed by a policeman who is pointing to the entrance of the Morgue. A distraught woman stands with her back to the viewer; a small girl weeps, her head in her hands. A crowd has gathered on the parapet of the *quai* to watch the drama that is taking place, and in the windows of buildings adjacent to the Morgue can be seen figures gazing curiously at the morbid scene below.[63]

Hyslop and Hyslop's conclusion to this description offers a framework for restoring Baudelaire's affective response to Paris under Haussmannization:

> In his depiction of a tragic event that was a familiar part of the teeming life of the city and in his concern for suffering humanity trapped in the depths of that city, Méryon is doing what Baudelaire has so often done in *Les Fleurs du mal* and in his *Petits Poémes en prose*.[64]

It was just such dramas and their settings that Baudelaire extolled in the *Salon of 1859* section on Méryon, drawing attention to "the wide vistas whose poetry resides in the dramas with which one endows them in one's imagination."[65] Considering this testimony, one can see that in Baudelaire's poems about Parisian inhabitants, the city recedes as a backdrop against which human figures and human dramas are magnified. This aggrandizement, in turn, affects both the poet's and the reader's embodiment of the figures, their (sensory) surroundings, and the dramas that move them. At the same time, the poet's ironic self-reflection mitigates these dramas.

When read with these close viewings of Méryon's plates, Baudelaire's "Of the Heroism of Modern Life" (*Salon of 1846*) foreshadows his affinity with Méryon's etchings. As previously noted, Baudelaire exhorted modern artists to find inspiration not only in "scenes of high life" but also in the scenes:

> of the thousands of uprooted lives that haunt the underworld of a great city, criminals and prostitutes; the *Gazette des Tribunaux* and the *Moniteur* all prove to us that we have only to open our eyes to see and know the heroism of our day.[66]

Analogously, the poems in *Tableaux Parisiens* not only register a Paris neglected by Haussmann's modernization but the people that populated the enduring medieval districts behind the new boulevards: old men, old women, the blind, prostitutes, and beggars. As the inhabitants of Paris in the late 1850s, the late 1840s "heroes of modern life" now metamorphose into the impoverished anti-heroes of modern life; they embody the moral decay of the high life and the physical distress of the uprooted lives of the underworld.

"The Seven Old Men" and "The Little Old Women" feature Paris's haggard, despised elders, who embody Paris *ancien* and become its metaphor. These two poems most directly align with Baudelaire's immediate association with Méryon and, therefore, are our focus here. In each poem, broken-boned and crippled people inhabit the darkened, narrow streets behind the glittering white facades of Haussmann's boulevards. "The Seven Old Men"[67] is set against the backdrop of the "teeming, swarming city, city full of dreams /… [Where] everywhere mysteries flow like the sap in a tree." It is just such a mystery that the poet-narrator happens upon while walking "One morning" along a "gloomy," "squalid" street where the houses are shrouded by a "dirty yellow fog" and "shaken by the heavy dump-carts." From this miasma, an old man, in "tattered yellow clothes" appears before the poet. Paris now becomes a city full of nightmares. The man's haggard and evil appearance is the focus of the poem.

This shocking apparition transmutes the poet's initial empathy into fear: "His aspect would have brought him showers of alms / If his eyes had not gleamed with so much wickedness [;] … / And his long shaggy beard, … / Projected from his chin as stiffly as a sword." The man's distorted and deformed body resonates with human and urban decay. "He was not bent over, but broken; his back-bone / Made with his legs a perfect right angle, / So that his stick, completing the picture, … / Of a lame quadruped[.]" Then the "sinister" old man's "likeness followed him: beard, eye, back, stick, tatters" sevenfold; seven identical specters appeared, "Who came from the same hell, … / Were walking with the same gait toward an unknown goal." The monstrous spectacle unnerves the poet:

> Let him who laughs at my disquietude,
> And who is not seized with a fraternal shudder,
> Realize that in spite of such decrepitude
> Those hideous monsters had an eternal look!

Disgusted and terrified, the poet "turned [his] back on that hellish procession" and returned home. Alone and detached from the haunting "underworld of a great city," the poet reflects upon the decrepitude he had not only witnessed but now embodied:

> Exasperated like a drunk who sees double,
> I went home; I locked the door, terrified,
> Chilled to the bone and ill, my mind fevered, confused,
> Hurt by that mysterious and absurd happening!

The poet's ordeal ends with his reason defeated, while "[his] soul, old sailing barge without masts, / Kept dancing, dancing, on a monstrous, shoreless sea!"

Although the poet is unhinged, his retreat to solitude and psychic interiority re-enacts the *dédoublement* of ironic self-reflection. In this case, he has reflected upon the sensory shocks the city and its inhabitants imparted, inducing a metaphoric recovery of his "soul" in the infinite realm of time and space. Therein, he recognizes his agedness, which allies him with the old men in tatters and the ancient city that envelops them. In this dialectic of dualities—of the self and non-self, the physical and psychic—that irony engenders, Baudelaire once again resolves his fear of madness, a fear that Méryon's madness had recently invoked. As noted, in his January 8, 1860 letter to Poulet-Malassis, he had thanked "heaven" for sparing him a similar fate.[68] In the March 4 letter to his mother regarding the delivery of the *Vues de Paris* albums, Baudelaire suggests Méryon's own ironic resolve. About the etching *Le Stryge* (*The Gargoyle*), which depicts a gargoyle overlooking Paris, Baudelaire muses: "How the devil does this man draw [so] peacefully on an abyss, I do not know."[69] Here, we can assume Baudelaire's elective affinity with Méryon (as with Guys, "the painter of modern life") resides in episodic retreats from the urban clamor and crowds to master and represent its resonances in creative work, whether in etching or poetry.

While in "The Seven Old Men" the poet is repulsed by the old men's specters and must retreat to his home to realize his affinity with them, in "The Little Old Women"[70] the irony resides in the poet-narrator's keenness to see beauty and poetry beneath the women's worn and broken features. Indeed, the poet embodies the women as they traverse the city and bear its sensory and societal assaults on their bodies. In doing so, Baudelaire releases the women from their decrepitude and restores their youth by remembering the women's enduring beauty and invoking their youthful memories. Watching the procession of the old women, the poet recalls Victor Hugo's Éponine and Lais from Greece, among other "Mothers with bleeding hearts, courtesans, saints." In this self-doubling and reversal of aging, irony permeates "the philosophical reveries of a Parisian flâneur."[71]

"The Little Old Women" begins with an ironic inversion in the first of four sections. Amidst the urban turmoil, the poet scours "the sinuous folds" of "teeming Paris," where, as in "the old capitals," "*even horror, becomes pleasant*" (emphasis added). As he roams the urban landscape, the poet-flâneur watches "For singular creatures, decrepit and charming. / These disjointed monsters were women long ago, / [...] Monsters, hunchbacked, broken / Or distorted, let us love them! they still have souls." The poet's account of his peripatetic course of watching, remembering, and imagining the lives of women, past and present, is the subject of the second and third sections of the poem. However, the poet's eulogy to the old women's youth and his deploration of the ridicule they suffer are summarized in the second quatrain of the fourth, final part of the poem. Here, he recounts the societal and physical assaults on the women's bodies:

> O you who were charming or who were glorious,
> None recognizes you! A drunken ruffian
> Passing by insults you with an obscene remark;
> A dirty, nasty child frisks about at your heels.

The poet also speaks directly to the heroines of the modern city and testifies to his embodiment of their psychic beings:

> I see your untried passions come into full bloom;
> I live your vanished days, gloomy or filled with light;
> My heart multiplied enjoys all of your vices!
> My soul is resplendent with all of your virtues!

That these old women are among the lives "uprooted" by modernization manifests in their deformity, threadbare clothes, and unsteady gait. Baudelaire foregrounds these attributes in the first and the fourth sections. In the first instance, the poet describes the women as "Clothed in tattered petticoats and flimsy dresses / They creep, lashed by the iniquitous wind, / Trembling at the clatter of the omnibuses[.]" Tattered clothing and instability are not only symbols of the passage of time but also symbols of a city transformed, wherein native inhabitants become refugees of their old habitats and aliens in their new surroundings. In the fourth section, Baudelaire inscribes these symbols into a pictorial vignette, captured when the poet-flâneur's roaming has stopped. This vignette registers Baudelaire's close looking at Méryon's *Vues de Paris* and his re-imagining of Méryon's fragile, ephemeral figures silhouetted against the massive walls of Paris *ancien*.

> Thus you trudge along, stoical, uncomplaining,
> Amid the confusion of cities full of life,
> Mothers with bleeding hearts, courtesans, saints,
> Whose names in years gone by were on everyone's lips.
> ....
> Wizened shadows, ashamed of existing,
> With bent backs, you timidly keep close to the walls
> And no person greets you, strange destinies!
> Human wreckage, ripe for eternity!

Just as these two stanzas evoke Méryon's etchings of *The Morgue* and *Le Pont-Neuf*, the following final stanza evokes the nostalgia for old Paris that the poet shared with the etcher. In this regard, the old women to whom the poet bids farewell becomes analogous with Paris in ruins:

> Ruins! my family! O kindred minds!
> I bid you each evening a solemn farewell!
> Octogenarian Eves, upon whom rests
> God's terrible claw, where will you be tomorrow?

More directly than "The Seven Old Men," "The Little Old Women" is a metaphor for Paris in ruin. In figuring the old women as Paris, Baudelaire reimagined the "poetry" of Méryon's views of Paris in which "none of the complex elements that compose the painful and glorious décor of civilization has been forgotten."[72] The poet's lament for the old women also resounds with his lament for old Paris, as voiced in "The Swan." In this poem, Baudelaire described the making of Haussmann's Paris as a "*bric-à-brac confus*" (confused bric-a-brac) to reference "the physical realities of demolition and rebuilding."[73] To affect this aural, temporal, and tectonic mash-up, Baudelaire juxtaposes the old and new and weights ephemeral memories with the heaviness of "rocks." Méryon's etchings have "endowed" the poet's "imagination" with ironic, contrastive doubling in each poetic instance.

If we accept these ironic inflections in *Tableaux Parisiens*, we can recover in Baudelaire's description of Méryon's etchings of Paris a modernity that differs from the etcher's more desolate representations. Indeed, a closer reading of Baudelaire's *Salon of 1859* excerpt reveals ironic contrasts that both underscore ambiguity and relieve the lamentations over Paris in ruins. Here, Baudelaire transforms spires into fingers, smokestacks into obelisks, smoke into vomit, scaffolds into spider webs, and closed vistas into open ones—transformations that evoke the "paradoxical beauty" of Paris in the throes of transformation from the old city into its modern incarnation. Baudelaire conceived the prose poems that comprise *Le Spleen de Paris* in the spirit of the city's reincarnation. In doing so, he transposed the dramatic vistas of *Tableaux Parisiens* into ironic snapshots of modern urban existence. In Chapter 4, I read Baudelaire's rationale-as-manifesto for writing prose poems, an introductory letter in *Le Spleen de Paris* interpreted in tandem with the photographic record of Paris in transition and transformation. This word-and-image reading further uncovers the collective ambiguity present in the poet's elegiac stance.

Notes

1 That Baudelaire became a signifier of modernized Paris is evident in his twentieth- and twenty-first-century critical legacy, a discourse marked by two opposing views of Baudelaire's modernity and his so-called ambiguous response to Paris transformed—at once nostalgic and hopeful. See, for example, T. J. Clark, "Introduction" and "The View from Notre Dame" in *The Painting of Modern Life*: 3–78; and Claude Pichois and Jean-Paul Avice, *Baudelaire-Paris* (Paris: Editions Paris-Musées/Quai Voltaire, 1993): 27–32. Clark also identifies "ambiguity" as a Parisian attitude toward modernity in general and, specifically, toward Haussmann's modernization of Paris.

2 See Oxford English Dictionary, s.v. "ambiguity (n.), sense 1.a," September 2024, https://doi.org/10.1093/OED/1070385339; and Merriam-Webster.com Dictionary, s.v. "ambiguity," accessed January 14, 2025, https://www.merriam-webster.com/dictionary/ambiguity.

3 The following synopsis of Haussmannization is derived from various predominantly English-language sources. For a detailed socio-historical account, see David Harvey, *Paris, Capital of Modernity* (New York: Routledge, 2006); for architectural and urban history, see Anthony Sutcliffe, "Paris as the Hub of French Industrialisation: Building a European Capital under the Second Empire, 1852–70," in *Paris: An Architectural History* (New Haven: Yale University Press, 1993): 83–104; for a broadly narrative account, including biographies of Napoléon III and George-Eugène Haussmann, see Stephane Kirkland, *Paris Reborn: Napoléon III, Baron Haussmann, and the Quest to Build a Modern City* (New York: Picador, 2014).

4 For a visual and analytical survey of Daumier's Haussmann-themed caricatures, see Jennifer Pride and Kingsborough Art Gallery, *Daumier's Paris: Caricature and Cultural Trauma in the Age of Haussmann* (Brooklyn: Kingsborough Community College, 2016).

5 Alain Corbin, "Charting the Cultural History of the Senses," in *Empire of the Senses: The Sensual Culture Reader*, ed. David Howes (Oxford: Berg, 2005): 129.

6 David Howes, "Olfaction and Transition: An Essay on the Ritual Uses of Smell," *The Canadian Review of Sociology and Anthropology* 24, no. 3 (August 1987): 398–416, as noted by Corbin, "Charting the Cultural History of the Senses," 129.

7 Jill Steward and Alexander Cowan, "Introduction," in *The City and the Senses: Urban Culture since 1500*, eds. A. Cowan and J. Steward (Aldershot, Hampshire: Ashgate, 2007): 2.

8 Steward and Cowan, "Introduction," 2.

9 See Aimée Boutin, "Rethinking the Flâneur: Flânerie and the Senses," the introduction to the special issue of *Dix-Neuf*, of the same title, 16, no. 2 (July 2012): 124–32. Boutin and other sensory study scholars seek not to displace the hierarchy of vision that had dominated twentieth-century scholarship but to reintegrate vision with the other senses (Boutin, 124). In so doing, they offer many other sensory experiences that inform the study of the visual arts.

10 Boutin argues that this hierarchy of vision is mainly due to Walter Benjamin's influential unfinished book *Charles Baudelaire: A Lyric Poet in the Era of High Capitalism* [Eng. trans. 1973]. In this and other works, Benjamin's "metaphysical concern with the flâneur's disembodied, alienated gaze [which] is absent from the earlier material, more clearly intent on describing how the flâneur senses the city," in "Aural Flânerie," *Dix-Neuf* 16, no. 2 (July 2012): 150.

11 Boutin, "Rethinking the Flâneur," 131.

12 Boutin, "Aural Flânerie," 149–61, 149.

13 Boutin, "Aural Flânerie," 160.

14 Baudelaire's personification of Constantin Guys epitomized the flâneur for his time and established the "avant-garde" archetype for future critical and creative writing. Boutin, "Rethinking the Flâneur," 128–9.

15 "Flâneur" is now the accepted English translation of "idler" commonly used by Baudelaire scholars. See the English translation of Baudelaire, "The Painter of Modern Life," in Charles

Baudelaire, *The Painter of Modern Life, And Other Essays*, ed. and trans. Jonathan Mayne (London: Phaidon, 1964): 9.

16 Baudelaire, "The Painter of Modern Life," in *Selected Writings*, 399–400.

17 Baudelaire, "The Painter of Modern Life," 400.

18 As Lois Boe Hyslop and Francis E. Hyslop explain, "In the creation of his cityscape, Baudelaire relies mainly on the power of suggestion. Paris is seldom seen in his poems. Picturesque descriptions and precise details are strangely lacking. Yet, if the city is not seen, it can be heard, and its presence can be felt or sensed." Hyslop and Hyslop assiduously catalog these sounds emitted from the people and places that comprise the Parisian cityscape, in "Baudelaire and Meryon: Painters of the Urban Landscape," *Symposium: A Quarterly Journal in Modern Literatures* 38, no. 3 (Fall 1984): 208–9.

19 Baudelaire, "The Loss of a Halo," in *Baudelaire: His Prose and Poetry*, ed. T. R. Smith, trans. Joseph T. Shipley, (New York: Boni and Liveright, 1919): 99.

20 See Christopher Prendergast, *Paris and the Nineteenth Century* (Oxford: Blackwell, 1992): 132.

21 Baudelaire, "The Generous Gambler," in *The Prose Poems and La Fanfarlo*, trans., intro, and notes Rosemary Lloyd (Oxford/New York: Oxford University Press, 1991): 74.

22 David Howes, "Charting the Sensorial Revolution," *The Senses and Society* 1, no. 1 (March 2006): 115.

23 I have borrowed the phrase "*les lieux de mémoire*" from Pierre Nora, "Introduction" to and the French title of *Realms of Memory: The Construction of the French Past, vol. 3: Symbols*, dir. Pierre Nora, ed. English edition Lawrence D. Kritzman, trans. Arthur Goldhammer (New York: Columbia University Press, 1998): ix–xii; original French title of Nora's four-volume series: *Les Lieux de Mémoire* (Paris: Gallimard, 1984–97).

24 Baudelaire, "Of the Heroism of Modern Life," *Salon of 1846*, in *Selected Writings*: 104–7.

25 Francis Scarfe, "Introduction" to *Baudelaire: The Poems in Prose and "La Fanfarlo,"* vol. II, trans. Francis Scarfe (London: Anvil Press Poetry, 1989): 11. The collection comprises 45 prose poems written between 1857 and 1861. These were published piecemeal in different reviews before Baudelaire died in 1867, while a posthumous volume with five additional poems was published in 1869 (Scarfe, 11–2). Pichois states that Baudelaire began writing prose poems in 1855 and initiated efforts to publish them in 1861; Pichois with Ziegler, *Baudelaire*, 287.

26 Baudelaire, *Salon of 1859*, in *Art in Paris 1845–1862* (Ithaca: Cornell University Press/Phaidon, 1981): 200–1. Note that the section on "landscape" was omitted from the *Salon* as published in Baudelaire, *Selected Writings*, 285–324. Baudelaire republished this tribute to Méryon in an essay on contemporary etchers, "Peintres et Aquafortistes" ("Painters and Etchers," 1862, in *Art in Paris 1845–1862*: 217–22); Méryon did not exhibit in the Salon of 1859. See Louis Auvray, *Exposition des Beaux-Arts. Salon de 1859* (Paris: A. Taride, 1859): 92. For Baudelaire's interpretation of etching as a modernist technique and aesthetic, see Michele Hannoosh, "Etching and Modern Art: Baudelaire's 'Peintres et Aquafortistes,'" *French Studies* 53, no. 1 (1989) : 47–60; and Baudelaire, *Salon de 1859*, in *Oeuvres Complètes de Charles Baudelaire. II*: 335.

27 In this context, I draw upon Benjamin's interpretation of Baudelaire's description of Méryon (which the former quoted from Abraham, Piere (ed), *Arts et Littératures Dans la Société Contemporaine*, II, in *Encyclopédia Française*, vol. 17 [Paris: Larousse, 1936]: 293), a version he transcribed in "The Paris of the Second Empire in Baudelaire," in *Charles Baudelaire: A Lyric Poet in the Era of High Capitalism*, trans. Harry Zohn (London: NLB, 1973): 88–9. As Benjamin notes, his (more gruesome) view of Méryon's depictions of Paris is informed by Gustave Geffroy's *Charles Méryon* (Paris, 1926): 88, n. 62.

28 Méryon had been institutionalized at Charenton from May 16, 1858 to August 25, 1859. After a brief return to the asylum at Charenton in late 1866, he was released to visit the Universal Exhibition of 1867, where several of his etchings were shown. Tragically, a violent storm broke out and finally shattered his reason. He did not recover from the relapse, believing himself to be Christ held captive by the Pharisees. He claimed there was not enough food in the world and that he would not wrong the poor by taking their sustenance. He refused to eat, and on February 14, 1868, Charles Méryon died of starvation. "Charles Méryon: Biography," https://www.campbell-fine-art.com/artists.php?id=80.

29 Baudelaire, *Salon de 1859*, in *Art in Paris 1845–1862*, 201.

30 Baudelaire, letter to his publisher Auguste Poulet-Malassis, January 8, 1860, in *Correspondance*, t. 1, eds. Claude Pichois and Jean Ziegler (Paris: Gallimard, 1973): 655. Notably, Baudelaire mentioned a reconciliation with Constantin Guys in the same letter. Here, he characterized (inadvertently?) Guys in stark contrast to Méryon. He wrote that the former is a "charming man, witty, and he is not ignorant, like all the writers" (656).

31 Benjamin, "The Paris of the Second Empire in Baudelaire," in *Charles Baudelaire: A Lyric Poet*, 87.

32 Benjamin, "The Paris of the Second Empire in Baudelaire," 82–3.

33 Benjamin, "The Paris of the Second Empire in Baudelaire," 87.

34 Benjamin, "The Paris of the Second Empire in Baudelaire," 82.

35 Benjamin, "The Paris of the Second Empire in Baudelaire," 82–3.

36 Jacques Damade, ed. *Paris, 1860: Eaux-Fortes sur Paris & "Les Tableaux Parisiens,"* texts Charles Baudelaire, ills. Charles Méryon, pref. and notes Jacques Damade (Paris: Editions la Bibliothèque, 2001). The exhibition, which I visited, displayed Baudelaire's poems, published initially as *Tableaux Parisiens*, with Méryon's *Vues de Paris*.

37 Here, we should distinguish between the unpublished album(s) of loose plates gathered in a portfolio and the proposed publication of plates bound together in a single volume.

38 For Méryon's engraving for the title page *Eaux-fortes sur Paris*, see Charles Méryon, *Eaux-fortes sur Paris* (Paris: [Méryon], 1852), http://catalogue.bnf.fr/ark:/12148/cb420674610; beginning with this link the Bibliothéque nationale de France identifies plates 1–12 from this series held in its collection. According to Méryon's biographer, Philippe Burty, only 12 of the 21 plates depicted Paris buildings and vistas; for a complete list of the *Eaux-fortes sur Paris* etchings, see Philippe Burty, *Charles Méryon, Sailor, Engraver, and Etcher: A Memoir and Complete Descriptive Catalogue of His Works*, trans. Marcus B. Huish (London: Fine Arts Society, 1879): 50. Burty identifies 23 etchings that comprise *Eaux-fortes sur Paris*. This count includes the title page, four verses, two frontispieces, twelve principal etchings [Paris], two minor etchings of the arts of Paris, and two tailpieces. Baudelaire lists Méryon's 12 etchings of Paris in his letter to his mother, Mme. Aupick (March 4, 1860), *Correspondance*, t. 2, 4.

39 One problem in identifying the exact images that Baudelaire would have viewed is that Méryon constantly reworked the original plates; for the multiple states of each of the 12 Paris engravings, see Burty, *Charles Méryon*, 50–71.

40 Here, Baudelaire refers to the ninth state (Fig. 2.2), which is not included in Burty, *Charles Méryon*.

41 Baudelaire, Letter to Poulet-Malassis, January 8, 1860, *Correspondance*, t. 1, 654–5. Méryon replaced the balloons with birds in the ninth state of the etching; see Charles Méryon, *Le Pont-au-Change* (Paris: [Méryon], 1854), http://catalogue.bnf.fr/ark:/12148/cb420738631. However, Baudelaire would have seen more than one state of this plate since he had compared it to the earlier version with the hot air balloon (which I discuss below); the hot air balloon appears in the eleventh state; Hyslop and Hyslop ("Baudelaire and Meryon") report that there are 12 states of this etching (209). They also give credence to the Emperor's release of eagles over Paris: "Since the eagle was the symbol of Napoleon III for whom the artist felt great antipathy, the sight of so many birds in the Paris sky seems less incongruous than it might first appear [.] Philippe Burty suggests this interpretation, basing his contention on Méryon's belief that Napoleon III released birds of prey from the Tuileries to threaten the citizens and to remind them of the *coup d'etat* of 1851." The authors add that Baudelaire's letter to Poulet-Malassis (January 8, 1860) "confirms his theory[.]" (206–7).

42 Baudelaire, Letter to Poulet-Malassis, January 8, 1860, in *Correspondance*, t. 1, 655. The "Prince" here refers to Napoléon III, née Charles-Louis Napoléon Bonaparte, the third son of Louis Bonaparte, the King of Holland and younger brother of Napoléon, and Hortense de Beauharnais. The "coup" refers to Bonaparte's coup d'état of December 2, 1851, at the time as President of France under the Second Republic, replacing the representative government with his dictatorship, and one year later as emperor of the Second French Empire. Napoléon III was caricatured as a sphinx in popular satire; see David Baguley, *Napoleon III and His Regime: An Extravaganza* (Baton Rouge: Louisiana State University Press, 2000): 1. See also John Hay's poem, "The Sphinx of the Tuileries," probably written while Hay resided in Paris during his diplomatic service between 1865–70 (https://www.poemhunter.com/john-hay/biography/).

43 "The Murders in the Rue Morgue" is a short story by Edgar Allan Poe published in *Graham's Magazine* in 1841. The story's location, personages, and drama match Méryon's depiction of the Paris morgue and the people who surround it. The story revolves around the double murder of Madame L'Espanaye and her daughter at their home in the Rue Morgue, a fictional street in Paris. The murders occurred in a fourth-floor room that was locked from the inside. Several witnesses reported hearing two voices at the time of the murder, but they could not identify the spoken language. Dupin, the narrator's friend and amateur detective, concludes that an "Ourang-Outang" (orangutan) killed the women. He places an advertisement in the local newspaper asking if anyone has lost such an animal; a sailor soon arrives looking for the creature, which he then sells.

44 Baudelaire, Letter to Poulet-Malassis, January 8, 1860, in *Correspondance*, t. 1, 655.

45 Baudelaire, *Exposition Universelle 1855*, qtd. in Pascal Pia, *Baudelaire*, trans. Patrick Gregory (New York: Grove Press, 1961): 88. This translation most closely follows the spirit of the original French text: "Le beau est toujours bizarre. Je ne veux pas dire qu'il soit volontairement, froidement bizarre, car dans ce cas il serait un monstre[.] Je dis qu'il contient toujours un peu de bizarrerie, de bizarrerie non voulue, inconsciente, et que c'est cette bizarrerie qui le fait être particulièrement le Beau," Baudelaire, *Exposition Universelle 1855*, in *Oeuvres complètes* (1999): 724.

46 The subject title evokes strong literary associations—the Pont-au-Change is featured in the novel *Les Misérables* by Victor Hugo.

47 As a district, the Île de la Cité was the most extensively modified during the second phase of Haussmann's renovations (1859–1867). Medieval streets and buildings were demolished, and government buildings were either demolished and rebuilt or modified to suit the new North-South boulevards that cut through the Île.

48 Baudelaire, Letter to Victor Hugo, December 13, 1859, in *Correspondance*, t. 1, 627. In a letter to Eugene Crepet, dated May 11 (?), 1860, Baudelaire mentions that he had received a letter from Hugo, in which the author was unusually cordial and was "very spiritual, which is even more peculiar (apropos Méryon);" "très spirituelle, ce qui est encore plus singulier (à propos de Méryon)," Baudelaire, *Correspondance*, t. 2, 41.

49 For this quotation and those following, Baudelaire, letter to Poulet-Malassis dated February 16, 1860; qtd. and trans. in Pichois with Ziegler, *Baudelaire*, 263.

50 Charles Méryon, Letter to Baudelaire, February 23, 1860, in Baudelaire, *Paris, 1860*, 11.

51 Baudelaire, Letter to Mme. Aupick, February 28, 1860, in *Correspondance*, t. 1, 683. Baudelaire also sent the set of Méryon's etchings to Mme. Sabatier, see Letter to Madame Sabatier, March 4, 1860, in *Correspondance*, t. 2, 6.

52 Baudelaire, Letter to Mme. Aupick, March 8, 1860, in *Correspondance*, t. 2, 4. In this letter, Baudelaire includes a list of the plates that comprise the album, noting the structures that have been demolished. What remained of his own copy is suggested by Baudelaire's list of paintings, drawings, and engravings composed in 1864–5 (after his arrival in Belgium); the list includes three Méryon etchings among items left "pour [Hippolyte] Lejosne." See *Correspondance*, t. 2, 430–1.

53 As catalogued in Burty, *Charles Méryon*, Méryon inscribed verses on the plates, and in one case, on a separate page for: *Le Petit Pont* (56); *Rue des Mauvais Garcons* (tailpiece for the first part of *Eaux-fortes sur Paris*) (59); *La Petite Pompe* (Pompe de Notre Dame; the frontispiece to the second part of "Paris Views") (63); *Le Pont-au-Change* (the verse titled "L'Esperance"—a reference to Poe's "Rue Morgue" text—was inscribed on a separate page) (67); and *L'Abside de Notre Dame de Paris* (70).

54 Baudelaire, Letter to Poulet-Malassis, *Correspondance*, March 11, 1860, t. 2, 8. According to Hyslop and Hyslop ("Baudelaire and Meryon"), Méryon also rejected Baudelaire's offer to write a manual, while Baudelaire succeeded in persuading the government to purchase several sets of prints (198).

55 See Baudelaire, *Paris, 1860*. Baudelaire recruited Champfleury and Banville to ask the Minister to buy Méryon's etchings of Paris (11).

56 Baudelaire, *Paris, 1860*, 16.

57 Baudelaire may have responded to Méryon, as implied in a letter to Philippe Burty dated 1861–2; that correspondence has not been found. *Correspondance*, t. 1, 215; n. 6 [764].

58 For chronology, see Pichois with Ziegler, *Baudelaire*, 936–8. The above chronology alters Hyslop and Hyslop's ("Baudelaire and Meryon") suggestion that the new poems in *Tableaux parisiens* were written, in part, in response to Méryon's *refusal* to accept Baudelaire's collaboration, concluding that some of these poems "may even have been partly inspired by Méryon's prints" (198).

59 Magnifying glasses were commonplace in private and public spaces. While no magnifying glass appears within the picture frames, Honoré Daumier's paintings of intimate gatherings of men viewing prints suggest how Baudelaire and Méryon may have examined *Les Eaux-fortes sur Paris*; see for example: *Les Amateurs D'estampes* (The Print Collectors; 1865; Kunsthalle Mannheim) and *Les Amateurs D'estampes* (The Print Collectors; 1860–3; Clark Art Institute, Williamstown, MA). Baudelaire's solitary viewing of Méryon's prints is suggested in Daumier, *L'Amateur d' Estampes* (The Print Collector; 1860–5; Museum Boijmans Van Beuningen, Rotterdam). See the use of magnifying glasses for examining artworks in public exhibitions in Daumier's graphic illustrations: *Un Véritable Amateur* (The True Art Lover; 1847); *Eh! bien en regardant ce tableau de près on finit par y découvrir des qualités, on voit que la couleur est bonne.* (Well, if you look very closely, you might end up finding some quality! The color seems to be good; 1865; and *Les Amateurs de Tableaux* (Painting Experts; 1865).

60 Hyslop and Hyslop, "Baudelaire and Meryon," 209.

61 The Paris Morgue was located on the eastern tip of the Île de la Cité behind Notre Dame, where the Mémorial des Martyrs de la Déportation is now located.

62 Baudelaire, Letter to Mme. Aupick, March 4, 1860, *Correspondance*, t. 2, 5.

63 Hyslop and Hyslop, "Baudelaire and Meryon," 209.

64 Hyslop and Hyslop, "Baudelaire and Meryon," 209.

65 Baudelaire, *Salon of 1859*, qtd. in Benjamin, "The Paris of the Second Empire in Baudelaire," 88–9.

66 Baudelaire, "Of the Heroism of Modern Life," *Selected Writings*, 104–7.

67 Baudelaire, "The Seven Old Men" ("Les Sept Vieillards"), in *Les Fleurs du Mal/Flowers of Evil* [2nd ed., 1861], trans. by William Aggeler, in *The Flowers of Evil* (Fresno, CA: Academy Library Guild, 1954), https://fleursdumal.org/poem/221.

68 Baudelaire, Letter to Poulet-Malassis, January 8, 1860, in *Correspondance*, t. 1, 665.

69 Baudelaire, Letter to Mme. Aupick, March 8, 1860, in *Correspondance*, t. 2, 4.

70 Baudelaire, "Little Old Women" ("Les Petites Vieilles"), in *Les Fleurs du Mal/Flowers of Evil* [2nd ed., 1861], trans. William Aggeler (Fresno, CA: Academy Library Guild, 1954), https://fleursdumal.org/poem/222.

71 Baudelaire, letter to Poulet-Malassis dated February 16, 1860; qtd. and trans. in Pichois with Ziegler, *Baudelaire*, 263.

72 Baudelaire, *Salon of 1859*, qtd. in Benjamin, "The Paris of the Second Empire in Baudelaire," 88.

73 As Christopher Prendergast claims, the poem reveals the paradox of modern Paris writ large: while this is a poem about a city in fragments and ruins, it exonerates "Paris as another Troy-to-be, with its prefiguring of modernity as antiquity not in order to dignify former with the aura of the latter, but rather to project a history as issuing in the city's death." Prendergast, *Paris and the Nineteenth Century*, 207–8.

3 On Baudelairean Modernity

Neuroaesthetic Modeling of Ironic Self-Reflection

This chapter initiates my extension of experiential aesthetics into "experiential" neuroaesthetics. Functioning as intervals between historiographic chapters, this chapter and Chapters 5, 7, and 10 will present various scientific models and empirical evidence to capture nineteenth-century modernity's embodied, lived experience in diverse artistic, verbal, and visual forms. As noted in the Introduction, I refer to these as empirical chapters that reflect on historical evidence. In general, *experiential aesthetics* pertains to the maker's and viewer's physical and emotional transitions as they navigate the temporal-spatial dimensions of an urban environment and re-experience it in the work of art. On the other hand, *experiential neuroaesthetics* pertains to the subject's visceral and affective transformation, including pre-conscious and cognitive processes. Unless otherwise noted, I use "visceral" in the metaphorical sense, not anatomical.[1]

In Chapter 2, I laid the groundwork for Baudelaire's self-doubling model, a pivotal concept in the development of ironic self-reflection. This model, along with the mirroring metaphor, is key to understanding the influence of philosophical and aesthetic writings of the German Romantic ironists. Furthermore, I demonstrated how E. T. A. Hoffmann's comic tales nurtured Baudelaire's ironic self-reflection. I asserted that the poet shifted *dédoublement* from comic laughter to philosophical introspection—the self-reflection he attributed to making and viewing paintings of modern life. In this first empirical chapter, I wish to align ironic self-reflection and its mirroring metaphor with twenty-first-century neuroaesthetics. Here, neuroaesthetics offers models to correlate the viewer's reception of formal and figural elements of ironic parody with the psychic and cognitive processing of non-art environments, such as changed cityscapes. From this perspective, we can hypothesize the nineteenth-century subject's neural activity—whether as art maker or viewer—in response to the urban environment, the maker's mediation of the sensory experiences, and the viewer's affective and cognitive reactions to that mediation.

The psychological and neuroscientific models employed in this chapter are groundbreaking for their investigations of the aesthetic experience of artwork with self-reflection and negative valence (adverse responses). For over a decade after the founding of neuroaesthetics, researchers limited their studies to "the neural bases of beauty perception in art"[2] and contingent pleasure and reward systems—that is, positive valence. The Baudelairean, and subsequently the modernist, aesthetic of rupture transgresses beauty, problematizes "reward," and compels a subjectivized interior gaze. Equally important, the studies used here focus on the viewer's transformative experiences (in terms of emotion and self-identity) during their engagement with artworks. While the transformative aesthetic experience has long been the domain of art psychology, it is now the subject of neuroscientific experiments.

DOI: 10.4324/9780429442315-4

Neuroaesthetic Model 1: Pelowski and Akiba's Transformative–Behavioral Approach

Matthew Pelowski and Fuminori Akiba's article (hereafter Pelowski and Akiba), "A Model of Art Perception, Evaluation and Emotion in Transformative Aesthetic Experience,"[3] offers a behavioral/valuation model with a focus on modern and postmodern art in which "discrepancy" produces deep aesthetic experience. Here, "discrepancy" pertains to viewers' understanding of modern and postmodern art, contrasting with viewers' "fluent" response to an artwork's beauty, as in traditional figural art. Here, the rubric of "discrepancy" correlates with rupture and ironic parody, characteristics that I ascribe to Baudelairean modernity. Pelowski and Akiba seek to go beyond "the analytic tradition [with its cognitive focus] of emphasizing successful assimilation of the art information." Their exemplar introduces "a five-stage model of art-perception, organized around initial disruption and subsequent meta-cognitive reflection and self-transformation."[4]

A psychological model that accommodates irony's philosophical internal gaze and its temporal dimension of self-transformation makes Pelowski and Akiba's scheme vital for my project. The model connects the meta-cognitive processes with the visceral effects of discrepancy processing, which align with the viewer's reaction to ironic parody. Significantly, Pelowski and Akiba's model is founded on the premise that the artist is motivated by "the goal of [inducing] deepened art experience"[5] in the viewer. This premise helps explain and substantiate how the viewer's direct engagement in evaluating artistic techniques involves an assessment of the artist's motivation. Their model merits more detailed scrutiny for my project because Pelowski and Akiba's understanding of the creative process and aesthetic experience continuum matches the Baudelairean paradigm.

Pelowski and Akiba approach the aesthetic experience from two opposing viewer responses to discrepancy. First, there is the viewer who either succumbs to "the human instinct of escape or to assimilation through surface evaluation or self-withdrawal." This viewer, for instance, would be either averse to Manet's painterly distortions or, when familiar with or knowledgeable about the artist's intentions, would fluently process the painting's overall harmony. (Note that Pelowski and Akiba did not use Manet's paintings in their experiment.) The second kind of viewer, the primary subject of their study, is open to "schema change." When confronted by unfamiliar (or defamiliarized) artworks, the viewer overcomes the escapist/withdrawal instinct and instead "enters an intractable position whereby one might use their disruptive encounter as a means of self-reflection and transformation, and therefore come to behave or see something new[.]" In this case, the viewer may also "feel pleasure, epiphany, and harmony."[6] As we have seen, this behavior matches Baudelaire's response to Hoffmann's ironic tales and, by extension, the "ideal" viewer of the paintings of modern life.

Pelowski and Akiba identify three "control" stages of art evaluation that precede the viewer's self-transformation. First, the formal (surface) elements in which discrepancy is "perceived" are directly evaluated. When discrepancy cannot be assimilated, evaluation is followed by a second control stage—"a period of meaninglessness and loss of understanding of the artist's motive." The art evaluation concludes with "a fundamental shift in assessment from the informational plane to a meta-cognitive mode of 'experiential' interpretation."[7] In the last two stages of Pelowski and Akiba's five-stage model, the direct evaluation of the artwork shifts to self-evaluation, wherein meta-cognitive [knowledge-based] reassessment switches from "outward 'subjective awareness' to inward 'self-awareness.'" During this phase, "the schema change appears to temporarily intensify the preceding tension and emotion." At this point, the change is psychological and physiological, affecting the autonomic nervous system (ANS; Figure 3.1).[8] Pelowski and Akiba argue that *discrepancy processing* engages both subsystems of the ANS as a "coordinated activation of parasympathetic [accelerated nervous] response and sympathetic [slowed down nervous] response [making it] such a potent psychological state."[9] Following this schema change, the final "aesthetic phase" of experience involves a "latency period" during which the viewer returns "to harmonious selfless contemplation of the new elements that allowed for [their] preceding experience." The viewer "re-engages with the final mastery/assimilation of the previously unfamiliar elements [which reflects] a relative adjustment in her/his self-image or relationship to the world."[10] This meta-cognitive process also involves "the arousal of 'aesthetic' harmony, pleasure and emotion itself,"[11] accounting for deep aesthetic experience.

As reviewed here, the continuum between artistic intentionality and the viewer's meta-cognitive aesthetic processing is vital for substantiating how the work of art mediated the lived experience of Haussmannized Paris. It is also crucial for confirming the cognitive and visceral effects of Haussmannization shared by artists and viewers. Pelowski and Akiba's model also applies to the embodied experience of non-art environments. They argue that a model that:

> joins self and schema change allows this model to expand... beyond art, providing a basic mechanism for self-transformations in relation to environmental stimuli, without needing a definition of 'artworks.' This also draws together both man-made and natural stimuli... as well as objects not intended to be art.[12]

This is what, by way of John Dewey,[13] Pelowski and Akiba refer to as the "aesthetic phase" of everyday life experience. When applied to Baudelaire's experience of the modern city, this model allows us to identify how the discrepancy between topological and social changes registered as meta-cognitive rupture and visceral effects. In Baudelaire's view, such daily experiences were the grounds for escape, self-reflection, and aesthetic reconciliation. Following the poet's mandate, the Baudelairean painters sought to translate the transformative experiences of everyday life, both psychic and physical, into the medium of their art to arouse deepened art experience through which the viewer might master urban encounters and return to life "renewed."

Neuroaesthetic Model 2: Brown and Colleagues' Transformative–Neural Approach

The self-transformative processes of the aesthetic experience are also the focus of Steve Brown and colleagues' study of the brain areas activated for aesthetic appraisal across sensory modalities (for example, light, sound, taste, temperature, pressure, and smell).[14] This neural/brain activity study supports Pelowski and Akiba's psychological model to further substantiate the physiological origins of ironic parody as an aesthetic response to Haussmannization. According to their efforts to "naturalize" aesthetics (from an evolutionary stance), Brown and colleagues' study is especially pertinent to restoring the embodied, aural, and olfactory responses to the nineteenth-century material environment. As they argue, "aesthetic processing is, at its core, the appraisal of the valence of perceived objects. This appraisal is in no way limited to artworks but is instead applicable to all types of perceived objects."[15] Brown and colleagues' study is equally essential for justifying a retrospective approach to Baudelaire's engagement with brain science: it

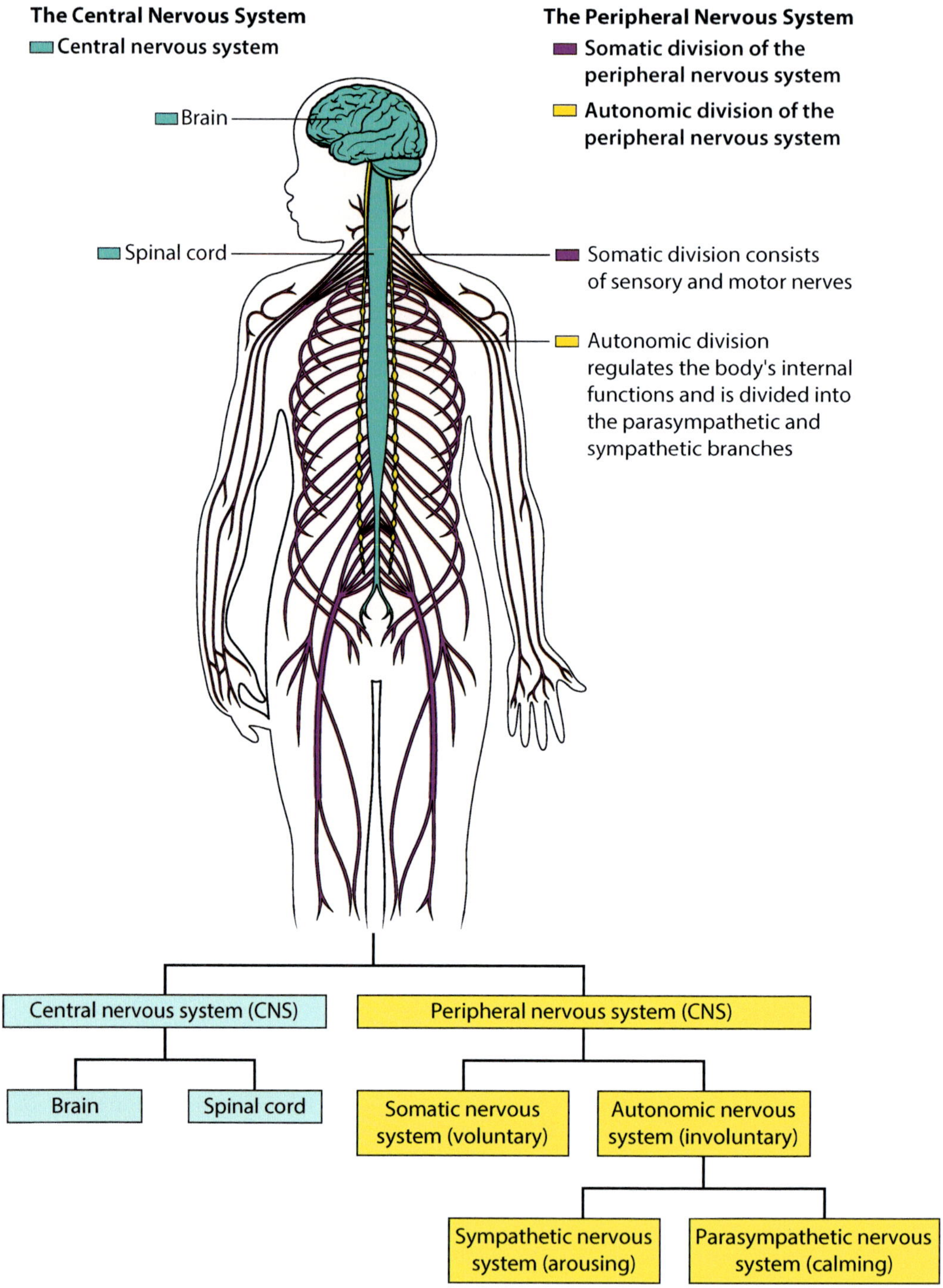

Figure 3.1 Central and peripheral nervous system. Illustration by Susan Tyler.

provides an updated anatomical mapping of the interaction between external and internally generated stimuli as first formulated by Johannes Müller (treated in Chapter 1).

Brown and colleagues state that "[t]he principal aim of the study was to examine zones of polysensory convergence for aesthetic processing" to hypothesize an aesthetic neuronal workspace.[16] To this end, they presented a "comprehensive analysis… of neuroaesthetic processing by reporting the results of voxel-based[17] meta-analyses of 93 neuroimaging studies of positive-valence aesthetic appraisal across four sensory modalities."[18] Based on these findings, they identified which brain region registers valence, in general, and which sub-regions register positive or negative valence.[19] Anatomically considered, the orbitofrontal cortex (OFC) and the anterior insula were identified as the most activated parts of the brain during aesthetic appraisal. Brown and colleagues concluded, however, that "[t]he right anterior insula… emerged as the strongest candidate for a supramodal area for positive-valence aesthetic processing."[20] (Figure 3.2)

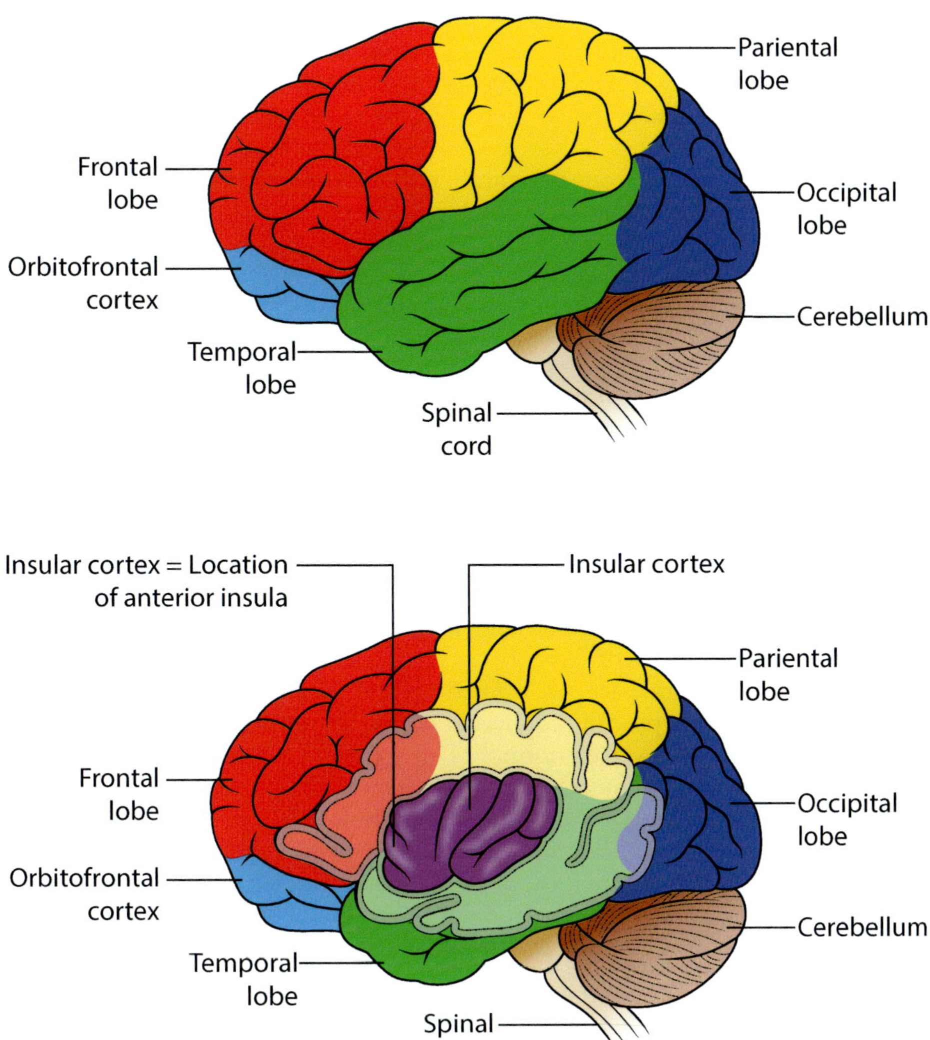

Figure 3.2 Brain anatomy. Illustration by Susan Tyler.

What is particularly compelling for Baudelaire's aesthetic of self-doubling is Brown and colleagues' evidence that the OFC and anterior insula are *bivalent*; each region can process both negative and positive valence. The team also identified each region's different functions for processing these valences: "The OFC is seen as an extension of 'what' sensory pathways that process object-recognition information, whereas the anterior insula contains a meta-representation of bodily responses from inputs coming from the interoceptive cortex in the posterior insula."[21] Although their findings are based on experiments that measured outcomes for localizing positive valence, Brown and colleagues' detailed analysis of these bivalent brain regions and their concordant processing provides the most viable model for mapping the dynamics of an aesthetic of psychic and physical rupture and ironic self-reflection. More broadly considered, this model is foundational to hypothesizing an experiential neuroaesthetics.

Based on Brown and colleagues' "naturalized" neural model, I would identify the anterior insula's functions as closely related to the self-reflection occurring during one's temporal-spatial immersion in an urban environment and one's embodied response to painted representations of the city. According to Brown and colleagues, during both the art and non-art experience:

> [t]he most strongly activated [neural] region corresponds with the anterior insula proper, a paralimbic region associated with interoceptive awareness/insight..., a sense of the 'feeling self'..., core affect..., the subjective experience of emotions such as pain, sadness, anxiety and disgust..., and the capacity to predictively anticipate the impact of emotional events on the body's responses.[22]

More specifically, Brown and colleagues identify the right anterior insula as the most activated sub-region for positive-valence aesthetic appeal: "the right anterior insula contains a meta-representation of the state of the body that is associated with the subjective awareness of the 'feeling self.'"[23] This analysis of the anterior insula's processing of external stimuli closely matches Baudelaire's description of Guys' immersion in the crowded Parisian boulevards and his subsequent self-reflections recorded in pen, ink, and watercolor renderings.

Brown and colleagues' research also identifies "a supramodal brain area that is active during the process of object appraisal," what they call a "core circuit for aesthetic processing." Their description of this core circuit merits our attention as it corresponds to an aesthetic neuronal workspace (as defined above). In this neural region, the OFC and anterior insula "are reciprocally connected with one another and show parallel functions when it comes to reward and emotion."[24] During aesthetic processing, the OFC and anterior insula co-function to process external sensory perceptions and internal visceral "self" feelings. As discussed in Chapter 1, Müller's binary nervous system and Dr. Joseph Moreau's "healthy" brain paradigm are echoed herein. In the latter case, just as the healthy brain functions free of hashish or madness, the core circuit for aesthetic processing manifests as neural equilibrium and self-controlled self-reflection. Baudelaire's notion regarding the healing effects of both the creative process and the aesthetic experience is affirmed in Brown and colleagues' overall view of the core circuit for aesthetic processing. They state:

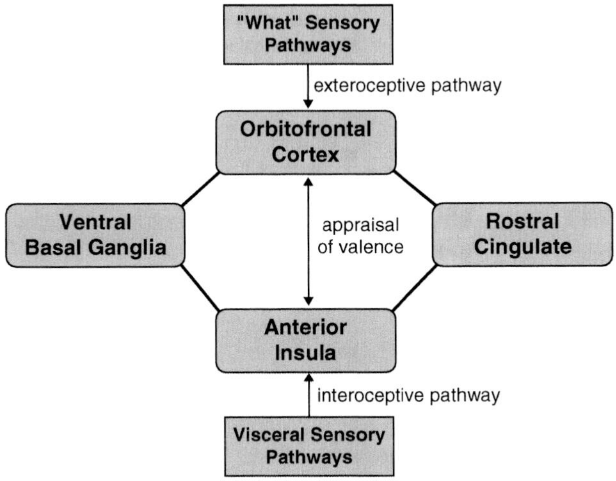

Figure 3.3 The Core Aesthetic Network of the Human Brain. In Brown et al., "Naturalizing Aesthetics: Brain Areas for Aesthetic Appraisal Across Sensory Modalities," *Neuroimage* 58.1 (Sept. 1, 2011): 256, figure 4. Courtesy Elsevier.

> An important way to naturalize aesthetic processing is to see it as the appraisal of valence of perceived objects, and to argue that this appraisal process comes about through a comparison between subjective awareness of current homeostatic state – as mediated by the anterior insula – and exteroceptive[25] perception of objects in the environment, as mediated by the sensory pathways leading up to the OFC. The basic model of aesthetic processing that we propose involves an interaction between interoceptive and exteroceptive processing via recurrent connectivity between anterior insula and OFC, respectively, as shown in Fig. 4. We call this a "core circuit for aesthetic processing"[.][26] (Figure 3.3)

In concluding this first empirical chapter, we should recall that Baudelaire's notion of the healing effects of artistic production and reception is grounded in ironic self-reflection. This being the case, we can more securely link Baudelaire's reception of German Romantic irony to his coincident reception of nineteenth-century brain science. The fusion of these two discourses in his aesthetic of *dédoublement* further justifies a neuroarthistory of the painter of modern life.

Notes

1 I use the term "visceral" figuratively, as defined here: visceral "a. Affecting the viscera or bowels regarded as the seat of emotion; pertaining to, or touching deeply, inward feelings. Obs. after 17th c. and revived in the 20th." "visceral, adj." OED Online. Oxford University Press. http://www.oed.com.proxy.lib.fsu.edu/view/Entry/223878?redirectedFrom=visceral+.

2 Di Dio Cinzia and Vittorio Gallese, "Neuroaesthetics: A review," *Current Opinion in Neurobiology* 19, no. 6 (2009): 682.

3 Matthew Pelowski and Fuminori Akiba, "A Model of Art Perception, Evaluation and Emotion in Transformative Aesthetic Experience," *New Ideas in Psychology* 29, no. 2 (2011): 80–97.

4 Pelowski and Akiba, "A Model of Art Perception," 80. "Metacognition is defined as the capacity to reflect on, evaluate, and control cognitive processes such as decision-making, memory, and perception. It involves the ability to assess one's own performance and is crucial for adaptive behavior in various contexts." L. Baker, "Metacognition," *ScienceDirect Topics*, 2010. https://www.sciencedirect.com/topics/neuroscience/metacognition.

5 Pelowski and Akiba, "A Model of Art Perception," 93.

6 Pelowski and Akiba, "A Model of Art Perception," 92.

7 Pelowski and Akiba, "A Model of Art Perception," 93.

8 Autonomic nervous system (ANS): "the portion of the nervous system innervating smooth muscle and glands, including the circulatory, digestive, respiratory, and reproductive organs. It is divided into the sympathetic nervous system and parasympathetic nervous system. Autonomic responses typically involve changes in involuntary bodily functions, such as heart rate, salivation, digestion, perspiration, pupil size, hormone secretion, bladder contraction, and engorgement of the penis and clitoris," in APA Dictionary of Psychology, 2023, https://dictionary.apa.org/autonomic-nervous-system. The "sympathetic nervous system is a network of nerves that helps your body activate its 'fight-or-flight' response. This system's activity increases when you're stressed, in danger or physically active," "Sympathetic Nervous System (SNS): What It Is & Function," Cleveland Clinic, 2022, https://my.clevelandclinic.org/health/body/23262-sympathetic-nervous-system-sns-fight-or-flight#:~:text=Your%20sympathetic%20nervous%20system%20is. The "parasympathetic nervous system is a network of nerves that relaxes your body after periods of stress or danger. It also helps run life-sustaining processes, like digestion, during times when you feel safe and relaxed," Cleveland Clinic, "Parasympathetic Nervous System (PSNS): What It Is & Function," Cleveland Clinic, 2022, https://my.clevelandclinic.org/health/body/23266-parasympathetic-nervous-system-psns#:~:text=Your%20parasympathetic%20nervous%20system%20is.

9 Pelowski and Akiba, "A Model of Art Perception," 94.

10 Pelowski and Akiba, "A Model of Art Perception," 90.

11 Pelowski and Akiba, "A Model of Art Perception," 90.

12 Pelowski and Akiba, "A Model of Art Perception," 91.

13 John Dewey was one of American pragmatism's early founders (active between 1882 and 1953). Dewey formulated an experiential philosophy of education. Believing that philosophy had become overly technical and intellectualistic, he argued that "philosophy needed to be reconnected with education-for-living." Within this system, "philosophy becomes an activity taken by interdependent organisms-in-environments." Individual (and then social) growth occurs when the human "organism" assesses everyday social conditions and values against their preconceptions, what Dewey called "active adaptation." See David Hildebrand, "John Dewey," in *Stanford Encyclopedia of Philosophy*, ed. Edward N. Zalta and Uri Nodelman (Summer 2024 Edition), https://plato.stanford.edu/archives/sum2024/entries/dewey.

14 Steven Brown, et al., "Naturalizing Aesthetics: Brain Areas for Aesthetic Appraisal across Sensory Modalities," *Neuroimage* 58, no. 1 (September 1, 2011): 250–8. Definition: "A sensory modality (also called a stimulus modality) is an aspect of a stimulus or what is perceived after a stimulus," "12.1C: Sensory Modalities," *Medicine LibreTexts*, July 20, 2018, https://med.libretexts.org/Bookshelves/Anatomy_and_Physiology/Anatomy_and_Physiology_(Boundless)/12%3A_Peripheral_Nervous_System/12.1%3A_Sensation/12.1C%3A_Sensory_Modalities.

15 Brown et al., "Naturalizing Aesthetics," 250.

16 The concept of an aesthetic neuronal workspace issues from Global Workspace Theory (GWT). Bernard J. Baars, who authored GWT, "hypothesized that conscious awareness stems from a discrete network of neurons that are widely distributed throughout the cortex. This network receives input from the various sensory regions of the brain and puts it all together – ... to create one unified picture of reality in a continuous stream that we experience," Steven Novella, MD, "The Global Workspace – Consciousness Explained?," *NeuroLogica Blog*, March 23, 2010, https://theness.com/neurologicablog/index.php/the-global-workspace-consciousness-explained/. Using functional magnetic resonance imaging (fMRI), neuroscience researchers have identified an aesthetic neuronal workspace that involves "activation in a distributed system including the occipital lobes, temporal lobe structures in the ventral stream involved in

object (fusiform gyrus) and scene (parahippocampal gyrus) perception, and the anterior insula—a key structure in experience of emotion. [Researchers] also observed activation in the posterior cingulate cortex bilaterally—part of the brain's default network. These results suggest that viewing paintings engages not only systems involved in visual representation and object recognition, but also structures underlying emotions and internalized cognitions;" Oshin Vartanian and Martin Skov, "Neural Correlates of Viewing Paintings: Evidence from a Quantitative Meta-Analysis of Functional Magnetic Resonance Imaging Data," *Brain and Cognition* 87 (June 2014): 52–6.

17 Voxels: "In 3D computer graphics, a voxel represents a value on a regular grid in three-dimensional space.... Voxels are frequently used in the visualization and analysis of medical and scientific data [...]," Wikipedia Contributors, "Voxel," Wikipedia (Wikimedia Foundation, December 9, 2019), https://en.wikipedia.org/wiki/Voxel. "Voxel-based morphometry (VBM) is a neuroimaging technique that investigates focal differences in brain anatomy. The core process of VBM is segmenting the brain into grey matter, white matter, and cerebrospinal fluid, warping the segmented images to a template space and smoothing;" Kiyotaka Nemoto, "[Understanding Voxel-Based Morphometry]," *Brain and Nerve = Shinkei Kenkyu No Shinpo* 69, no. 5 (May 1, 2017): 505–11, https://doi.org/10.11477/mf.1416200776.

18 Brown et al., "Naturalizing Aesthetics," 250.

19 The researchers excluded negative-valence appraisals due to the scarcity of research in this area, a gap I will address in Chapter 10.

20 Brown et al., "Naturalizing Aesthetics," 252. Supramodal: "That [which] transcends sensory modalities, such as vision and hearing. Usually, it refers to an area of the brain that implements abstract functionality common to more than one source of sensory data. For instance, the areas of the brain concerned with language processing can integrate data from the visual, auditory, and tactile areas of the brain," "Supramodal," Wiktionary, December 12, 2022, https://en.wiktionary.org/wiki/supramodal#:~:text=supramodal%20(comparative%20more%20supramodal%2C%20superlative).

21 Brown et al., "Naturalizing Aesthetics," 256. "Interoception refers to sensing the physiological condition of the body; primary interoceptive cortex in dorsal posterior insula receives sensory input from the body's internal milieu," Christine D. Wilson-Mendenhall et al., "Primary Interoceptive Cortex Activity During Simulated Experiences of the Body," *Journal of Cognitive Neuroscience* 31, no. 2 (February 2019): 221–35, https://doi.org/10.1162/jocn_a_01346.

22 Brown et al., "Naturalizing Aesthetics," 254.

23 Brown et al., "Naturalizing Aesthetics," 254.

24 Brown et al., "Naturalizing Aesthetics," 255.

25 Exteroceptive stimulus: "a stimulus that arises in the external world and is sensed by an organism through any of the five senses of sight, smell, hearing, touch, or taste;" "Exteroceptive Stimuli," in *APA Dictionary of Psychology*, American Psychological Association, 2023, https://dictionary.apa.org/exteroceptive-stimulus.

26 Brown et al., "Naturalizing Aesthetics," 256.

4 Mirroring Haussmannization in Baudelairean Word and Photographic Image

Introduction

This chapter traces the two parallel paths introduced in Chapter 2, which are etched on the cultural and sensorial map of Haussmannized Paris and mirrored in Baudelaire's ambiguity toward Paris *ancien* (old) and Paris *moderne*. The first path manifests in Baudelaire's ambiguity-cum-nostalgia, articulated in his praise for and poetic emulation of Charles Méryon's engravings of "*vieux Paris*" (old Paris).[1] The second path recovers a collective ambiguity by aligning Baudelaire's art criticism and *Le Spleen de Paris*, comprising his prose poems, with the photographic record of "old" and "new" Paris. On the one hand, I explore how the photographic archive influenced and validated Baudelaire's modernist aesthetic in "The Painter of Modern Life." This perspective underscores the interplay between word and photographic image, which fixed modernized Paris as a national *lieu de mémoire* (memory space),[2] a symbol of technological progress, cultural continuity, and orderly social reform. However, the new photographic technology also mirrors the cultural ambiguity to which Baudelaire gave voice. Photographers captured the disappearing city and its transformation, mediating nostalgia for its past and celebrating its contemporaneity. To elucidate this paradox, I will focus here on Charles Marville, a preeminent photographer of Paris in his own time and ours.[3] The photographic archive comprised of his work also marks the capital city as *un lieu d'expérience* (an experiential space), the embodiment of physical, sensorial, and affective changes that arise from paradoxical encounters with the old and the new and the demolitions that joined the two.

While only summarily noted in this chapter, the neuroaesthetic models reviewed in the previous chapter underlie my interpretation of Marville's photographs and the embodied and affective responses to Haussmannization they convey. Here Pelowski and Akiba's model for self-reflection and transformation during the aesthetic process pertain to creators' and viewers' ambiguity (resistance/acceptance) toward the city's stasis and mutations that Marville's photographs record. Likewise, as Pelowski and Akiba assert, the psychic and cognitive processing of non-art environments, such as changed cityscapes, mirror multivalent aesthetic processing. I thus maintain that the physical and material ruptures of the urban fabric would have compelled Parisians' ongoing self-reflection and transformation. In Chapter 5, the empirical chapter that follows this one, I discuss the data derived from an experiment of the experiential embodied and metacognitive[4] effects of Marville's photographs on present-day viewers.

DOI: 10.4324/9780429442315-5

Photography: The Mirror as Metaphor and Mnemonic Device

To view Baudelaire's texts in tandem with photography, I extend the mirror trope, introduced in Chapter 1, to the new medium and review "The Painter of Modern Life" in this context. There are two ways in which I use the mirror as an interpretive device in this word-and-photographic-image exegesis.

1 Taking the position of the nineteenth-century viewer, I regard the photograph as a mirror image of reality, as a record of Parisians' daily encounters while traversing Haussmann's newly widened streets and boulevards and those narrow, medieval streets left untouched by Haussmann's ruler and wrecking ball.
2 On a more theoretical level, the mirror is a metaphor for *memory* and the *cognitive experience* of modernity.[5] As I see it, nostalgia and hopefulness form a mirroring relationship between a pre-existing order and the present, a mirroring shaped by memory. However, ambiguity constitutes a third cognitive response since the past and the present that memory reflects are constantly in flux. Claude Pichois and Jean-Paul Avice in *Baudelaire-Paris* identify *ambiguity* as the dominant characteristic of Baudelaire's poems, expressing his love-hate relationship with Paris, both old and new.[6] As I will show, each of these facets is reflected in Marville's Old and New Paris series, dating from 1866–77.

Within the broader scope of this book, posing photography as a metaphorical mirror of cultural ambiguity and as a mnemonic mirror of Haussmannization serves two purposes. First, the photographic interplay of nostalgia and modernity corroborates Baudelaire's ironic response to Haussmannization and its subsequent visual translation as ironic parody. As stated in Chapter 2, I treat ambiguity as an analogue of irony and *dédoublement* (doubling), which includes ironic parody. Second, this approach substantiates the visceral experience of rupture and fragmentation and its subsequent manifestation as a pictorial technique practiced by Baudelairean painters of modern life.

To argue that Baudelaire was creatively engaged with contemporary photography is contrary to most art historical scholarship.[7] It is generally held that Baudelaire opposed photography because its mechanicity robbed the artist of his imaginative faculties. This interpretation is based on Baudelaire's *Salon of 1859*, where he chastised artists who appropriated the mechanical realism that photography offered. From this perspective, scholars have overlooked Baudelaire's advocacy of photography as an *aide de mémoire* (memory aid) and the impact that photography had on his formulation of a modern aesthetic. My focus on Baudelaire and Paris photography does fit within a smaller corpus of literary scholarship that interprets Baudelaire's *Les Fleurs du Mal* and *Le Spleen de Paris* as the verbal counterparts of the photographs, representing "living" records of Baudelaire's melancholy and disaffection with the moral and social decadence of modern Paris.[8] My approach also differs from these melancholic interpretations: I interpret this word-and-photographic-image dynamic as more nuanced, incorporating Baudelaire's nostalgia and his celebration of modernity and modernist tenets for artistic practice.

A Nuanced Celebration of Modernity: Baudelaire, Gautier, and *Le Spleen de Paris*

My restoration of the celebratory side of Baudelairean modernity and its intersection with photography opens with a verbal dialogue between the poet and his mentor,

Théophile Gautier (1822–1909). This dialogue is framed by their shared cognitive experiences of mid-century Paris and begins with two versions of Baudelaire's dedication of *Les Fleurs du Mal* to Gautier. The second version was published at the time and abbreviated an earlier draft that Baudelaire sent to Gautier for approval. In the published dedication, Baudelaire addressed Gautier as the "impeccable poet" and "perfect magician of French letters."[9] In the first draft, Baudelaire beseeched Gautier to support *Les Fleurs du Mal*, which he called "these sickly flowers," adding, "I know that in the ethereal realms of true Poetry, Evil does not exist any more than Good, and that wretched dictionary of melancholy and crime may legitimize the reactions of morality, as the blasphemer confirms religion."[10] Of importance here is that, in response, Gautier faulted the poet for "drawing attention to scabrous aspects ["le côté scabreux"] of the book and denouncing them."[11] Gautier's caution against excessive evil, extolled in Baudelaire's original dedication, makes this private correspondence more critical than the published dedication. By suggesting that Baudelaire downplay the decadent aspects of the collected poems and, by extension, of the dedication, Gautier helps us restore the writer's double-sided response (qua ambiguity) to Paris transforming and transformed.

For both writers, Paris provided a wellspring of paradox and compelled, in turn, the paradoxical nature of their works. Here, as never before, multifarious contrasts between the old and new, the rich and poor, the magnificent and base defined the quotidian reality and generated a modern paradigm. Thus, in *Les Fleurs du Mal*, the poet narrates voyages between the external world of Paris and the world of the imagination, knowing that one cannot exist without the other. As argued in the previous chapter, in the prose poems of *Le Spleen de Paris*, Baudelaire's narrative vignettes and their protagonists are based on the poet-flâneur's lived experiences of the Parisian streets and boulevards. As such, they mirror the city's dualism of good and evil as equal and necessary terms of cultural progress, propelling its dynamic ruptures and transient rhythms.[12] It is just such an embodiment of Paris as *un lieu d'expérience* that shaped the prose poems and marked their modernity.

As the poet stated in a letter to Arsène Houssaye,[13] the (future) publisher of *Le Spleen de Paris*, the lived experience of a modern city motivated both the subject and style of the prose poems. He extolled the disparate collection of poems as "fragments," resonating with the broken urban fabric of Haussmannization. For Baudelaire, the prose poem best suits "the description of modern life, or rather of *a* modern and more abstract life" (original emphasis). He explains, "This obsessive ideal [of the prose poem] springs above all from frequent contact with enormous cities, from the junction of their innumerable connections." While Baudelaire conceived the poems' subjects as mirrors of modern life, their abstract and fragmented forms are inextricably linked to the sensorial rhythms of the modern city. As he proclaims:

> Who among us has not... dreamt of the miracle of a form of poetic prose, musical but without rhythm and rhyme, both supple and staccato enough to adapt itself to the lyrical movements of our souls, the undulating movements of our reveries, and the convulsive movements of our consciences?

Similarly, Gautier celebrated this complex, lived experience in his journalistic essays. In his arts and literary reviews, reports on social events, popular fashions, and cultural customs written between 1832 and 1872, Gautier registered the spirited, cheerful voice of a Parisian boulevardier.[14] Viewed within Gautier's practice as a journalist, his 1854 review of Édouard Fournier's guidebook *Paris Démoli. Mosaïque de Ruines* (1853; Paris Demolished. Mosaic of Ruins) encapsulates his excitement with the rapidly changing urban terrain:

> *Paris démoli*... could not have arrived at a better moment. The deep trenches, several already splendid streets, crisscross the city in all directions; the small islands of houses disappear as if by magic, new prospects open, unexpected views take shape, [so] that anyone who believed to know his way is misled by streets born yesterday. In so many places, the face of Paris is changed from ground to rooftop.

He continues:

> It is a curious spectacle, these open houses with their floors suspended over the abyss, their colored or floral wallpapers still marking the shape of the rooms, their staircases that lead to nowhere,... their bizarre ornaments and their violent ruins[.]... This upheaval is not without beauty; shade and light play picturesque effects on this debris,... but here the grounds are cleared and are leveled, new buildings shoot up in their youthful whiteness, and the old city returns in a palatial gown, completely embroidered with radiant sculpture.[15]

The pictoriality of Gautier's description is noteworthy since Fournier's book was not illustrated. In each chapter of *Paris Démoli*, Fournier describes buildings and districts lost to war and destruction and narrates their attendant historical events and personages. Gautier thereby enhanced Fournier's historical account of demolished streets and buildings with picturesque descriptions that record his direct encounters with the rapidly changing urban terrain. As I later discuss, Gautier's picturesque view of demolition issues from an enduring romantic discourse on the "poetics of the ruins," a phrase coined by Daryl Lee regarding the aesthetic response to the discourse engendered by the later (1870s) "Commune ruins picturesque," referring to the 1871 Paris Commune.[16] While Gautier contributed to this discourse,[17] he inscribed the picturesque aesthetic markers of the poetics of the ruins in his earlier introduction to *Paris Démoli*. Here, I adapt Lee's structural analysis of the process that transforms the trauma of destruction into a "ruinist aesthetic." According to Lee, "The aim [of the ruinist aesthetic]: reconcile the shock of destruction with the pleasure of ruins. The means: dissociate laborer from product, ruin from [displacement and deprivation]." This dual process mirrors the cultural ambiguity engendered by Haussmannization. Extending Lee's Commune model backward in time, we can assert that while Haussmann was condemned for his destructive acts—for uprooting households, livelihoods, and trades—"at the same time his 'creations' were separated and celebrated as picturesque ruins."

Issuing from this picturesque discourse on the ruinist aesthetic, Gautier's eyewitness account also accords with some of the earliest photographs of Paris and its demolition sites. Take, for example, Henri Le Secq's photographs dating from 1852–3 that expose open rooms and dangling stairways of buildings demolished along the Rue de Rivoli or the "*ruines violentes*" of the former royal stables in the Place du Carrousel (Figures 4.1 and 4.2).[18] In the latter series, piles of disassembled masonry blocks and timber frames litter the Louvre forecourt. Taken from different vantage points and points in time, these photographs record the laborious stages of the demolition process (but without the laborers' presence) and its temporal dimensions. Gautier would have been both witness and

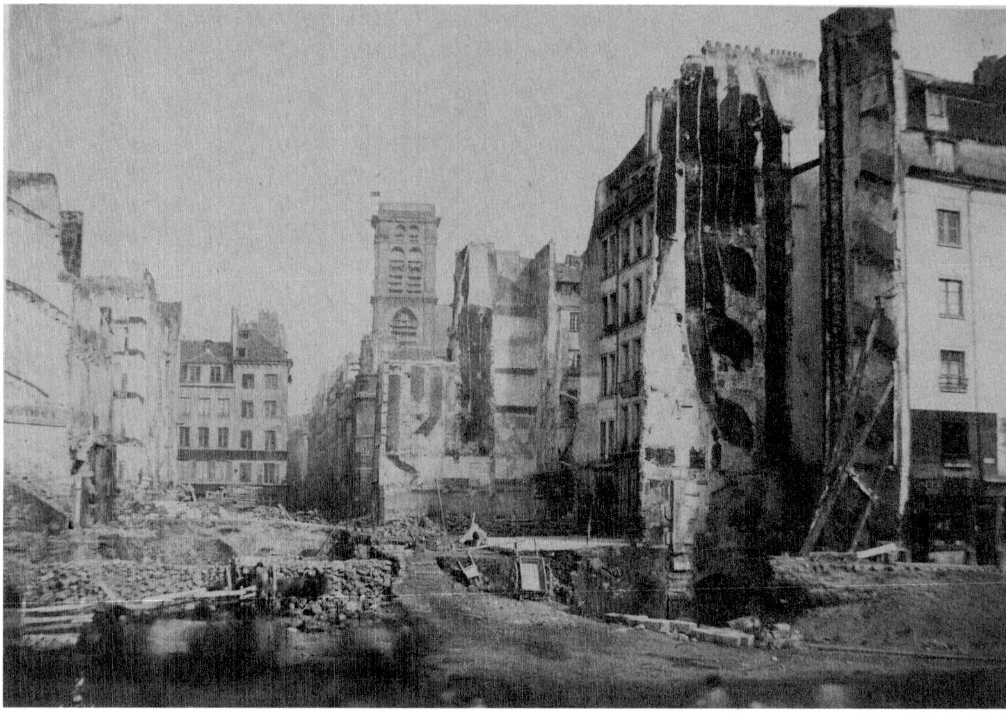

Figure 4.1 Henri Le Secq, *Rue Saint-Martin During the Construction of Rue de Rivoli*, 1848/1858, salted paper print, 7 5/8 x 10 5/6 in. Paris Musées/Musée Carnavalet, Histoire de Paris.

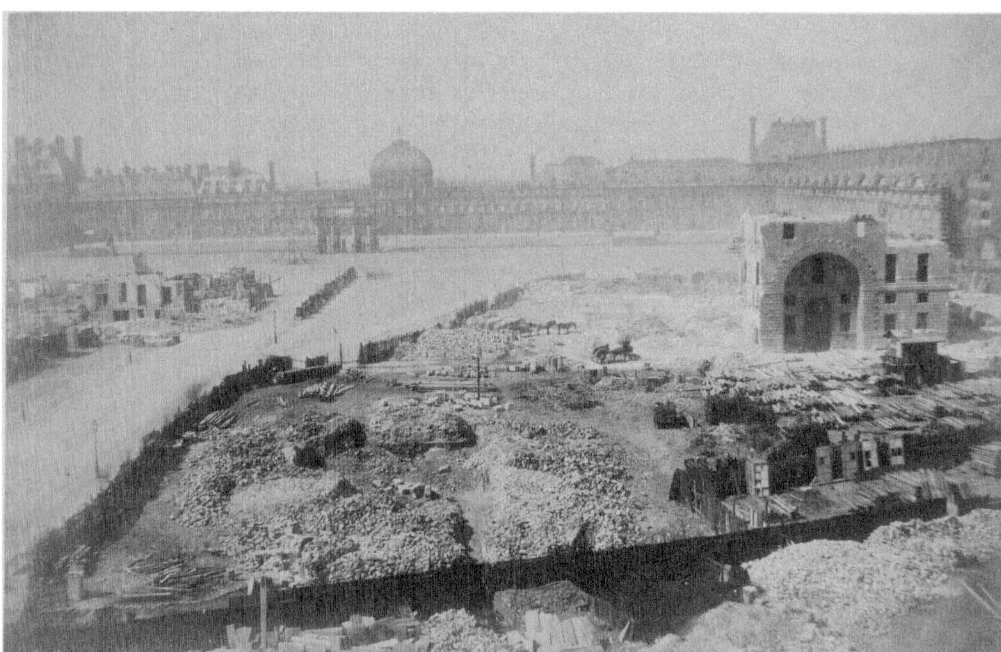

Figure 4.2 Henri Le Secq, *Place du Carrousel, 1er Arrondissement, Paris*, 1852, salted paper print, 7 5/6 x 12 in. Paris Musées/Musée Carnavalet, Histoire de Paris.

victim of this prolonged "upheaval." Hitherto, he had lived in the site of the former royal stables, among an assemblage of makeshift structures that formed a virtual shantytown.[19] His dual perspective of the ruins, both displacement and celebration, further accords with the poetics of ruins.

This word-and-photographic-image reciprocity, mediated by Gautier's pictorial prose, provides an equally viable portrait of Baudelaire's Paris in 1846, transformed by sporadic, urban renewal schemes at the city's heart.[20] While living on the Île Saint-Louis since 1842, Baudelaire wrote "Of the Heroism of Modern Life" as part of his *Salon of 1846* and first defined an aesthetic of dualities. He announced:

> All forms of beauty... have within them *something eternal* and *something transitory*—an absolute and a particular element.... The specific element of each type of beauty comes from the passions, and just as we have our particular passions, so we have our own type of beauty.[21]
>
> (emphasis added)

In 1855, when Haussmannization was well underway, the poet began writing the prose poems and realized a poetic form for capturing such urban "beauty."[22] At this point, Baudelaire may have been inspired by the paradoxical beauty that Gautier's review registered among the ruins since it was republished as the preface to Fournier's 1855 second edition of *Paris Démoli.*

Given these correspondences between Gautier's and Baudelaire's celebratory verbal images of modernity as sites for artistic and creative renewal, we can return to Baudelaire's "The Salon of 1859: The Modern Public and Photography," (hereafter "The Modern Public and Photography"), in his *Salon of 1859*, where he ostensibly condemned photography. Here, I want to suggest that he used photography as both a *memory aid* and an anathema to his more fully developed aesthetic of modernity when, in 1859, he began "The Painter of Modern Life." In the 1859 Salon review, Baudelaire considered photography a deterrence to the creator's imagination.

Photography and "Mnemonic Art"

Baudelaire opens the 1859 Salon review by attacking the smooth finish and detailed realism of the Salon exhibitors' paintings. This criticism aligns with Baudelaire's eschewal of academic conventions fostered by the official École des Beaux-Arts. In the next section, titled "The Modern Public and Photography," he proceeds to chastise those (academic) painters who use photography to perfect pictorial realism; they have "badly applied [the] advances of photography, [which] like all purely material progress, [has] greatly contributed to the impoverishment of French artistic genius." Baudelaire thus warns that if "photography is allowed to deputize for art..., it will not be long before it has supplanted or corrupted art altogether, thanks to the stupidity of the masses, its natural ally." This moral-aesthetic condemnation, however, allows for Baudelaire's endorsement of photography's "true duty"—"that of handmaid to the arts and sciences, but their very humble handmaid[.]" In this capacity:

> [l]et photography quickly enrich the traveler's album, and restore to his eyes the precision his memory may lack; let it adorn the library of the naturalist, magnify microscopic insects, even strengthen... the hypotheses of the astronomer; let it, in short, be

> the secretary and record-keeper of whomsoever needs absolute material accuracy for professional reasons.

Baudelaire's appraisal of photography allows us to identify him—and Gautier—as travelers among the ruins and renovations of their beloved city, and for whom photographs served as "record-keepers." Baudelaire's elaboration on the benefits of photography provides further evidence of this practice:

> Let [photography] save crumbling ruins from oblivion,... all those precious things, vowed to dissolution, which crave a place in the archives of our memories; in all these things, photography will deserve our thanks and applause.[23]

Baudelaire may have recalled using photography as a handmaid in writing these lines. Since Le Secq's works were in the public domain, we can assume that Baudelaire had the Place du Carrousel 'photographic record' in mind or within reach when he wrote the verse poem "The Swan" (1859). This poem records the cerebral processing of memories, qua self-reflections, as the poet-narrator traverses demolition ruins and new building construction. As such, the poem accords with the poetics of ruins and the ameliorative effects of the ruinous aesthetic. Baudelaire begins "The Swan" with a mirror reflection of the grieving widow Andromaque on a river surface that evokes the now industrialized Seine River[24]—"[t]hat mirror, poor and sad, which glittered long ago." From here, the poet "walked across the new [Place du] Carrousel" to lament that "Old Paris is no more (the form of a city / Changes more quickly, alas! than the human heart)." Standing in the now emptied *place* in front of the Louvre, the poet reflects upon the vanquished buildings of the shantytown:

> I see only in memory that camp of stalls,
> Those piles of shafts, of rough-hewn cornices, the grass,
> The huge stone blocks stained green in puddles of water,
> And in the windows shine the jumbled bric-a-brac.

At this site, the poet identifies with the titular swan that escaped from his cage at the demolished menagerie nearby. Like the swan, the poet feels both "ridiculous" and "sublime," "like a man in exile." However, he reframes the jarring contrasts of old and new as an internal mirror of ephemeral and transitory memories: "Paris changes! but naught in my melancholy / Has stirred! New palaces, scaffolding, blocks of stone, / Old quarters, all become for me an allegory, / And my dear memories are heavier than rocks." [25] Unlike the swan, the poet mitigates his alienation with an ironic reflection on his modern condition weighted by memories more burdensome than "blocks of stone."

As exemplified in "The Swan," Paris *démoli* becomes the source of artistic regeneration and modernity. Published on January 22, 1860, shortly after Baudelaire and engraver Méryon's first meeting (discussed in the previous chapter), "The Swan" was included in *Tableaux Parisiens*, the new section added to the second (1861) edition of *Les Fleurs du Mal*. As such, the titular swan re-embodies the alienated "heroes of modern life" (extolled in the *Salon of 1846*) who figured in "The Seven Old Men" and "The Little Old Women." At the same time, the text can be viewed in tandem with Baudelaire's contemporaneous critical writing, "The Painter of Modern Life," first drafted in 1859. In this light, the *tableau*-poem figures are the nostalgic counterpoint to the celebration of modernity, which shapes the *dédoublement* of a collective ambiguity and Baudelaire's ironic self-reflection. Within this dynamic, memory functions as a conceptual and creative agent, aligned with the mnemonic (record-keeping) function Baudelaire assigned to photography.

Baudelaire's notion of "mnemonic art" provides a point of convergence not only between the verse poem and "The Painter of Modern Life," but more importantly, between the latter text and "The Modern Public and Photography" from the *Salon of 1859*. These two critical writings converge as mirroring texts in two complementary ways. First, the "faculty of memory" endows the "painter of modern life" with the photographer's duty as a "record-keeper." In "The Painter of Modern Life," under the heading "Mnemonic Art," Baudelaire explains the onset of the creative process, wherein the artist engages in a "struggle between the determination to see everything, to forget nothing, and the faculty of memory[.]" While I treat this text in Chapter 1, regarding ironic self-reflection, here the faculty of memory performs like a camera, given its "habit of registering in a flash the general tones and shape, the outline pattern." In the same section on mnemonic art, Baudelaire reasserts memory's photographic function, as witnessed in Guys' painterly "execution," which issues from "the absorbed intenseness of a resurrecting and evocative memory, a memory that says to every object: 'Lazarus, arise.'"[26] However, memory's photographic, mimetic realism must be held in check. The creative process advances when the artist translates his impressions into "the culminating features of an object or its main characteristic." This translation, in turn, activates "the imagination of the viewer, [who] conjures up... the impression produced by objects on the [artist's] mind. In this case, the viewer becomes the translator of a translation[.]"[27] Although Baudelaire argues for an "abbreviation" of the distracting details that memory records, his cerebral understanding of mnemonic art also informs his validation of photography as a "record keeper."

The viewer-as-translator also operates in the reception of photographic images. For Baudelaire, photographs of Paris *ancien* and Paris *démoli* function as a mnemonic art, capturing the fleeting and ethereal cognitive experiences that memory fosters. Here, too, "the spectator becomes a translator," but in reverse order of the viewer-as-translator of the painterly record. In the case of photography, precise details trigger "stored" images in the viewer's memories. At this point, the viewer "exercise[s] their memory" to complete the "ethereal" valence projected from the photographic image.

In the second textual correspondence, the modernist aesthetic of transitoriness, upheld in "The Painter of Modern Life," becomes the antidote for (academic) painting's photographic realism, derided in "The Modern Public and Photography." As we have seen in both Guys' techniques and those of the later Baudelairean painters, the lack of finish, broken contours, and spatial distortions make modern life's ephemeral and transitory experiences tangible. Thus, Baudelaire's final words in his essay on photography become contiguous with "the transient, the fugitive, the contingent" elements of modernity that he seeks to advance:

> But if [photography] once be allowed to impinge on the sphere of the *intangible and imaginary*,... then woe betide us!... [Will not] a people, *whose eyes get used to accepting the results of a material science as products of the beautiful*,... singularly diminish its capacity for judging and feeling those things that are most *ethereal and immaterial*?[28]
>
> (emphasis added)

Despite Baudelaire's cautionary words, these last sentences are foundational to exposing the reciprocal side of the word-and-photographic-image correspondences: that of the urban photographers' response to Haussmannization as travelers among the ruins and renovations and, subsequently, their alignment with Baudelaire's painter of modern life.

Baudelaire, Photography, and Antiquarianism

As historian Paul-Louis Roubert has noted, Baudelaire's rhetoric about photography's "true duty" did not occur in a vacuum. His words strongly resonate with François Arago's argument for advancing the scientific, artistic, and touristic use of daguerreotype, which Arago presented to the Chamber of Deputies in 1839.[29] Although Arago's argument precedes Baudelaire's essay by twenty years, the earlier endorsement was often repeated in the following decades.[30] Baudelaire's censure of painting's dependency on photography was likewise derivative. The argument against using photography in the fine arts, resulting in the public's contaminated taste for exact realism, endured from photography's birth and was mandated in *Revue des Deux Mondes* just two years before Baudelaire's essay. Here, art critic Gustave Planche wrote "Le Paysage et les Paysagistes" (The Landscape and Landscapists; May 1, 1857) to denounce both the photographic "doctrines that today dominate landscape painting," which corrupt public taste and the artists who appeal to that taste. Extending Arago's tenet, Planche praises photography's "merits" as "a very faithful document in the absolute sense of the word;" however, he maintains, it "cannot succeed in… reveal[ing] anything imaginary."[31]

What are the consequences of reassessing the originality of Baudelaire's idea of photography as a record keeper and his shaming the public for fostering photographic realism as the standard for beauty in art? As we have seen in the case of Baudelaire's appropriations from Jacques-Joseph Moreau and Johannes Müller, Baudelaire extends these timeworn ideas to formulate an aesthetic program for advancing art beyond mimetic realism and prescribing an art that mirrors the ephemeral lived experience of modernity. His prescriptions, in turn, compelled painters of modern life to break with the conventions of academic painting to capture the fleetingness and fragmentation of their modern urban setting.

This modernist trajectory notwithstanding, Baudelaire's appropriations from Arago and Planche also expose his connection with an antiquarianist discourse that sanctioned photography as a record-keeper when faced with Haussmannization. Marville was central to the practitioners of this discourse. The following section reviews the antiquarianist practices during Haussmannization and Marville's photographs within this context. By examining Marville in dialogue with Baudelaire, we can address the question: to what degree did he insert the "ethereal and immaterial" in his photographic compositions and subjects to assuage the aesthetic arguments against photography and, thereby, to become both modern and aesthetic? In answering this question, I seek to re-embody this ethereal dimension—as both an aesthetic experience and photographic property—by viewing Marville's photography through the double lenses of memory and modernity.

Antiquarianism: A Shared Perspective

A review of antiquarian practices during the second half of the nineteenth century helps to establish Marville's artistic motivations and the civic initiative of the institutions that collected his works. While recent scholarship on Marville has brought Haussmann-era antiquarianism to light, this discourse also substantiates the photographer's shared perspective with Gautier and Baudelaire. Here, the antiquarian trope of "old Paris" signifies the collaborations and exchanges among artists, writers, and civic leaders. As witnessed in the previous chapter, Baudelaire's admiration for and collection of Charles Méryon's quasi-fantastic etchings of *vieux* Paris resonate with the antiquarianist discourse. Méryon's printmaking coincides with the emergence of antiquarianism in book and print production.

Not surprisingly, Fournier's *Paris Démoli* is a prominent marker of the burgeoning interests in old Paris during the mid-century. First published in 1853, when Haussmannization began, *Paris Démoli* was revised and updated in 1855 and 1883. In addition to adding Gautier's review as the "Preface" of the latter two editions, Fournier added his introductory essay, "Introduction: Voyage à Travers les Nouvelles Ruines" (Introduction: Journey Through the New Ruins). While Fournier's title resonates in Baudelaire's touristic wanderings through demolition sites, the *Paris Démoli* texts also mirror the collective ambiguity of nostalgia and hope. Gautier and Fournier marvel at the newly built boulevards and edifices but lament the loss of historically meaningful buildings.[32] For example, while the first two paragraphs of Fournier's sixty-five-page "Introduction" echo Gautier's admiration for the newly built boulevards and edifices (adding beauty and comfort to the city), the remaining pages follow the format of the main text, recording the history of recently demolished buildings and streets. Adolphe Martial Potémont's *Ancien Paris* (Old Paris; 1864) extended Fournier's antiquarianist project with an album of 300 engravings. However, Potémont never acknowledged Haussmannized Paris; he only presented those edifices that had been demolished. In doing so, he portrayed a quintessential "picturesque Paris, anecdotal and vibrant with history."[33] His project was a success. In 1867, he completed the series *Ancien Paris* with the addition of 48 new plates, and then in 1877, he published *Les Boulevards de Paris* (The Boulevards of Paris), a series of 45 plates depicting old thoroughfares known as the "*Boulevards*."[34] In this volume, the variously authored commentaries focus on the disappearing life and culture of the boulevards and the necessity of preserving their memory in word and image.

Picturesque Practices and Public Institutions

Fournier's and Potémont's nostalgic approach to historical sites, ruins, and vanished buildings conform to picturesque travel writing, which emerged at the end of the eighteenth century and extended to Romantic landscape art and poetry.[35] When travel writing focused on ruins, these texts brought the human drama and histories associated with ancient buildings and sites to life.[36] When illustrated, historical sites were framed within Claudian landscape vignettes[37] in which the reader was invited to embody the tourist's excursive (roving) movement and (exploring) gaze.[38] Here, too, picturesque travel and texts engaged the spectator/reader in a metacognitive aesthetic experience of self-reflection and transformation. While Marville practiced the picturesque style in his early career, he rid himself of the outwardly Romantic motif of Claudian vignettes. Even so, as in the picturesque mode, his photographs are equally "anecdotal and vibrant with history." For this reason, they served private collectors, illustrators, and public institutions who were heavily invested in saving old Paris in the collective memory. Thus, picturesque technical and aesthetic practices were at the heart of Haussmann-era antiquarianism.

As Arago, and then Baudelaire, prescribed, from mid-century photography became the handmaid to the record-keeping industry, with photographers supplying images for illustrators and writers of antiquarianist publications. However, photography had a more significant role in establishing and supporting historical societies, museums, and archival collections. Curator Sarah Kennel's observation about Marville's photographs of old Paris (begun in 1862) broadly applies to institutional and private collecting of photography to preserve the memory of pre-Haussmann Paris: "Marville's pictures did not emerge sui generis, but were produced within a wide field of competing representations of 'Vieux Paris' (Old Paris), a city that had existed both physically and conceptually in the art and literature of the mid-nineteenth century." To recognize this cultural practice, she adds, "is to recognize that representing Paris in the 1860s was a highly charged endeavor."[39]

Marville was at the center of this "highly charged endeavor." The photographer sold his works to or received commissions from public institutions founded on antiquarianism. For example, Kennel suggests that Marville's Old Paris series may have been acquired by "the Service des Travaux historiques [Historic Works Service] to sponsor important works on the history of Paris and to gather material, tangible documents that could provide a 'demonstrative history' of the city." That project, initiated in 1867, was described by a municipal counselor and committee member for Historic Works, Charles Poisson. Even so, Poisson's description of the Service's project invokes the collective ambiguity, fostering nostalgia and pride in the city's modernization. As he notes, in establishing the Service des Travaux historiques:

> the current municipal administration... pays a debt to the past that it proudly incurred by transforming the city; and... it reconstitutes its history by showing, in the most striking and easy-to-grasp form, what existed in place of the boulevards, promenades, and squares with which it has endowed the Parisian population.

We can assume, with Kennel, that photography played a crucial role in this mission:

> Though Poisson does not specifically mention photography, his description of the museum's ongoing accumulation of 'views of streets, courtyards, bridges, and quays' in the 'most striking and easy-to-grasp' form leaves open the strong possibility that Marville's Old Paris photographs—even if they were originally commissioned for another purpose—found a home in the [Carnavalet] museum's topographic collection.[40]

Marville: From Picturesque Aesthetic to Precisionist Sublime

Marville's work forms a photographic archive that spans the heyday of Haussmann's transformations of Paris. In the early 1860s, Marville began taking photographs for the *Album du Vieux Paris* (Old Paris Album). By 1864, he received an official city commission to take 425 views of existing streets and structures tagged for demolition, a project he completed in 1869.[41] In late 1874, Marville was commissioned to record Haussmann's new streets, a project he completed in 1877. Marville received the later 1874 commission from the city's Travaux de Paris (Public Works) in preparation for an exhibition at the Paris 1878 Exposition Universelle, where his photographs of old Paris were exhibited next to the new—arranged street by street—to present to the world Haussmann's *grands travaux* (major works) and the city as a model of modern urbanism.[42]

While this before-and-after comparison was not intended in the original commissions, we can interrogate each of Marville's Paris projects as responses to, and documents of, Gautier's and Baudelaire's written descriptions of the city transformed. In this light, Marville's corpus evokes the collective ambiguity to which these writers gave voice. Whether viewed separately or together, the photographic images of newly built streets and buildings and those yet to disappear trigger the collective memory (and individual self-reflection) of the rapidity and inevitability of change. Here, I examine how Marville's official portraits of Paris addressed issues of urban modernism that align him with Baudelaire and, subsequently, the Baudelairean painters of modern life and the ways they channeled subjective experiences of Haussmannization into their art.[43] This experiential approach corroborates recent scholarship on Marville, published in the exhibition catalogue *Charles Marville: Photographer of Paris* (2013). As Kennel, the exhibition's curator, states, "Outside of whatever instrumental purposes sparked the commission[s], we must also attend to the particularities of the images themselves and explore how Marville's relationship to the city he was charged to represent may have shaped them."[44]

Marville: Artistic Training and Early Photography

To forge this Baudelairean alignment, it is essential to establish Marville's artistic training and contacts before becoming Paris's official photographer. In the early 1830s, Marville began his artistic training as a painter in the Académie Suisse, considered an alternative to the official École des Beaux-Arts.[45] In 1834, Marville embarked upon a career as an illustrator, and in 1838, he studied at the Athénée des Beaux-Arts to hone his skills in the illustration trade. Through the 1840s, Marville's main activity was producing illustrations of landscapes, cityscapes, and architectural views for wood engravings published in travelogues and mass-market news media. As we shall see, Marville's practice as an illustrator highly impacted his photographic framing and composition techniques.[46] Likewise, the mutual working relations between illustrated books, graphics publishing, and photography "may [also] account for Marville's fluid transition from illustration to photography in 1850."[47]

We should also consider how Marville may have engaged Baudelaire's writings and art criticism. For one thing, Marville was not only closely involved with artist-painters as an illustrator in the publishing industry, but he also would have come into contact with literary writers.[48] For another, his connection with artists was further secured in his professional practices—first as a copyist and then as a photographer of paintings in the Louvre and at the Salon.[49] As a photographer for contemporary artists, Marville was most frequently employed by Salon luminary Jean-Auguste-Dominique Ingres (1870–1867).[50] Although Ingres was a leader and protector of academic practices, he was a subject of Baudelaire's criticism in his *Salon of 1846* and his review of the 1855 Exposition Universelle. Along this trajectory, spanning his early training as an illustrator and then a photographer, Marville would become familiar with artistic debates and a critical discourse spawned by Baudelaire's Salon and exhibition reviews.[51]

As Kennel reports, "By 1850 [Marville] had all but abandoned illustration in favor of the relatively new practice of photography on paper." By the end of 1851, he had become the chief contributor to Louis-Désiré Blanquart-Evrard's photographic publications, lasting until 1855.[52] This professional affiliation was important since Blanquart-Evrard pioneered photographic publishing, technical developments, and critical and technical writing on the new medium. Blanquart-Evrard even advanced techniques to ensure photography's artistic practices and effects.[53] Given this affiliation, Marville changed his professional status from an "*artiste-peintre*" (artist-painter) in 1851 to an "*artiste-photographe* (artist-photographer)," a term he used for the first time in the *annuaire* (directory) of 1852.[54]

While employed by Blanquart-Evrard, Marville photographed public buildings and landscapes. In this capacity, he gained access to more progressive artist-painters and picturesque photographic practices while traveling with Barbizon School landscape painters to the forests of Fontainebleau.[55] During these excursions, he translated the painterly qualities of plein-air landscape painting into the photographic medium. In doing so, he captured "the slight movement of leaves and render[ed] the dappled sunlight filtering through the branches as a flickering patchwork of light and dark that enlivens the entire image."[56]

Back in Paris, Marville mastered the urban subjects and commercial side of photographic production. His commercial activities had begun before Blanquart-Evrard's business closure in 1855, making photographs for architects, artists, and museums.[57] For these clients, "Marville photographed different views of the quays of the Seine, bridges, market squares, monuments, buildings, and quiet gardens, capturing both the public spaces and the private corners of the city."[58] From the late 1850s to the early 1860s, Marville received city commissions that prepared him for his appointment as official photographer of Paris. These commissions included the documentation of several construction and restoration projects.[59] Significant was the 1858 commission to photograph the Bois de Bologne and its transformation from royal hunting grounds to public pleasure gardens. Here, he collaborated with Adolph Alphand, who redesigned the Bois in 1853 and, under Haussmann, was appointed department head of Promenades et Plantations (Department of Promenades and Plantings) in 1856 and director of Travaux de Paris in 1870.[60] Marville's professional affiliation with Alphand finally secured his appointment as the official photographer of Paris and his long-term collaboration with the architects and engineers who worked under the directing engineer.[61] Providing photographic documentation for a wide variety of municipal subjects, Marville realized his mature style, which Kennel describes as a shift from picturesque lighting and atmosphere to "a lucid and expansive manner that expertly conjoined vivid description with a rigorously balanced composition and at times downright dazzling depiction of space."[62] We can observe Marville's stylistic maturity in the lucid quality of the Old Paris photographs, his first commission as the city's official photographer.

Marville as Photographer of Old Paris: The Picturesque and Mnemonic Art

My analysis of Marville's Old Paris photographs is framed by two related discourses discussed earlier in this chapter. First, I align him with Baudelaire's sanction of photography's capacity for "sav[ing] crumbling ruins from oblivion" and their shared antiquarianist motives amidst "Haussmann's creative destruction of Paris in the name of modernization."[63] Second, I align Marville with the picturesque aesthetic of the ruin that Parisian antiquarianists helped to sustain. While Kennel argues that Marville had, by the 1860s, dispensed with "the last vestiges of a picturesque sensibility," I disagree with this assertion. Indeed, her vivid descriptions allow the recovery of enduring picturesque elements, particularly the compositional devices that facilitate the excursive view of the picturesque traveler and the mnemonic function of the picturesque ruin. By tracing the tenets of the ruinist aesthetic from Baudelaire and Gautier to the photographic corpus of

Marville's Old Paris, we can better understand how this archive signifies Paris as *un lieu de mémoire*, transforming its whole and its parts into mnemonic monuments.

As previously noted, Marville's Old Paris photographs were likely purchased for the topographic collection of the Musée Carnavalet, the city's historical museum, and as guides for the engraved illustrations in the city's book publications of old Paris. Illustrators also used them for collectible antiquarian books, like Potémont's *Ancien Paris*. In his translation from photograph to etching, Potémont teased out and heightened the picturesque details associated with old Paris, skillfully multiplying lights against darks and mid-tones as defined by building surfaces and shadows cast on cobblestone streets (Figure 4.3). Crumbling plaster veneers and irregular projectiles of streetlamps and window shutters also add picturesque hallmarks of variety and roughness. As Kennel observes, these details could be read as both condemnatory and nostalgic regarding Haussmannization and old Paris as these terms coalesced in popular culture. As a countermodel to negative readings of old Paris, I follow Potémont's picturesque interpretation of Marville's photographs to establish a mnemonic reading of his Old Paris photographs. This reading suspends the pejorative connotations of "the claustrophobic constriction of the streets" and damp, sewage-laden cobblestones.[64]

Potémont's attraction to the picturesque elements in Marville's Old Paris images was an "informed reading."[65] These signifiers were already embedded in his photographic images' subject matter and composition thanks to Marville's training as an illustrator in the picturesque tradition. As I have argued elsewhere, the nineteenth-century understanding of the "picturesque" exceeded its etymological definition of "like a painting," especially in associating ideas with ruins.[66] In this discursive context, illustrators and viewers attached historical and moral meaning to ruinous buildings and places ravaged by time, endowing them with palimpsest-like layers of signification.[67] These associations are also endemic to the "*pittoresque*" (picturesque) as a moniker for the topographic views of old Paris. Here, the poetics of the ruin is based on the contingent association of ideas attached to a place. While Marville's images of narrow streets, irregular passages, and tottering, uneven buildings are not historical monuments per se, they represent neighborhoods demolished and yet to be demolished whose names hold locational and generational heritage and historical trajectories of lived experiences. As such, photography becomes a mnemonic art and transforms the lived experience into the experiential aesthetics of the ruin.

Marville's photographic technique enhances the tangible presence of these lived spaces and facilitates the viewer's spatial immersion (Figure 4.4). As curator Françoise Reynaud explains, Marville provided specific references to locations where a viewer might be standing with map-like precision: "[Marville] often positioned his camera so that the signs located to the left or right of an intersection, which gave the names of the streets, squares, and other thoroughfares, were visible. As a result, it is generally easy to identify the position of the camera simply by consulting a Paris map of the period." This topographic specificity adds to the viewer's spatial memory of the place, as does Marville's method of photographing a location from multiple perspectives, "moving his camera to record the scene from opposite directions."[68]

Marville's pictorial techniques also compel the viewer's immersive experience in his Old Paris photographs. Kennel observes that "the distinctive pictorial strategies Marville deployed—high-keyed description, clarity, and rigor—were designed to provide maximum visual data." These techniques served the archival functions of the city commissions and mediated the viewer's embodied experience within these storied "picturesque" spaces. This excursive view was a hallmark of picturesque composition. By adopting it,

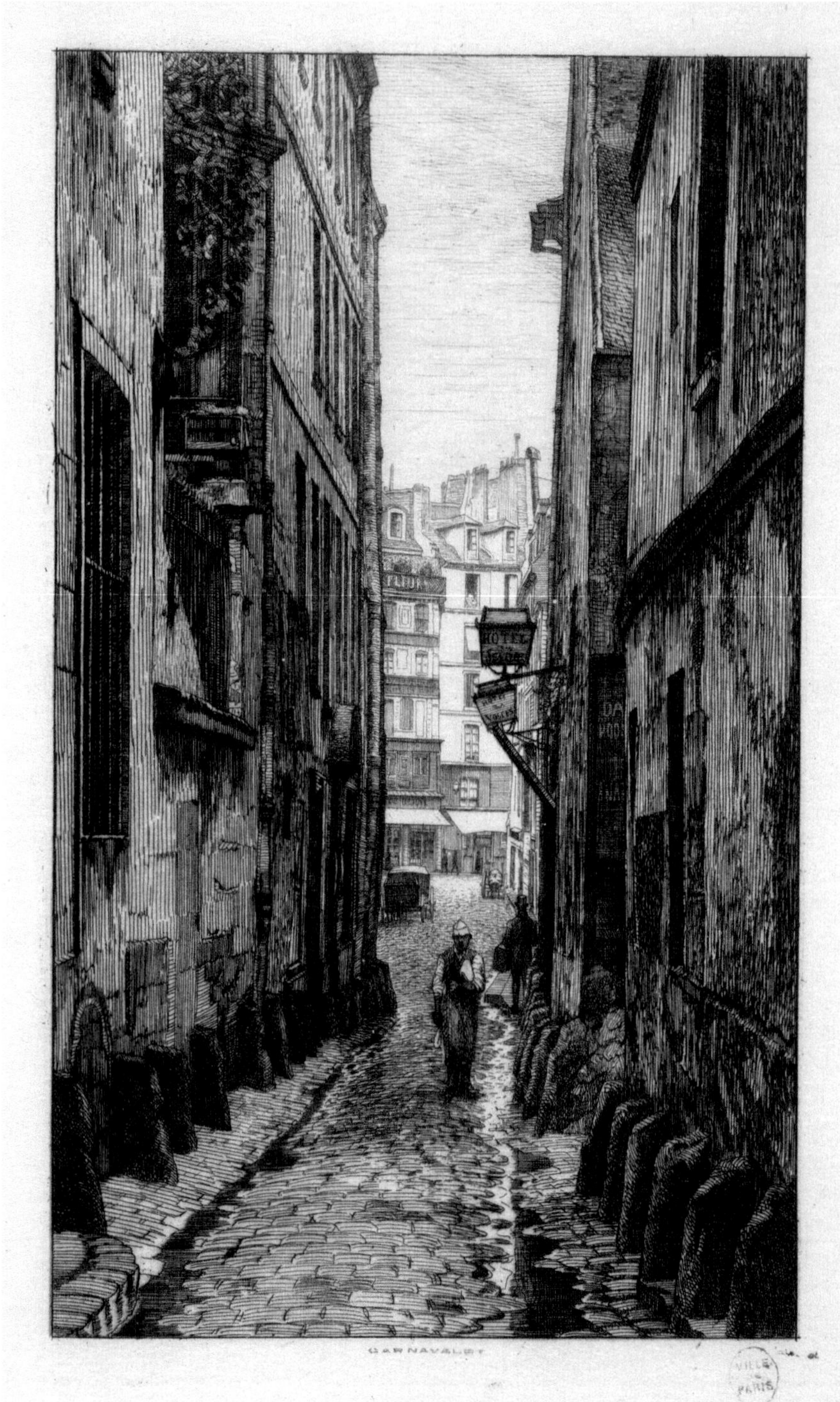

Figure 4.3 A.-P. Martial (called Adolphe Martial Potémont), *Rue du Croissant*. Published in *Ancien Paris* [Paris: Potémont, c. 1863], etching, 10 5/6 x 14 1/8 in. Paris Musées/ Musée Carnavalet, Histoire de Paris.

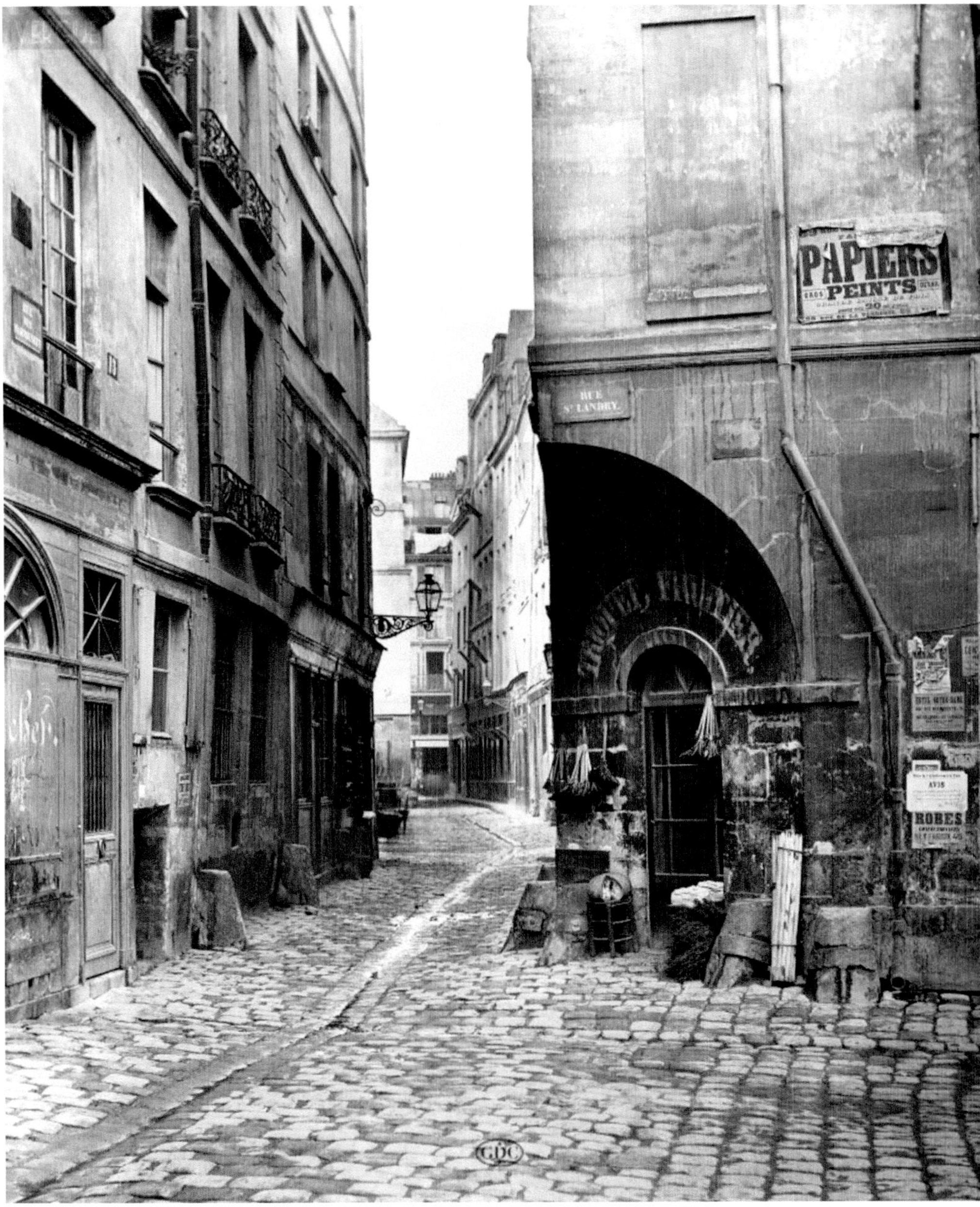

Figure 4.4 Charles Marville, *Rue des Marmousets, from Rue Saint Landry, Île de la Cité, Paris IVe*, 1865, 2015 print from digitally restored negative (original albumen print), 11 x 13 3/8 in. Bibliothèque Historique de la Ville de Paris. Credit: Laurent Gloaguen, *Vergue*.

Marville invites viewers to enter the spaces of his photographs. Kennel's description of an excursive, wandering perspective, integral to Marville's Old Paris compositions, supports a picturesque reading of these images: "Marville frequently provides an imaginative pathway for the viewer: space is haptic, human-scale, and the photographs, with a few exceptions, are made from the perspective of someone on the ground."[69]

Marville developed compositional devices to achieve this excursive effect: pathways framed by adjacent cross-streets and building flanks, deeply shadowed middle grounds relieved by illuminated spaces in the background, and spatial depth accentuated by pathways and building elevations aligned along orthogonals leading to a distant vanishing point. Marville's Old Paris photographs often offer vertical, close-up views instead of the new boulevards' expansive, horizontal, distant views to emphasize irregular pathways, buildings, and surfaces (Figures 4.5 and 4.6). In these ways, Marville staged an excursive narrative, a tactile history of streets and inhabitants. For the private collector of Marville's photographs, the inhabitants' lived experience becomes an aestheticized poetics of ruins. For the professional and institutional collectors, Marville's photographs serve as the *aide de memoire* for "whomsoever needs absolute material accuracy" to "restore to his eyes the precision his memory may lack" (Baudelaire). Here the poetics of ruins correlates with a metacognitive, embodied aesthetic experience. Whether physically traversing these streets or viewing their photographic images, the stroller or spectator engages in streams of memories and self-reflection.

Marville's photographic practices and the aesthetic functions they transmit are exemplified by his survey of the once-medieval district marked for demolition to widen and extend the Rue des Écoles, for example, in *Rue du Bon Puits from Rue Traversine. Paris Ve. Avril 1866* (Figure 4.5). As we enter the Rue du Bon Puits, a raking horizontal light shaft in the foreground initiates our approach from a passageway to the right. Not only does the light lead us into and along the cobblestone street, but it also highlights the wall that frames our vista to the left. Here, we immediately see a building elevation whose plaster facing has worn away, leaving bare the cement and rubble of the load-bearing wall. This ruinous building reminds the viewer of the medieval origin of the street, dating to the mid-thirteenth century.[70] Framing our entrance, the crumbling wall stands metonymically for the edict authorizing the street's demolition in May 1866—the date of Marville's photograph. Demolition took place four months later, in September 1866.

As in other photographs, the high-key clarity of details captures the street's temporal fragility. Traces of the stenciled name of a business are visible on the one-story façade next to the crumbling building. Marville's camera also seizes figurative details. In front of the adjacent building, a man sits still on a *chasse-roue* (guard stone) outside a doorway. (The *chasse-roue* and the absence of sidewalks were markers of old Paris.)[71] In Marville's Old Paris photographs, working-class inhabitants hold still, waiting for the camera to save them, like their streets, from oblivion.[72] However, while standing still, some figures appear to have budged, as traces of the seated man's companions—a man standing to his right and a man seated to his left—bear witness. Here, as elsewhere in Marville's photographs, technical limitations blur moving figures—the slower shutter speed and longer exposure time needed for fixing the impression on the glass plate negative. These technical "faults" function as temporal details. The moving figures register as shadows or streaks that contrast with the buildings' solid materials and surfaces and the vivid light-and-shadow patterns that define them. Caught in a doorway or street crossing, the figures' blurred traces capture their tenuous existence in collective memory and the liminal spaces of their habitat.

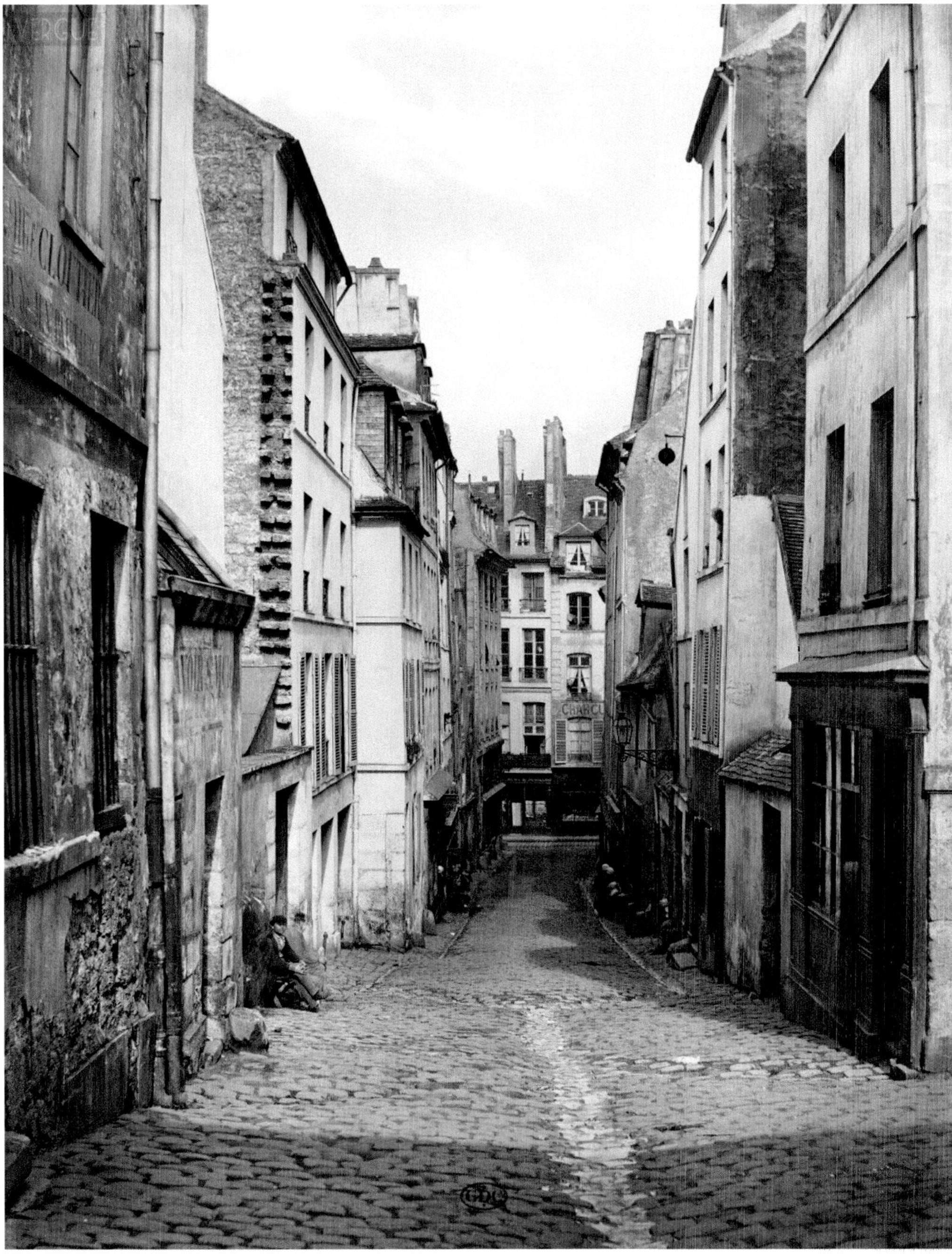

Figure 4.5 Charles Marville, *Rue du Bon Puits from Rue Traversine. Paris Ve. Avril 1866*, 1866, 2019 print from digitally restored negative (original albumen print), 9 3/4 x 13 5/8 in. Bibliothèque Historique de la Ville de Paris. Credit: Laurent Gloaguen, *Vergue*.

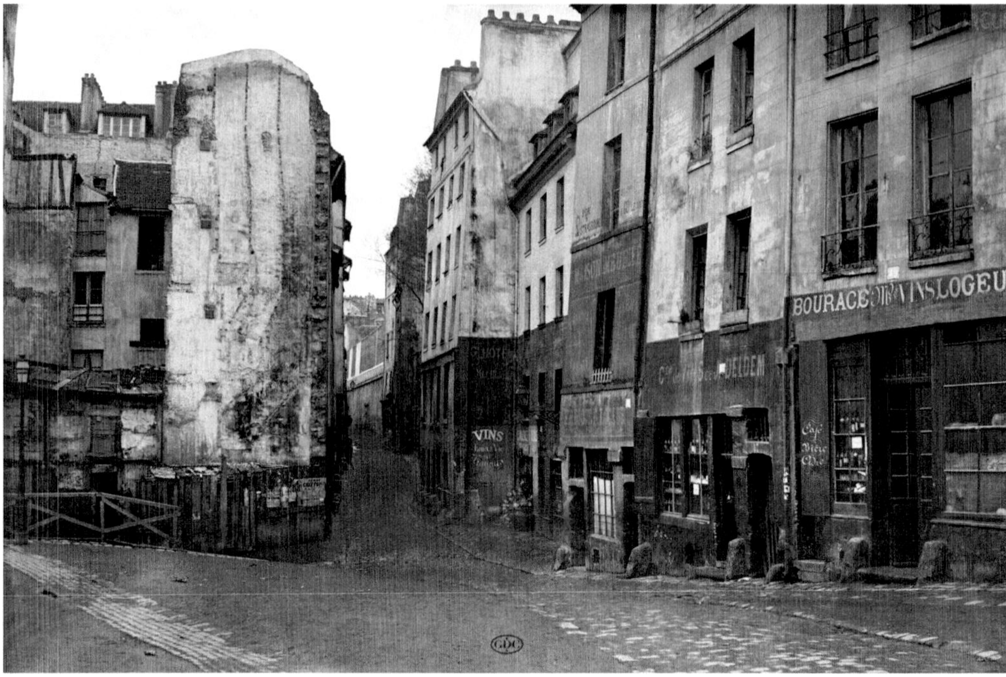

Figure 4.6 Charles Marville, *Rue Traversine, from Rue des Écoles. Paris Ve [1866]*, c. 1866, 2018 print from digitally restored negative (original albumen print), 14 1/2 x 9 5/8 in. Bibliothèque Historique de la Ville de Paris. Credit: Laurent Gloaguen, *Vergue*.

On the right side of the Rue du Bon Puit, we first encounter a shaded building whose windows gleam with reflected light from the sunlit buildings across the street. Upon closer inspection, a woman appears in the doorway; she has budged, rendering her a blur, more fleeting than the reflected light in the windows. Ironically, Marville left these technical errors intact as reminders of the edict just passed and the human displacement that followed. Assuming the antiquarian's view, we recall Baudelaire's "The Swan," where "in the windows shine the jumbled bric-a-brac," and here Marville makes tangible how "Paris changes! but naught in [our] melancholy,… Old quarters, all become for [us] an allegory."

The play of light and shadow continues to pull us into the photographed space, much like a picturesque landscape vista. The shaded buildings on the right sharply contrast with the sunlit buildings on the left. At the same time, highlights and shadows carry tonal gradations that clearly define varied elevations, rooflines, projections and recesses, and material textures. The play of lights and shadows also vividly exposes uneven cobblestones, another marker of old Paris, which Marville's photograph casts as a central motif. The cobblestone path, with its slightly depressed center gutter, defines the central axis of the composition and leads the eye down the street corridor. Halfway down the right side, the sloping roof over a two-story building reifies the irregularity of the built and paved fabric, as does the street lantern extending from a bracket. Beyond this marker, both streets are cast in shadow; the darkness is relieved at the vista's end. The orthogonal lines, reiterated by the building elevations, recede to the end of Rue du Bon Puits. At that point, the street intersects with the Rue Saint-Victor, lined by buildings illuminated by the same raking light that initiated and now terminates our excursion. Details on that street suggest signs of enduring domestic and commercial life—curtains frame the window above the *charcuterie* sign, while a poster announces a theatrical event for May 1866.[73] These details also tell a story; the Rue Saint-Victor was saved from the wrecking ball.

This good fortune was not the case for the five adjacent streets marked for demolition to make way for the extension of Rue des Écoles. Marville's photograph of the *Rue Traversine, from Rue des Écoles. Paris Ve [1866]* records the demolition work in progress (Figure 4.6). As the full title of *Rue du Bon Puits* indicates, while taking the photograph, Marville had been standing on, with his back towards, the Rue Traversine, testifying to the photographer's physical immersion in the demolition process and the temporal dimension of his images. In the Rue Traversine photograph, Marville captures his lived experience of the transient state of the built environment. From this prospect, he records a moment when the demolition process stood still due to a hiatus in the expropriation process in November 1865 (restarted in May 1866).[74] Here, transition details coalesce in the poetics of ruins; the photographer frames the vista with contrasting but consonant signs of demolition and lingering commercial life. On the left, remnants of the Rue Saint Nicolas-du-Chardonnet show buildings shorn of their abutments, enclosures, and adjoining walls. An interior courtyard of shanty structures is exposed to the street, as are the blackened voids of windows shorn of their glass panes. The fragile edifice and its debris are held back by a makeshift wooden guardrail and picket fence, barely containing the massive weight of masonry ready to fall.

Marville frames the right side of the photograph with buildings that converge with the viewer's space. He set his camera at a thirty-degree angle to the spine of Rue des Écoles to capture the road's curvature, natural to the medieval topography but soon to disappear under Haussmann's *percements* ("piercings" for street openings)[75] of straight-edged, ruled lines. On this side, a partially open door to a wine shop suggests the persistence of daily life in this ravaged street. Faded stenciled names of businesses unfold along the building facades, reminders of the waning street culture and the dismantling of commercial life. Wine bottles line two shop-front windows. As the curving Rue Travertine recedes into perspectival depth, barren tree branches emerge from a recess, silhouetted against a wall and then rising against the sky, yet another marker of temporal decay. Like the Rue du Bon Puits photograph, the orthogonals end in a field of light, highlighting the enclosed walls of the École Polytechnique, for which the demolished streets along the Rue des Écoles made way. That Marville highlighted this terminus speaks to Haussmannization as well; it was from this institution that engineering students honed their skills to realize Haussmann's *grands travaux*.

Demolition and an Enduring Picturesque Ruinist Aesthetic

With the completion of the Old Paris commission in 1869, Marville transferred his residual picturesque practices to Haussmann's demolition sites, a series in which he directly engaged with ruinist aesthetics.[76] Among the most spectacular views are the demolitions, excavations, and construction at the Avenue de l'Opéra. Among Haussmann's *percements*, the Avenue de l'Opéra was designed as an orthogonal avenue to frame architect

Charles Garnier's masterpiece opera building (1861–75), the jewel in Emperor Napoléon III's urban crown.[77]

In his Old Paris photographs, Marville's up-close compositions give the viewer direct access to the narrow passageways by cropping buildings that border intersections or the street's opening. In this way, the viewer follows the orthogonal lines embedded in the buildings, gutters, and curbs, allowing a gradual descent into the picture space. In contrast, Marville's close-up views of demolition sites plunge the viewer into the ground-level rubble. Marville employed a horizontal format as in the new boulevard photographs. However, close-up points of entry produce different results in these two series. In the boulevard photographs, the shift from vertical to horizontal format emphasizes the symmetry of buildings and streets, the expanse of light and air, and the focus on distant vistas (Figures 4.9 and 4.10). Conversely, in the demolition photographs, the horizontal format emphasizes a panoramic, immersive topography of rubble and buildings in various states of destruction. For example, in *Opening of Avenue de l'Opéra: Butte des Moulins from Rue Saint-Roch*, the photographer places himself, and the viewer, at the bottom of the pit of rubble (Figure 4.7). Disassembled building parts—stone slabs and blocks, timber, and dirt—expand across the foreground and middle ground and recede toward the background of a hollowed-out building.

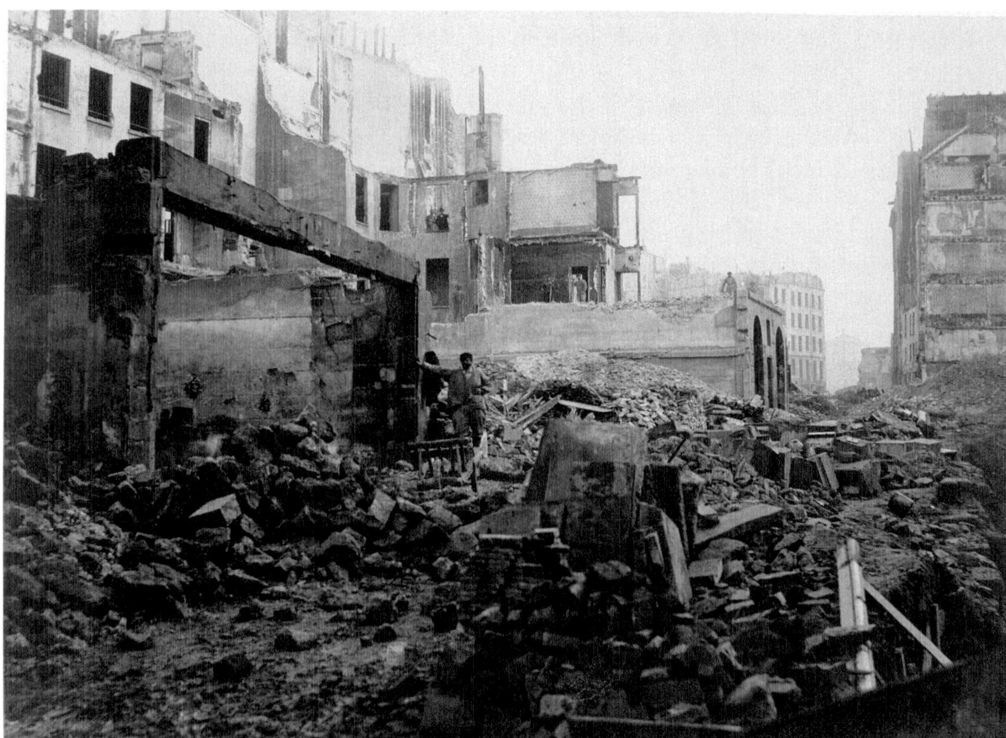

Figure 4.7 Charles Marville, *Opening of Avenue de l'Opéra: Butte des Moulins from Rue Saint-Roch*, 1876–7, albumen print, 10 x 14 in. Paris Musées/Musée Carnavalet, Histoire de Paris.

By viewing Marville's photographs within the picturesque aesthetics of the ruin, the conflation of photographic documentation and the poetics of ruins comes into view. Ironically, Marville has centered on a still-wallpapered wall, frontally framed against the cloudless sky (visible in the original print). This detail and the skeletal building invite viewers to recall Gautier's paradoxical gaze while traveling among Haussmannian ruins as "a curious spectacle," as noted earlier.[78] Marville's foregrounding of the ruins and our immersion therein also mirrors Baudelaire's encounter, in "The Swan," with the Louvre courtyard demolition site. Like the poet, the viewer is a traveler among "Those piles of shafts, of rough-hewn cornices, the grass, / The huge stone blocks stained green in puddles of water[.]"

Marville's pictorial devices compound this picturesque ruinist aesthetic. He arranges perspectival vistas of buildings in ruin, which lead the viewer's gaze into distant space. As in the picturesque landscape, perspectival depth is marked by receding planes parallel to the frontal picture plane and the orthogonal lines embedded within, each of which frames the focal point in the middle ground. The darker foreground offsets the more fully illuminated middle ground to highlight the vista's subject.

In *Opening of Avenue de l'Opéra: Butte des Moulins from Rue Saint-Roch*, Marville delineates pictorial recession with the planar volumes of hollowed-out building frames on the left and solid building blocks on the right, stripped of their facing materials and ornaments. To convey the perilous effects of falling and fallen rubble, Marville has aligned the orthogonals on an angle, veering to the right of the center point, occupied by a large stone boulder. A fallen timber directs our attention to the orthogonals' central axis, as does a stone slab to the left and a piles of stone blocks that align with the timber plank. The orthogonals' receding point ends in a luminous vista as in the Old Paris series. At this end, the newly built l'Opéra, faced in sparkling white Lutetian limestone and silhouetted by its crown-like dome, dissolves in a distant halo of atmosphere and light.

Given Marville's framing of the old and the new, we should consider how his photographs are inflected with his reading of the textual poetics of picturesque ruins.[79] As earlier noted, Gautier viewed Haussmannian demolitions as "not without beauty" given their "picturesque effects" of light and shade."[80] Marville's vista and the contrast between old and new again evoke Baudelaire's "The Swan," where "new palaces, scaffolding, blocks of stone" replace the "old quarters" that weigh in his memories. As the city photographer, Marville may have held similar memories when photographing *Opening of Avenue de l'Opéra: Butte des Moulins from Rue Saint-Roch*. As the location of the full title implies, the area cleared for the Avenue de l'Opéra featured two small hills (*buttes*), the Butte aux Moulins and Butte Saint Roch, partially leveled in 1615: "The Butte des Moulins (literally 'windmill hill') was a small hill—part natural, part built up—that had existed in Paris since ancient times. Dotted with windmills, it was a working-class neighborhood dominated by small trades until the mid-1870s."[81] Perhaps it was in homage to this now-displaced population that Marville filled his demolition photographs with demolition workers.

From the picturesque perspectives reviewed here, we can assume that Marville adhered to the ruinist aesthetic that, following Lee's model, aimed to "reconcile the shock of destruction with the pleasure of ruins." On the one hand, Marville's compositional, lighting, and photographic methods provided the means to this end. However, Marville subverted the ideological means to this end; he did not "dissociate laborer from product, ruin from [displacement and deprivation]."[82] As noted in the photographs reviewed, Marville included the workers engaged in Haussmann's "creative" destruction of old

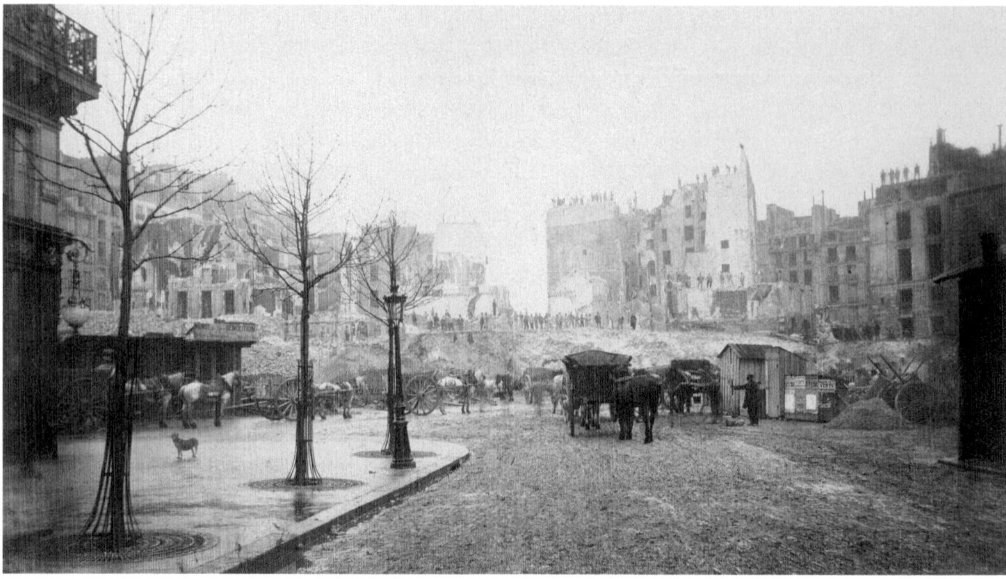

Figure 4.8 Charles Marville, *Opening of Avenue de l'Opéra: Buttes des Moulins, View Taken from the Rue Saint-Honoré*, December 1876–February 1877, albumen print, 10 x 14 1/4 in. Paris Musées/Musée Carnavalet, Histoire de Paris.

Paris, particularly in the Butte des Moulins district. In the former photograph of *Opening of Avenue de l'Opéra: Buttes des Moulins*, laborers stand amidst the rubble and are framed above by hollowed-out windows and doors. Another photograph, *Opening of Avenue de l'Opéra: Butte des Moulins, View Taken from the Rue Saint-Honoré* (December 1876–February 1877), more directly signifies Marville's celebration of the human labor force (Figure 4.8). In this image, horse-driven carts line the foreground, while the receding planes of the middle ground are lined with workers spread across the horizontal expansion of the picture frame: some stand atop a plateau of rubble, and others have scaled partially demolished buildings and rooftops of multistoried structures yet to be felled to ground level. Here, too, Marville has captured the demolition process, which is frozen still. Yet, he records the process as it unfolds in time and space, showing different levels of demolition and the men who wield the pickaxes, mallets, and shovels.

Perhaps, in 1876–7, Marville reflected upon and transposed Baudelaire's 1846 rallying cry for artists to discover the "heroism of modern life" in "the scenes... of the thousands of uprooted lives that haunt the underworld of a great city."[83] Indeed, these laborers mirrored the more recently uprooted lives, displaced from their rural villages and demolished Parisian *quartiers*. At the end of the day, they migrated to the city's outskirts of shantytowns, the newly established "underworld of a great city."

Marville's Double Vision: Paris Transformed

While Marville developed a picturesque aesthetic of ruins, he simultaneously developed a precisionist aesthetic to promote the Haussmannian vision of modernized Paris, at once sanitized, illuminated, spacious, and regulated. This difference notwithstanding, for the New Paris series, Marville sharpens the pictorial perspective to evoke the classical

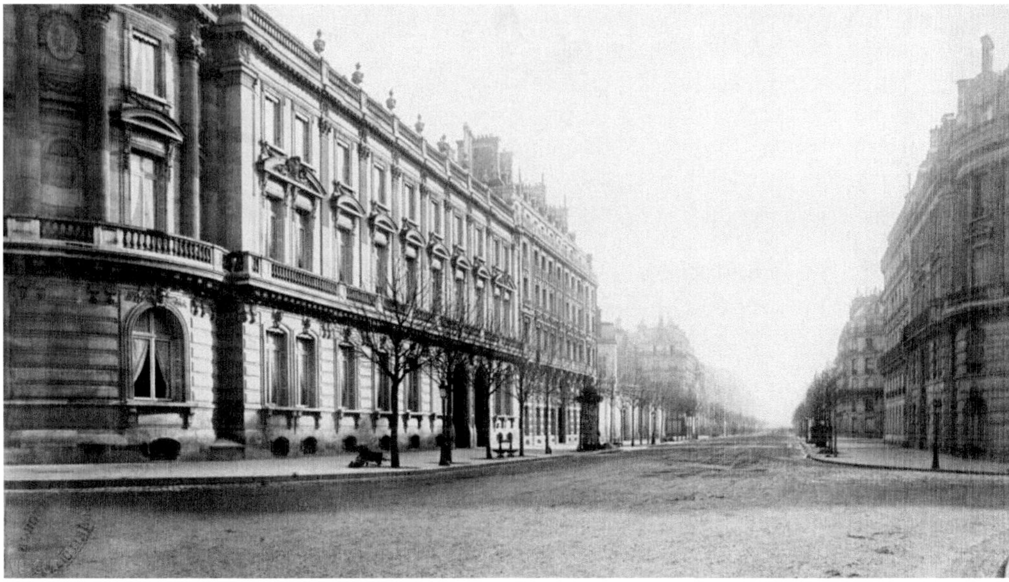

Figure 4.9 Charles Marville, *Boulevard Saint-Germain. Paris VIIe*, c. 1877, albumen print, 14 x 8 1/4 in. State Library of Victoria, Melbourne. Credit: Laurent Gloaguen, *Vergue*.

landscape features of tranquility and harmony. This pictorial strategy, in turn, harmonizes with the classical-style architecture, urban layout, and perspectival vistas implemented in Haussmann's axial and geometricized schemes. In the *Boulevard Saint-Germain. Paris VIIe* (c. 1877) Marville positions his camera on an angle to the street's axis to capture the Haussmann-era building on the left, the Cercle Agricole (designed by Henri Blondel and built from 1866–7) (Figure 4.9). Standing cattycorner to the building, Marville faced southeast, looking down at the recently extended Boulevard Saint-Germain, the most significant part of Haussmann's renovation of Paris on the Left Bank. (Indeed, this boulevard incorporated many of the streets Marville had photographed in the Old Paris series.) Marville's skewed angle allowed him to falsify the harmony of the buildings' alignment and regularity down each side of the street. While the Cercle Agricole appears to be a mirror image of the building across the street, cropped but for its rounded corner, a birds-eye view (1878)[84] of the intersection shows that the older building on the right occupied a squarish plat, as opposed to the new building's rectangular plat and elongated façade. Marville further obscured the irregular elevations that extend beyond the Cercle Agricole by focusing on the alignment of newly planted trees, creating a continuous orthogonal line, even as they blur into the distance.

Marville's photographs of the Boulevard Haussmann demonstrate his pictorial elaboration of Haussmann's perspectival alignment of newly built edifices. In this series, Marville also skews his camera at an angle, allowing an alignment of repeated details to carry our perception along the orthogonals to their convergence. These details include cornices, balconies, mansard rooflines, windows, and stringcourses. These architectural elements were standardized according to Haussmann-era building codes, such as planting trees at equidistance.[85] In the *Boulevard Haussmann, from Rue du Havre* (c. 1877), Marville captures the alignment of single trees along the diagonal course of the orthogonal line on the left side, while on the right side he restates the row of trees as a composite silhouette. This asymmetrical, angled view of the boulevard introduces variety and surprise to an otherwise static stage set (Figure 4.10).

While Marville exploits linear perspective to advance Haussmannian precision and order, he also adapts the painter's device of atmospheric perspective to invoke an element of the sublime.[86] The *Boulevard Haussmann, from Rue du Havre* exemplifies how Marville translates atmospheric perspective from classical landscape painting. As in painting, the orthogonal lines converge and dissolve into the light and haze that register the immeasurability of spatial depth. The infinite depth, combined with the countless repetitions of architectural details, compels astonishment and wonder. No less astonishing, Marville establishes that this sublimity is a modern phenomenon: he captures the blurred traces of figures moving along the boulevard as part of the transient space of Haussmann's design. By injecting the sublime into Haussmann's rationalist scheme, Marville betrays his double vision—he celebrates the city as a modern monument and mitigates modernity with the romantic quest for the infinite. Marville's lingering romanticism resonates with the wonder that Gautier enjoys among Haussmannian ruins.[87] As earlier noted, he wrote in the *Paris Démoli* review, "The deep trenches… crisscross the city in all directions;… [and] new prospects open [and] unexpected views take shape[.]"[88] Perhaps, when Marville 'romanticized' the scintillating new boulevards, he also reflected upon Baudelaire's double vision of modern beauty in "the scenes of high life"[89] that existed alongside the city's "underworld."

Ironically, among the ruins of old Paris, we can recover the flâneur and, perhaps, Baudelaire's and Gautier's enduring symbolic presences. In *Opening of Avenue de l'Opéra: Butte des Moulins, View Taken from the Rue Saint-Honoré*, Marville has staged the workers and the top-hatted bourgeois men (spectators, entrepreneurs, civil servants, engineers), side-by-side, along the horizontal axis of the middle ground littered with rubble. Here, too, he records a Baudelairean duality of the city's underworld and

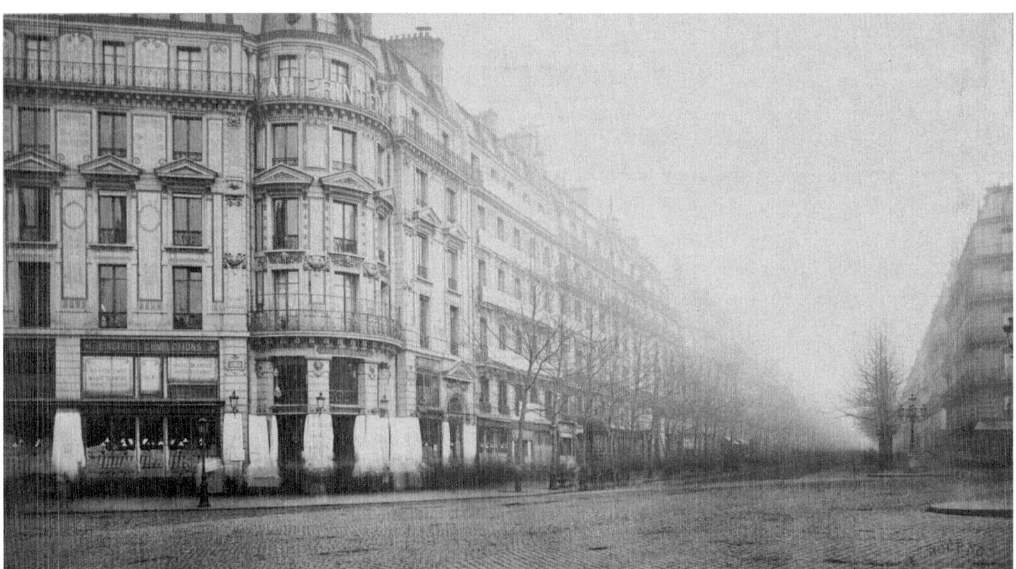

Figure 4.10 Charles Marville, *Boulevard Haussmann, from Rue du Havre*, c. 1877, albumen print, 8 x 14 1/4 in. State Library of Victoria, Melbourne. Image: Wikimedia Commons.

the high life (Figure 4.8). Behind the men, the slightly angled buildings converge toward the central vista of the Opéra building. The viewer looks across the proscenium-like foreground of shacks, horse, and carts. The cropped and shadowed shack on the right and the newly planted trees on the left complete the proscenium effect. Yet, even as viewers, we, too, stand amidst the rubble and mirror/embody the men at the center of the image. Which of the social classes might we identify? We can take our cue from the particularly prominent black coats and top hats—flâneur fashion hallmarks—silhouetted against the shorn, hollowed-out buildings on the left side.[90] If this is so, then we, like Baudelaire and Gautier, are both spectators of and travelers among the ruins. We, too, witness Paris's liminal, transitory space, old and new.

Conclusion

Given Marville's early training as a painter and illustrator and later affiliation with painters, we can assume he was familiar with Baudelaire's mandate against photography's status as a fine art. With this in mind, we can also assume that he took the critic's mandate as a challenge to translate the "ethereal and immaterial" (Baudelaire) associated with the fine arts into the photographic medium. If this is so, then Marville's pictorial devices—raking background light and tonal contrasts in the Old Paris and demolition series; the blurred traces of urban inhabitants, atmospheric perspective, and material dissolution in the New Paris series—register as the painter-turned-photographer's translations of the ethereal and transitory and his intervention in the concept of the Baudelairean painter of modern life.

Baudelaire's engagement with Paris photographers completes this circle of word-and-photographic image reciprocities. Friends with several photographers, with whom he collaborated in choosing various poses, settings, and attire for his portraits, Baudelaire would have witnessed their technical experiments and challenged their artistic aspirations. These liaisons made possible Baudelaire's eventual, although private, endorsement of a photograph that was more than just a memory aid but one that could stimulate or simulate the imagination. Writing to his mother in December 1865, he mused about a portrait photograph in which *flou* or blurring was the primary attribute, replacing an exact portrayal of the sitter's "wrinkles, warts, and defects."[91] Although Baudelaire claimed that "there is hardly anyone in Paris who knows how to do this," a recently discovered photograph taken by his friend Félix Nadar, dating from 1855–7, suggests otherwise (Figure 4.11).[92] In this photograph, one pose from a series, Baudelaire budged. The resulting image is blurred, especially on the left side. Nadar considered it an accident, and it was never issued during his lifetime.[93] For Baudelaire, this blurring would have approached a representation of the *ephemeral*, as it may have for Marville. Although Marville may have considered this a technical "fault" in his Old Paris series, he foregrounded the *flou* in his New Paris series. What is important is that as a witness to Baudelairean modernity, Marville adapted the paradoxical solution that the maligned photograph reflected—the transitory and the absolute, as in material reality—a solution that also recorded the ongoing Haussmannian transformations of Paris during the Second Empire and into the Third Republic.

In the next chapter, I reverse the historical perspective of viewing Marville's photographs in the context of Baudelairean modernity and, instead, focus on present-day viewer responses to Marville's Old and New Paris series. In collaboration with empirical

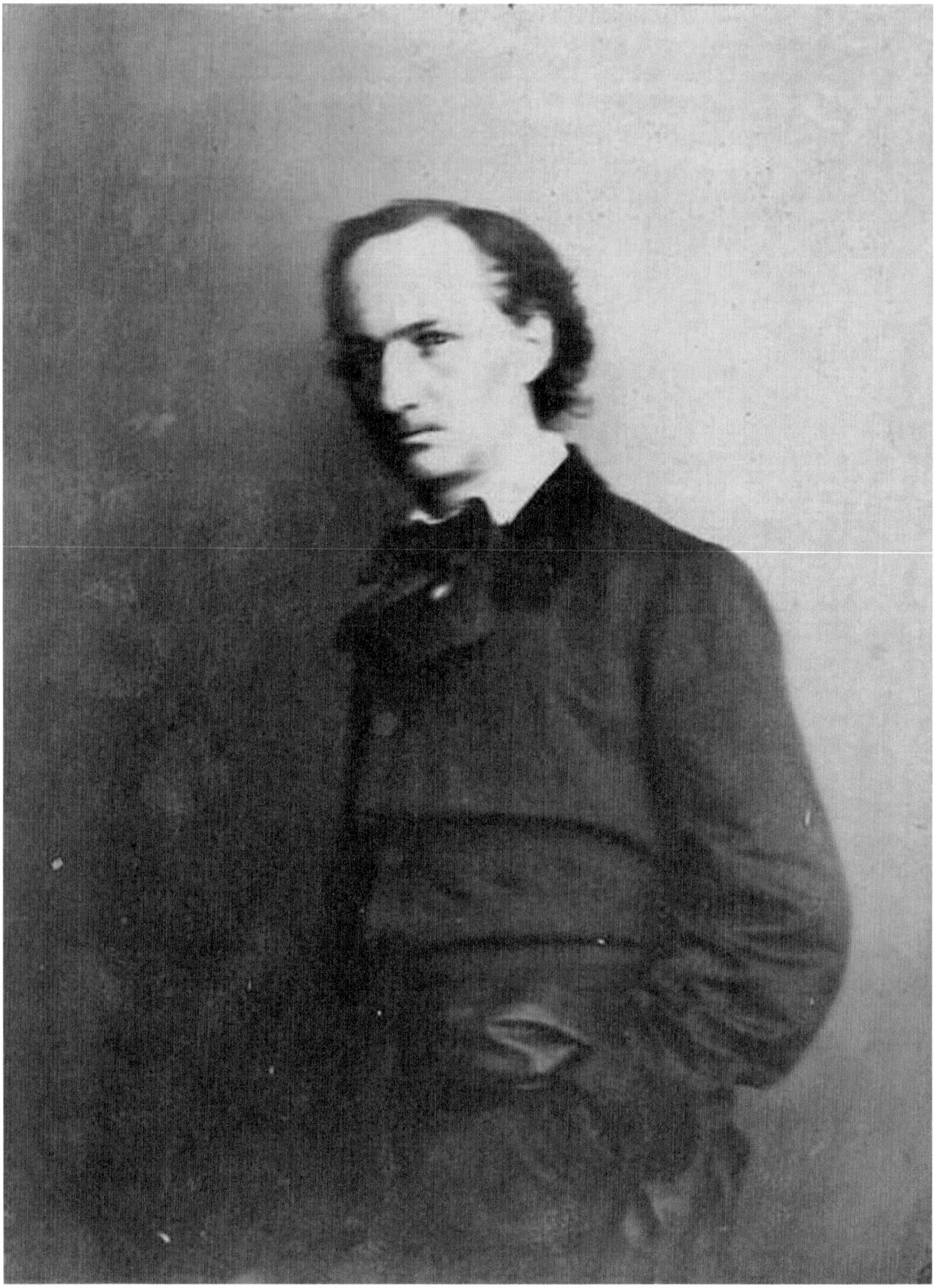

Figure 4.11 Félix Nadar, *Charles Baudelaire*, 1855/7, salted paper print from a collodion glass negative, 9 1/2 x 7 in. Musée d'Orsay, Paris, France. © RMN-Grand Palais/Art Resource, NY. Repro-photo: Hervé Lewandowski.

psychologists, I ran an experiment to determine whether Marville's formal devices triggered aesthetic, emotive, and embodied effects that art historians have assumed to be integral to photographs in the series. The experiment's premises, procedures, and results are examined therein.

Notes

1 In nineteenth-century texts, "*ancien*" and "*vieux*" are interchangeable descriptors of "old" Paris. I use the lowercase "o" and "n" when referring to "old" and "new" Paris, in general, and the uppercase "O" and "N" when referring to Marville's Old and New Paris series.

2 I have extended Pierre Nora's phrase "*les lieux de mémoire*" and here use the rubric "memory" to describe the formation of a verbal and visual discourse and, ultimately, symbolic. See Nora, "Introduction," in *Realms of Memory*, ix–xii.

3 Much of my interpretation of Charles Marville in the context of Baudelairean modernity is amplified in the exhibition catalogue, Sarah Kennel, *Charles Marville: Photographer of Paris* (Washington, DC: National Gallery of Art, 2013), which accompanied the exhibition of the same name and was organized by the National Gallery of Art, Washington, in association with the Metropolitan Museum of Art, New York. While I use this publication to support my thesis, my findings were first presented in conference papers and publications dating from the early and mid-2000s and subsequently published as "The Mirror as a Metaphor of Baudelairean Modernity," in *Orientations: Space/Time/Image/Word 5*, ed. Claus Clüver et al. (Amsterdam/ New York: Editions Rodopi, 2005): 16–36; and as "Re-Viewing Baudelaire's Paris: Photography, Modernity and Parody," in *Writing and Seeing: Essays on Word and Image*, ed. Rui Carvalho Homem and Maria de Fátima Lambert (Leiden: Brill, 2005): 145–56.

4 As noted in Chapter 3 (notes 3 and 4), I have borrowed the term "metacognitive" from Pelowski and Akiba's "A Model of Art Perception."

5 In *The Painting of Modern Life*, T. J. Clark identifies two responses to Paris's transformation: nostalgia for what has passed and optimism toward the changed (see "Introduction" and "The View from Notre Dame").

6 Pichois and Avice, *Baudelaire-Paris*, 27–32. Clark also identifies "ambiguity" as a Parisian attitude toward modernity in general and specifically toward Haussmann's modernization of Paris; *The Painting of Modern Life*, 47.

7 Conversely, literary historian Marit Grøtta argues that Baudelaire incorporated photographic ("semi-transparent and illuminated") views and framing in his prose poems, specifically in "Les Fenêtres" and "Mademoiselle Bistouri;" see "Reading/Developing Images: Baudelaire, Benjamin, and the Advent of Photography," *Nineteenth-Century French Studies* 41, no. 1–2 (2012): 80–90.

8 In his introduction to Marie de Thézy's *Charles Marville, Paris Disparu* (Paris: Paris Musées, 1995), Jean-Paul Avice acknowledges Baudelaire's endorsement of photography as a means of saving from oblivion "those things that time dissolves" (5). However, in his introduction above and in Pichois and Avice's *Baudelaire-Paris*, biographic accounts of Baudelaire's activities in Paris and his poems about Paris are illustrated with Charles Marville's photographs of old Paris and thus reinforce the pessimistic attitude attributed to Baudelaire's *Fleurs du Mal* and *Spleen de Paris*.

9 Baudelaire, "Dedication/To the impeccable poet/To the perfect magician of French letters/To my very dear and very revered/Master and friend/Théophile Gautier/With sentiments/Of the most profound humility/I dedicate/These unhealthy flowers"; "Dédicace/Dedication," in *Les Fleurs du Mal/Flowers of Evil* [2nd ed., 1861], https://fleursdumal.org/poem/098.

10 Baudelaire, "Premier Version de la Dédicace" (First Version of the Dedication), [March 8, 1855], Pichois with Ziegler, *Baudelaire*, 219. In the first draft of the Dedication, Baudelaire's adulatory prose is more florid, and he cites specific works by Gautier that he admired. Gautier's response is only known through Baudelaire's letter to his publisher, Auguste Poulet-Malassis, March 9, 1857; see Pichois with Ziegler, *Baudelaire*, 401, n. 7; and "À Auguste Poulet-Malassis," March 9, 1857, in Baudelaire, *Correspondance*, 379–80.

11 Quoted from Pichois with Ziegler, *Baudelaire*, 219; and "À Auguste Poulet-Malassis," March 9, 1857, in Baudelaire, *Correspondance*, 380. In the nineteenth century, the word 'scabreux'

may have had a more general meaning (rude, rugged, difficult, dangerous) than the current association with the obscene; see ARTFL Project, The University of Chicago, "SCABREUX, EUSE," in *Dictionnaire de L'Académie Française*, 6th Edition (1832–5), http://portail.atilf.fr/cgi-bin/dico1look.pl?strippedhw=scabreux&headword=&docyear=1800-1899&dicoid=ALL&articletype=1.

12 As translator Francis Scarfe has observed, "only seventeen of the fifty poems have anything to do with spleen [or ennui], and as a result, the dominant mood in the book is one which shows a positive approach to existence," cited from "Introduction" in Baudelaire, *Baudelaire: The Poems in Prose*, 13.

13 The letter to Arsène Houssaye, the editor of *La Presse*, appeared as the preface to his prose poems when published in *La Presse*, August 26, 1862. The following translation is from Baudelaire, *The Prose Poems and La Fanfarlo*, 30–1; from "À Arsène Houssaye," in Baudelaire, *Oeuvres Complètes*, vol. 1 (1975): 275–6.

14 See Claudine Lacoste-Veysseyre's introduction to the posthumously published collection of Théophile Gautier's journalistic oeuvre, *Paris et les Parisiens* (Paris: Boite à documents, 1996): 5–14. See also Gautier's introduction to *Paris et les Parisiens au XIXe siècle: Moeurs Arts et Monuments*. (Paris: Morizot, 1856) : i–iv. *Paris et les Parisiens au XIXe siècle* is a 465-page volume, consisting of texts by Alexandre Dumas, Gautier, Arsène Houssaye, Paul de Musset, Louis Enault and Du Fayl, with illustrations by Eugène Lami, Gavarni, and Rouargue.

15 Théophile Gautier, "Paris Démoli," originally published in *Moniteur Universel*, January 21, 1854; reprinted. "Préface," in *Paris Démoli. Mosaïque de Ruines* (Paris: A. Aubry, 1855): i–iv.

16 Daryl Lee, "Baudelaire's 'Paysage' and the Commune Ruins Picturesque," *Romance Notes* 45, no. 1 (2004): 107. The Paris Commune (1871) was an insurrection against the French government that established a revolutionary government from March 18 to May 28, 1871. It occurred after France's defeat in the Franco-Prussian War and the collapse of Napoleon III's Second Empire (1852–70). Communes that arose elsewhere in France were quickly suppressed, while the Commune of Paris alone faced the opposition of the royalist-dominated government located in Versailles. However, the insurgents could not organize themselves militarily, and on May 21 government troops entered an undefended section of Paris. During *la semaine sanglante*, or "bloody week," that followed, the regular troops crushed the Communards. The insurgents, in their defense, set up barricades in the streets and burned public buildings (among them the Tuileries Palace and the City Hall [Hôtel de Ville]). Large sections of Paris were reduced to rubble. About 20,000 insurrectionists were killed, along with about 750 government troops. See "Commune of Paris," *Encyclopedia Britannica*, June 28, 2024. https://www.britannica.com/event/Commune-of-Paris-1871.

17 Lee, "Baudelaire's 'Paysage,'" 110, and for the quotes that follow.

18 Several of LeSecq's photographs of demolition sites are published in Shelley Rice, *Parisian Views* (Cambridge: MIT Press, 1999), from a private photographic collection known as the *Album Berger* at the Bibliothèque Historique de la Ville de Paris, which I personally viewed in situ. An inscription inside the *Album Berger* states: "Don de M. Amedi Berger, président de la Cour des comptes, fils de M. J. Berger ancien préfet de la Seine, mort le 27 janvier 1881."

19 See Moncan and Heurteux, *Le Paris d'Haussmann*, 43.

20 The Île de la Cité was a focus of these changes under the July Monarchy, with the construction of a new street that ran along the length of the Île, three new streets that crossed it, and the construction of the Pont Louis-Philippe that connected the Right Bank with the adjacent Île Saint-Louis. Other improvements included clearing the area around the Hôtel de Ville, clearing the quays along the Right Bank, and demolishing the Hôtel Dieu annex buildings. For these and other urban renewal projects in the city center, see Hervé Maneglier, *Paris Impérial: La Vie Quotidienne Sous le Second Empire* (Paris: Armand Colin, 1990): 16–7. The Île de la Cite streets were completed in 1837.

21 Baudelaire, "Of the Heroism of Modern Life," in *Selected Writings*, 104–5.

22 According to Pichois, Baudelaire began writing prose poems in 1855; he turned to writing solely prose poems after publishing the 1861 edition of *Fleurs du Mal* and started efforts to publish the poems in 1861. Although 20 poems were published in Houssaye's *La Presse* in August and September 1862, the collection of 50 poems was published posthumously in 1869; Pichois with Ziegler, *Baudelaire*, 287–8.

23 Baudelaire, "Salon of 1859: The Modern Public and Photography," in *Selected Writings*, 297–8.
24 On the industrialization of the Seine, see Auguste Vitu, *Paris: 450 Dessins D'après Nature* (Paris [n.p.], 1890): 23–6, http://catalogue.bnf.fr/ark:/12148/cb365853927.
25 Baudelaire, "The Swan," in *Les Fleurs du Mal/Flowers of Evil* [2nd ed., 1861], trans. William Aggeler, in *The Flowers of Evil* (Fresno, CA: Academy Library Guild, 1954), https://fleursdumal.org/poem/220.
26 Baudelaire, "The Painter of Modern Life," *Selected Writings*, 407–8.
27 Baudelaire, "The Painter of Modern Life," *Selected Writings*, 406–7.
28 Baudelaire, "Salon of 1859: The Modern Public and Photography," *Selected Writings*, 297.
29 Baudelaire, "Le Public Moderne et la Photographie," notes Paul-Louis Roubert, *Études Photographiques* 6 (May 1999), online, http://journals.openedition.org/etudesphotographiques/185. Arago's text first appeared in English as [François] Arago, "Chamber of Deputies–Bill Presented to the Minister of Interior [France], June 15, 1839," in *An Historical and Descriptive Account of the Various Processes of the Daguerréotype and the Diorama* (London/New York: McLean Kraus, 1839): 2. I have modified the text with standard modern English. "Draughtsmen and painters, even the most skillful, will find a constant subject of observation in this most perfect reproduction of nature.... the process will afford them a quick and easy method of forming collections of sketches and drawings[.] ... In fact, to the traveler, to the archaeologist and, also to the naturalist, the apparatus of M. Daguerre will become an object of continual and indispensable use[.]" For a discussion of Arago's text, see Anne McCauley, "François Arago and the Politics of the French Invention of Photography," in Daniel P. Younger, ed., *Multiple Views: Logan Grant Essays on Photography, 1983–89* (Albuquerque: University of New Mexico Press, 1991): 43–69.
30 Baudelaire, "Le Public Moderne et la Photographie," n. 25 (Roubert).
31 Roubert, n. 29. Planche's text was originally published in *Revue des Deux-mondes* 9 (May 1, 1857): 763–4. Also in 1857, Charles Monselet, in *Les Ruines de Paris*, favorably translates the need to use photography to save the ruins as precious documents of old Paris: "For a long time we will not see a similar period, which, by its transient nature, was bound to escape painting, and to which only rare photographic prints are devoted," trans. and qtd. in Guillaume Le Gall, "Paris, Objet d'Histoire: Sauvegarde, Representation," *Le Vieux Paris*, online exh. (Paris: Bibliothèque nationale de France, 2007), http://expositions.bnf.fr/atget/arret/17.htm.
32 Sima Godfrey provides a comparable interpretation of Fournier's text in "From Memory Lane to Memory Boulevard: Paris Change!" in Mary Ann Caws, ed., *City Images* (New York: Gordon and Breach, 1991): 168–9, n. 1.
33 Monique Moulène, "L'Ancien Paris de Martial Potémont," Le Blog Gallica, Bibliothèque nationale de France, published online January 1, 2013, https://gallica.bnf.fr/blog/01012013/lancien-paris-de-martial-potemont?mode=desktop.
34 Moulène, "L'Ancien Paris de Martial Potémont."
35 I have elsewhere treated the picturesque discourse and its British origins in Lauren S. Weingarden, "John Ruskin: The Picturesque Discourse and the Language of Architectural Naturalism," in *Louis H. Sullivan and a 19th-Century Poetics of Naturalized Architecture* (Farnham, England: Ashgate, 2009): 39–70.
36 These writers depended on associationism theories. The earliest theory was based on landscape design and modeled the reader/viewer who recalls works of art that the picturesque landscape evokes. As David Marshall argues, later forms of associationism "were more inclusive, [and] they look[ed] forward to the writings of Romanticism," with these forms "appealing to emotions, moods, and more general acts of imagination, ... depending on correspondent ideas"; "The Problem of the Picturesque," *Eighteenth-Century Studies* 35, no. 3 (2002): 430. It is this later tradition to which I refer in the following argument for picturesque photography.
37 Features of Claude Lorrain's (1600–82) classical landscapes were adapted to the picturesque aesthetic, including ancient edifices and ruins located in the middle ground and surrounded by hills, mountains, and trees, arranged along orthogonal lines, leading to a vanishing point in the background of diffused atmospheric light. These features are often framed as a vignette composition, with indefinite or muted borders, to suggest a self-contained vista within an infinitely expansive landscape.
38 I have adapted the "excursive" moniker from Elizabeth K. Helsinger, who developed the trope of "excursive sight" in *Ruskin and the Art of the Beholder* (Cambridge: Harvard University

Press, 1982). As she explains, "excursive sight—the play of eye or mind moving leisurely through a scene or subject—produces the pleasures that earlier writers associated with the picturesque and Ruskin defined as the particular experience of the ordinary beholder" (69).

39 Kennel, "Charles Marville: Hidden in Plain Sight," in Kennel, *Charles Marville*, 29.

40 Kennel adds that we cannot know with certainty if this was the case: "Because the Hotel Carnavalet [a private sixteenth-century mansion] was undergoing renovations (it would open, as the Paris historical museum, in 1880), these early collections were stored at the Hotel de Ville, which burned down during the Commune on May 24, 1871" (30). Notably, the Musée Carnavalet was established "on the initiative of Seine Prefect Haussmann, perhaps as a way of compensating for the partial destruction of Paris. It was meant to house a new institution designed to document Paris, while paying particular attention to how the collections would be presented," "The Musée Carnavalet." Carnavalet. www.carnavalet.paris.fr, n.d., https://www.carnavalet.paris.fr/musee-carnavalet.

41 There is no extant document to establish the exact date and agency of the city commission. See Françoise Reynaud, "Marville and Old Paris," in Kennel, *Charles Marville*, 195. Reynaud proposes the commission issued from either the Service du Plan (the city planning department that surveyed, mapped, and studied the leveling of the streets in preparation for the *grands travaux*) or the Service des Travaux historiques (Historic Works Department).

42 Joke de Wolf, "Paris on Display: Marville's Photographs at the Universal Exhibitions," in Kennel, *Charles Marville*, 219. Travaux de Paris was an extension of Promenades et Plantations (Walkways and Plantings).

43 As I have done in this chapter, Kennel also associates Marville with "the art of Edouard Manet, Charles Baudelaire, and many others we have come to associate with modernism," yet she does not develop these associations (29). While Kennel's approach entails a monographic study of Marville, I view Marville more broadly, within a discursive matrix of Baudelairean modernity. Kennel's study provides valuable historical and biographical evidence to substantiate my argument.

44 Kennel, *Charles Marville*, 29.

45 For Marville's professional training, see Kennel, *Charles Marville*, 5–6. Unlike the official École des Beaux-Arts training, the Académie Suisse "was open to all men, without entrance examination, and offered drawing classes from live models for a modest fee." The Académie Suisse was considered an alternative to the official art school; many artists who would be identified as anti-academic or "avant-garde" studied and socialized with other artists here, including Marville's contemporaries Eugéne Delacroix, Jean-Baptiste Corot, and Honoré Daumier.

46 See also Thézy, *Charles Marville.*

47 Kennel, *Charles Marville*, 6, 7.

48 Marville's career as an illustrator began with the publication of his drawings in two important books published by Henri-Leon Curmer: Le Maistre de Sacy's *Le Histoire de L'ancien et du Nouveau Testament* and Prudence-Guillaume de Roujoux's *Histoire Pittoresque de l'Angleterre*, published under the direction of Charles Nodier and Baron Taylor (Kennel, *Charles Marville*, 6). Curmer was also the publisher of the quasi-satiric physiologies, *Les Français Peints par Eux-mêmes* (1840–42), and Marville was friends with many of the artists working in Curmer's publishing house; see Kennel, *Charles Marville*, 5.

49 Marville registered as a copyist at the Louvre in July 1844; Kennel, *Charles Marville*, 9.

50 Kennel, *Charles Marville*, 22.

51 Although Marville professionally aligned with the conservative Ingres, his political alliance pointed in the opposite direction. Marville illustrated for left-leaning publishers following the revolution-induced political crisis of 1848; Kennel, *Charles Marville*, 10.

52 Kennel, *Charles Marville*, 10; see her discussion of Marville's work for Blanquart-Evrard (12–3). Blanquart-Evrard's Imprimérie Photographique was the first successful photographic publishing house, founded in 1851 and folded in 1855.

53 Marville was a prominent presence in Blanquart-Evrard's series *Album Photographique de L'artiste et de L'amateur*, first published in 1851 (Kennel, *Charles Marville*, 10, 12). For Blanquart-Evrard's technical and theoretical writings see "On the Intervention of Art in the Practice of Photography," *British Journal of Photography*, August 1, 1863; Part II–Practical Means–Manipulations," *British Journal of Photography*, August 15, 1863: 326 (http://albumen.conservation-us.org/library/c19/blanq2.html); and *La Photographie, ses Origines, ses*

Progrès, ses Transformations (Lille, France, 1869), https://gallica.bnf.fr/ark:/12148/bpt6k1231104.

54 Kennel, *Charles Marville*, 13, 17. Marville also claimed this identity in a letter to the French Ministry of Public Works requesting permission to photograph public buildings.

55 Barbizon School painters included the precursors of Impressionism, such as Camille Corot, Théodore Rousseau, and Jean-François Millet. Gustave Le Gray, who developed painterly tonal effects, is among the earliest photographers of Fontainebleau Forest, beginning his repertoire in 1851. See Heilbrunn Timeline of Art History, ed., "Gustave Le Gray (1820–84)," Metmuseum.org, 2020, https://www.metmuseum.org/toah/hd/gray/hd_gray.htm.

56 Kennel, *Charles Marville*, 14. These picturesque practices conform with Blanquart-Evrard's essay on making photography artistic via light effects in "On the Intervention of Art in the Practice of Photography." When Marville exhibited his photographs at the 1862 London Universal Exhibition, critics praised the picturesque effects he had mastered under Blanquart-Evrard's purview (Kennel, *Charles Marville*, 26).

57 Kennel, *Charles Marville*, 17–8. Given these "widely varied and increasingly commercial uses for his photography, [...] in the 1862 annuaire, [Marville] abandoned the description 'artist-photographer,' touting himself instead as 'Photographer of the imperial museum of the Louvre and of the city of Paris, Gold Medal of Italy, drawings, collections of the museums of Italy and France, reproductions[.]'" (20).

58 Kennel, *Charles Marville*, 15.

59 Kennel, *Charles Marville*, 23–4. Marville also produced utilitarian images of finished buildings for the architects responsible for the building projects of Second Empire Paris.

60 Kennel, *Charles Marville*, 25.

61 Kennel, *Charles Marville*, 26.

62 Kennel, *Charles Marville*, 26.

63 Kennel, *Charles Marville*, 28.

64 Kennel, *Charles Marville*, 33.

65 As explained in the Introduction, I have adapted the informed reader model from literary reader-response theory, particularly Wolfgang Iser in "The Reading Process: A Phenomenological Approach," (1972), and Stanley Fish, *Is There a Text in This Class?: The Authority of Interpretive Communities* (1980). They argue that any individual interpretation of a text is created in an interpretive community of minds comprised of participants who share a specific reading and interpretation strategy.

66 See Lauren S. Weingarden, "William Gilpin: The Poetics of Picturesque (Painterly) Perception and Architectural Representation," in Ulla-Britta Lagerroth et al., eds., *Interart Poetics: Essays on the Interrelations of the Arts and Media* (Amsterdam/Atlanta: Editions Rodopi B.V., 1997):179–93; James Ackerman, "The Photographic Picturesque," *Artibus et Historiae* 24, no. 48 (January 1, 2003): 73–94; and Wolfgang Kemp and Joyce Rheuban, "Images of Decay: Photography in the Picturesque Tradition," *October* 54 (Autumn 1990): 102–33.

67 The association of ideas attached to ruins was especially amenable to Gothic ruins and their picturesque features. Marville's self-portrait in the Gothic ruins of the *Church of Bacharach, Southern End* (1854; https://www.nga.gov/features/marville.html) indicates his immersion in this discourse. This photograph was published in an album titled *Les Bords du Rhin*, based on his travels throughout Germany in 1853–54.

68 Reynaud, "Marville and Old Paris," in Kennel, *Charles Marville*, 199.

69 Kennel, *Charles Marville*, 31.

70 Rue du Bon Puits was inhabited by 1245 and was named in 1250, as recorded in Félix Lazare et Louis Lazare, *Dictionnaire Administratif et Historique des Rues et Monuments de Paris* (Paris: Bureau de la Revue Municipale, 1855), http://vergue.com/post/760/Rue-du-Bon-Puits. In the following discussion, I draw upon the online collection of Marville's photographs, *Vergue* (http://vergue.com/), created by Laurent Gloaguen and first published online September 25, 2013. Each photograph is annotated with pre- and post-Haussmann maps inscribed with old, demolished, and new streets and delineations of Marville's locations and lens angles when photographing the sites. The photographic information includes histories of the street, derived from both archival records and the contemporaneous guidebooks that Marville would have used.

71 "The chasse-roues were stone posts which served to protect the plinth (a building's slab base) from the wheels of carts and cars (hence their name). They also served as retreats for pedestrians

in narrow lanes. Sidewalks made these posts useless, so they are only seen in Marville's photographs in front of houses with no sidewalks. Almost all wheel-carriages disappeared from Paris public roads; they were systematically removed when a sidewalk was built." ("Rue des Prêtres Saint Séverin, c. 1866," *Vergue*, https://vergue.com/post/724/Rue-des-Pretres-Saint-Severin).

72 As Kennel explains, "Occasionally we glimpse inhabitants of the city who stood still long enough during the exposure (roughly from three to fifteen seconds) to register on the negative" (*Charles Marville, 33*).

73 For a transcription of the poster and associated chronology of Marville's photograph of the street, see "Rue du Bon Puits, 1866," *Vergue*, http://vergue.com/post/760/Rue-du-Bon-Puits.

74 "Rue Traversine, c. 1866," *Vergue*, http://vergue.com/post/754/Rue-Traversine.

75 Translation from Kennel, *Charles Marville*, 28. For contemporaneous definitions of "percement," see *CNRTL*, n.d., https://www.cnrtl.fr/definition/percement. In the context of Haussmannization, "percement" acquires a pejorative meaning, such as bulldozing through old streets to make way for the new. The Metropolitan Museum of Art more positively translates "percement" as "construction" in the caption "Charles Marville: Percement de l'avenue de l'Opéra (Construction of the avenue de l'Opéra)," The Metropolitan Museum of Art, https://www.metmuseum.org/art/collection/search/307390, a demolition photograph discussed below. Here I have translated "Percement" as "Opening" to accord with the demolitions, rather than "construction," recorded in the photographs.

76 Marville adapted his ruinist aesthetic to the city of Paris commission to photograph the ruins of the Hotel de Ville in June or early July 1871 ("Timeline," in Kennel, *Charles Marville*, 225). According to Kennel, Marville "was one of the few members of the first generation of French 'primitive' photographers also to photograph the aftermath of the Paris Commune[.]" In 1877, Marville offered to reprint negatives of the Old Paris series, the positives of which had been destroyed in the Hotel de Ville fire during the Commune. (Kennel, *Charles Marville*, 3, 30).

77 Marville's earliest photographs of the area designated for Charles Garnier's l'Opéra date from 1864. Marville returned in 1877 to photograph the Avenue de l'Opéra still under construction; see Thézy, *Charles Marville*, 16.

78 Gautier, "Paris Démoli," 39.

79 Marville took advantage of the wet collodion process "that gave the negative a transparency that effected a great precision in the highlights and shadow" to achieve picturesque effects in the demolition series, such as sharply defined light and dark contrasts and precisely captured fragmented and rough surfaces. Albumen prints, made from these negatives, provided the finished work's precision, tonality, and smoothness. Ackerman, "The Photographic Picturesque," 91.

80 Gautier, "Paris Démoli," 39.

81 "Construction of the Avenue de l'Opéra: The Butte des Moulins (From the Rue Saint-Roch), December 1876," National Gallery of Art, "Charles Marville: Photographer of Paris/Paris Transformed," 2013, https://www.nga.gov/features/marville/paris-transformed.html.

82 Lee, "Baudelaire's 'Paysage,'" 110.

83 Baudelaire, "Of the Heroism of Modern Life," in *Selected Writings*, 106.

84 See "Le Paris de L'exposition. Le Nouveau Boulevard Saint-Germain dans Toute son Étendue." (1878). Paris: Capital of the 19th Century. Brown Digital Repository. Brown University Library. https://repository.library.brown.edu/studio/item/bdr:87058/.

85 See Marville's *Paris. Boulevard Saint-Michel. Déplacement d'un Arbre Sur une Plate-Forme* (Removing a Tree From a Platform; 1876), the view taken from the Rue Auguste-Comte toward the north, showing the transport and planting of trees along the boulevard (Bibliothèque Historique de la Ville de Paris, https://bibliotheques-specialisees.paris.fr/ark:/73873/pf0001016880/v0001.simple.selectedTab=record).

86 As used here, the sublime refers to the Kantian mathematical sublime wherein "our ability to intuit is overwhelmed by size (the huge building)," inducing a feeling of awe and, in turn, a pleasurable experience. See Douglas Burnham, "Kant, Immanuel: Aesthetics," *The Internet Encyclopedia of Philosophy (IEP): A Peer-Reviewed Academic Resource*, 2019, https://www.iep.utm.edu/kantaest/.

87 See also Daryl Lee, "The Ambivalent Picturesque of the Paris Commune Ruins," *Nineteenth-Century Prose* 29, no. 2 (2002): 144.

88 Gautier, "Paris Démoli," 39.

89 Baudelaire, "Of the Heroism of Modern Life," in *Selected Writings*, 106.

90 On the left, it is the rue d'Antin. The large building (11, rue d'Antin and 47, rue neuve Saint-Augustin) is one of two buildings built in 1703 by Jean de La Motte, King's Buildings Clerk, on the Rue Neuve Saint-Augustin. To the right, the stone arch is probably a remnant of the hotel's courtyard frontage, which was at 10, rue d'Antin (the tree was planted in the courtyard of this mansion); "Percement de L'avenue de l'Opéra, 1877," *Vergue*, http://vergue.com/post/216/Percement-avenue-Opera. See also Marville's assembly of top-hatted men and laborers in Percement de l'avenue de l'Opéra: chantier de la rue d'Argenteuil près de la rue Saint-Honoré, https://www.carnavalet.paris.fr/collections/percement-de-lavenue-de-lopera-chantier-de-la-rue-dargenteuil-pres-de-la-rue-saint.

91 Baudelaire, *Correspondance*, vol. II, 554. Jean Adhémar discusses Baudelaire's collaborations with photographers: "Baudelaire et les Photographes: Paris et Bruxelles," *Bulletin: Musées Royaux des Beaux-Arts de Belgique, Bruxelles*, 1–3 (1985–88): 258–9. For the nineteenth-century etymology of *flou* and its different meanings in painting as opposed to photography, see Pauline Martin, "Le Flou du Peintre ne Peut Être le Flou du Photographe," *Études Photographiques* 25 (May 2010), document 2512, posted online May 5, 2010, http://journals.openedition.org/etudesphotographiques/3060.

92 My conclusion regarding Baudelaire's positive response to the photographic *flou*, as recorded in his letter to his mother and the "faulty" Nadar portrait photograph, was first published in Lauren S. Weingarden, "The Mirror as a Metaphor of Baudelairean Modernity," 16–36. See also Dana MacFarlane, "'Waiting Still:' Baudelaire and the Temporality of the Photographic Portrait," *History of Photography* 36, no. 1 (February 2012): 3–14.

93 Nadar's photograph was first published in 1993 in Pichois and Avice, *Baudelaire-Paris*, 208. Jérôme Thélot uses this portrait as a departure point for arguing that Baudelaire and Nadar were good friends and that their debates were mutually stimulating for their respective arts. Thélot also contends that Baudelaire wrote his poem "Le Rêve d'un Curieux," dedicated to Nadar, as homage to their friendship. However, Thélot maintains that Baudelaire remained resistant to photography, just as the poet of "Le Rêve" regrets that the photographer only records "cold truth." Jérôme Thélot, "'Le Rêve d'un curieux,'" *Études photographiques* 6 (May 1999), online, http://journals.openedition.org/etudesphotographiques/186. See MacFarlane, "'Waiting Still'" for a Benjaminian interpretation of Baudelaire's "ambivalent assessment of the merits of photography," 10–4.

5 A Collaborative Investigation into Affective and Embodied Aesthetic Responses to Charles Marville's Old Paris and New Paris Series

Introduction

In this chapter I investigate viewers' affective and embodied responses to Charles Marville's photographs to reify the lived experience of nineteenth-century subjects who navigated the Paris streets before and after Haussmannization. Under this inquiry, I initiated a collaborative project with two empirical psychologists, Matthew Pelowski[1] and Corinna Kühnapfel,[2] to survey present-day viewers' responses to Marville's photographs of old Paris and new Paris. The objective of this investigation was to answer the question, "Can present-day photographic viewing and, by extension, can empirical aesthetics be used to corroborate a multisensory past?" This question, in fact, undergirds my entire book project. By using a transdisciplinary method that combines art history and neuroaesthetics, can we identify and understand the mobilized spectator's affective and cognitive responses to the urban and cultural disruptions incurred by Haussmannization?

In our collaborative experiment, we selected 56 photographs from Marville's Old Paris and New Paris series that were presented to participants on their computer/device screens. For each photograph shown, viewers responded to questions regarding three factors: their subjective feelings of nostalgia, optimism, or fear; their haptic responses to spatial and tactile conditions in the photographs and the depicted contents; and their cognitive projections of embodied sensations of moving through spaces defined by architecture, paving stones, surface textures, and light and shadows. We then assessed the viability of the outcomes concerning my art historical objectives.

The outcomes of this empirical study were two-fold. First, we co-authored an article titled "Embodying the Photographic Image: Charles Marville's Old Paris and New Paris Series," written for an anthology on embodied aesthetics.[3] Here, we presented the methodologies and results of the investigation in three authorial voices, comprising the three-part format of our discussion. In the first section, I wrote from the art historian's perspective. Psychologists Pelowski and Kühnapfel introduced our survey's empirical method and data in the second section. In the third section, we jointly wrote concluding statements and suggestions for further research. In the present chapter, I summarize the empirical data to focus on the second outcome of our study: how to reconcile the misalignment between our empirical evidence, derived from present-day viewers, and the art historical evidence and historiographic record of my original objective.

DOI: 10.4324/9780429442315-6

The Art Historical Problem and Methodology

In Chapter 4, I presented the historical connection between photography and modern urbanization as the backdrop for Marville's commissions. This connection remains foundational to the purpose of the present chapter, as is the nineteenth-century critical debate on photography as a documentary medium or a fine art production. In what follows, I extend the parameters of the debate discussed in Chapter 4 to provide a context for identifying Marville's photographic techniques that transformed photographic documentation into a locus for viewers' emotive and embodied responses. I also focus more directly on these techniques as they impacted Marville's approach to his photographic commissions for the city of Paris. As discussed in the previous chapter, Marville modified his techniques to conform, first, to the archival purposes of the Old Paris series (1864–9) and, subsequently, to the promotional purposes of the New Paris series (1876–7).[4] Ideally, at this point, we should determine nineteenth-century responses to and critical reception of Marville's photographs in his own time. However, substantive textual assessments do not exist. I thus rely on art historical descriptive and interpretive methods gleaned primarily from exhibition catalogues dating from the late-twentieth and early twenty-first centuries.[5] These contemporary sources are productive for two kinds of analysis: first, to identify Marville's formal and technical devices that compel viewers' imagined (embodied) movement through space; second, to tease out the photographs' associative (mnemonic) and emotional effects. However, these sources also reveal the writers' phenomenological biases, what I call a *curatorial gaze*, focusing on and favoring Old Paris over New Paris photographs.[6] As constructed here, the curatorial gaze is based in reader reception theory and its corollary, "the ideal reader," concepts laid out in this volume's Introduction.[7] The curatorial gaze is pivotal for countering the misalignment between our empirical evidence and my art historical objectives. As presented later in this chapter, the curatorial gaze registers the responses of informed (trained) viewers, who employ historical analyses, and can be distinguished from those of the uninformed (untrained) viewers who participated in our study.[8]

A Critical Context for Marville's Photographic Techniques: Photography-as-Document or Photography-as-Fine-Art

As established in Chapter 4, the Second Empire (1852–70) coincided with the spectacular rise of photography. A shared certainty united photographers and society at that time—a faith in the power of resemblance and the value of empirical evidence inherent in the photographic image. From photography's inception in the 1830s and through the 1850s, rapidly developed technological advancements contributed to the public's focus on photography's mechanistic features inherent in its documentary functions.[9] These connections between technology and public faith in empirical evidence are foundational to the documentary function of photography during Haussmannization. As previously noted, Marville's Old Paris series and New Paris series demonstrate how photography served a dual documentary role—to preserve a record of the past and to promote the newly ordered built environment.[10] These dual missions coincided with the photographer's technical and formal changes for each series, involving lighting, contrast, camera height, formatting, and framing. As the contemporaneous discourse on photography reveals, the medium exceeded its strictly documentary function and intruded upon the fine arts. This discourse also reveals Marville's photographic practice as a slippage between photography-as-document and photography-as-fine-art. This slippage triggers emotive and embodied responses intrinsic to certain viewers' aesthetic experience. In this regard, photography-as-fine-art aligns with Méryon's graphic renditions of old Paris, as discussed in Chapter 2. Here I contend that Marville developed pictorial skills and techniques that exceed photographic documentation, from which emotive and embodied responses arise. In the following discussion I demonstrate that he developed these skills and techniques in the context of a nineteenth-century discourse on photography as fine art.

Photography's Mimetic and Mnemonic Functions

As discussed in Chapter 4, Charles Baudelaire's "The Modern Public and Photography," a section of his 1859 Salon review, encapsulates the photography-as-documentation side of the debate.[11] While his argument was already established in critical discourse beginning with the birth of photography in 1839, his critique of photography and its mimetic function was more far-reaching than his predecessors'. Baudelaire penned "The Modern Public and Photography" to chastise painters who used photography to perfect pictorial realism. His reprimand issues from the dichotomy he promoted between photography's mechanical reproduction of reality and the importance of the imagination in the fine arts.[12] Given this contrast, Baudelaire soundly endorsed photography's documentary mimetic function. Asserting that photography's "true duty" is that of a "record-keeper," he further endorsed photography's mnemonic function: "Let [photography] save crumbling ruins from oblivion,... all those precious things, vowed to dissolution, which crave a place in the archives of our memories[.]"[13] Although the photography-as-fine-art perspective has received less historical attention than its counterpart, it is essential to the emotive and mobilizing stimuli comprising the embodied aesthetic of Marville's photographs.

Photography's Aesthetic Function and the Photographer's Transformative Skills

The photography-as-documentation side of the debate provided the basis upon which advocates of photography-as-fine-art formed their counterargument. Nineteenth-century critics who promoted photography's fine art status pointed out the photographer's role in transforming the real into an aesthetic ideal, a transformation achieved by managing the camera apparatus, the developing process, and compositional devices. For example, in 1851, photography critic Francis Wey stated that "Photography... must not only translate but interpret," as exemplified by the various ways a photographer could alter a photographic image by manipulating the negative.[14]

Photography-as-fine-art advocates also promoted photography's transformative functions. In his 1851 article, photographer Charles Négre explained the continuum between the photographer's interpretation of his photographed subject and the viewer's affective responses to the photographic image:

> When the chemist prepares the paper, the artist directs the lens, and by means of the three beacons that guide him ceaselessly in the study of nature–observation, feeling, and reasoning–he reproduces the effects that make us dream, the simple patterns that move us, and the sites with powerful and bold silhouettes that surprise us and frighten us.[15]

In photographer Eugène Durieu's review of an 1855 photography exhibition, he enforced photography's aesthetic function by focusing on the photographer's technical alterations of his subject:

> When one wishes to obtain a picture rather than a reproduction, when one renders an artistic effect in the view of a landscape or a portrait; that is the moment when the photographer becomes a true creator and must meet very specific requirements.[16]

As the previous chapter explored, these artistic precepts were foundational to Marville's early training as a topographical (picturesque) illustrator, skills he honed as an urban photographer of old Paris and new Paris. In both series, and following the two commissions, Marville controlled the camera's position in relation to the street and horizon, the angle of the camera lens, the framing of his subject, the orientation of the glass negative (horizontal or vertical), and the light, shadow, and tonality of the urban setting.

Formal Analysis of Marville's Photographs

Bridging the Gap: Formal Analysis and Empirical Methodology

At the level of formal analysis, we collaborators could bridge art historical methods with quantitative methods of empirical research. We developed survey questions for our experiment to contrast the formal features of Marville's Old Paris and New Paris photographs.[17] As earlier noted, each viewer responded to questions regarding three factors: their subjective feelings of nostalgia, optimism, or fear; their haptic responses to spatial and tactile conditions in the photographs; and their cognitive projections of embodied sensations of moving through spaces defined by architecture, paving stones, surface textures, and light and shadows. Our contrastive method was validated by the 1878 Universal Exposition hosted by the city of Paris. At that exposition, Marville's Old Paris series and New Paris series were displayed side by side for the first time, endorsing the civic administrators' intention of celebrating the city's modernity and, coincidently, condemning old Paris.[18] These ideological motivations aside, I identified eight formal and technical features that accorded with our survey questions and the contrasting visual elements of the Old Paris and New Paris series. These features are vertical and horizontal format, movement or stasis, pedestrian-level viewpoint, asymmetry, spatial compression or expansion, lighting effects, haptic effects, and low-angle shots and wide-angle effects. I condense the formal categories below and present sample photographs to demonstrate the contrastive method used in our survey.

Old Paris/New Paris: Vertical and Horizontal Formats

Marville's vertical and horizontal formats were central to our analysis and survey samples. While Marville used both vertical and horizontal formats for the Old Paris series, he exclusively employed the horizontal format for the New Paris series.[19] Marville used the vertical format to portray up close the narrow passageways that characterized the medieval origins of the old Paris streets. He chose the horizontal format to frame intersections or squares from which the old Paris streets irregularly trajected. Conversely, Marville solely used the horizontal format for the New Paris series to emphasize the breadth, continuity, and orderly extension of Haussmann's boulevards.

Old Paris: Movement; Lighting; Asymmetry

Marville's Old Paris photographs are inscribed with the photographer's and, by implication, the viewer's embodied movements. As examined in Chapter 4, Marville translated the experience of embodied mobility, central to the picturesque landscape aesthetic, into urban photography. According to archivist Marie de Thézy, Marville was commissioned to record the progression of the demolitions of the streets. He therefore adopted a topographic approach, photographing a single street from one end, and then the other, and from various intersections.[20] The title of the vertically framed photograph *Rue de l'École de Médecine, from Cour du Commerce* (1866) designates such intersections and reflects the photographer's traversal along a meandering pathway (Figure 5.1). The horizontal formats of the Old Paris series more often capture entire street crossings, such as in *Rue Taranne, from Saint Benoît Intersection* (1866) (Figure 5.2).[21]

Marville also inscribed an excursive view, analogous to the topographic approach, within the photographic composition, inviting the viewer to imagine traversing the narrow, winding streets that lead into pictorial depth. The excursive view is inscribed by obliquely receding planes at angles to the frontal plane. These planes are further defined by the light and dark contrasts of the building masses as sunlight and shadows play across their surfaces. Marville's asymmetrical compositions also contribute to the viewer's dynamic spatial experience. As seen in *Rue des Marmousets, from Rue St. Landry* (Figure 4.4) and *Rue de l'École de Médecine, from Cour du Commerce*, by placing his camera at a slightly off-center position from the street's axis, Marville emphasized the irregular patterns that force the viewer's gaze around an abrupt bend in the street or a corner wall that juts out in the foreground or middle ground.

New Paris: Movement; Lighting; Asymmetry

Given that Haussmann designed the boulevards using a single-point perspective comprised of orthogonals and a vanishing point, the new avenues provided Marville with ready-made compositions of perspectival harmony. As a result, he positions himself and his camera in a single spot, abandoning the topographical approach and the excursive view. Both the photographer and viewer assume a static position. Mirroring the camera lens, the viewer's gaze is activated from a single vantage point, propelled along the uninterrupted orthogonal lines of the uniformly scaled and detailed buildings and by trees and lampposts spaced at regular intervals. In these images, such as *Boulevard Haussmann, from Rue du Havre* (c. 1877) and *Boulevard St. Marcel, from Avenue des Gobelins*, a misty or translucent light uniformly spreads across building elevations and evenly paved streets (Figures 4.10 and 5.3). As in pictorial perspective, the photographer shrouds the converging lines in an atmospheric haze to emphasize the boulevard's ostensibly infinite dimensions in space and its newly imposed order on the cityscape. Marville also arranges asymmetrical compositions for the New Paris series by placing himself and his camera slightly off-center from the streets' central axes. While this diagonal sight line mitigates the boulevards' rigid symmetry, it also emphasizes the unbroken orthogonal lines of Haussmannian perspective.

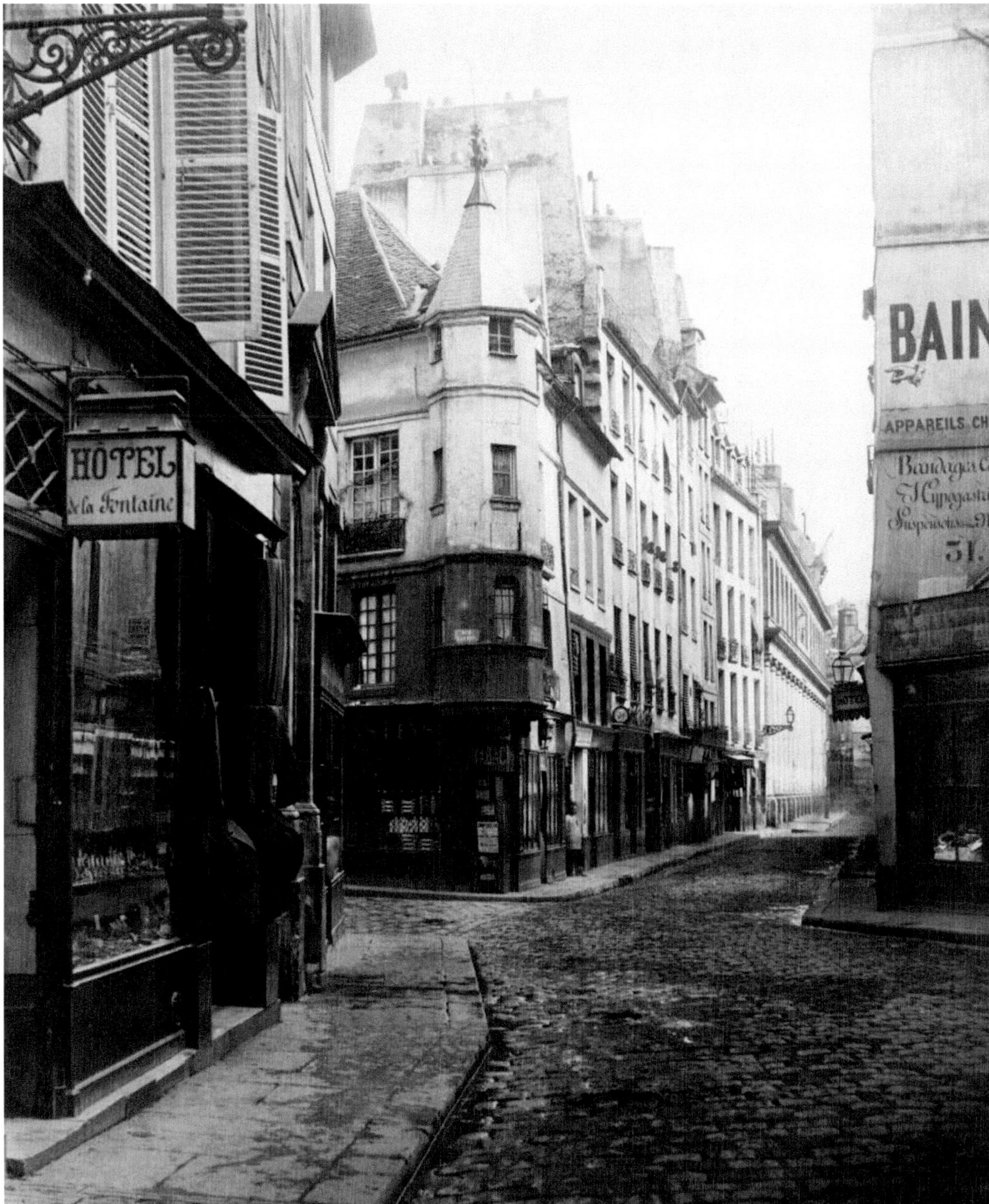

Figure 5.1 Charles Marville, *Rue de l'École de Médecine, from Cour du Commerce, Paris VIe*, c. 1866, albumen print, 14 x 10 5/8 in. Bibliothèque Historique de la Ville de Paris. Credit: Laurent Gloaguen, *Vergue*.

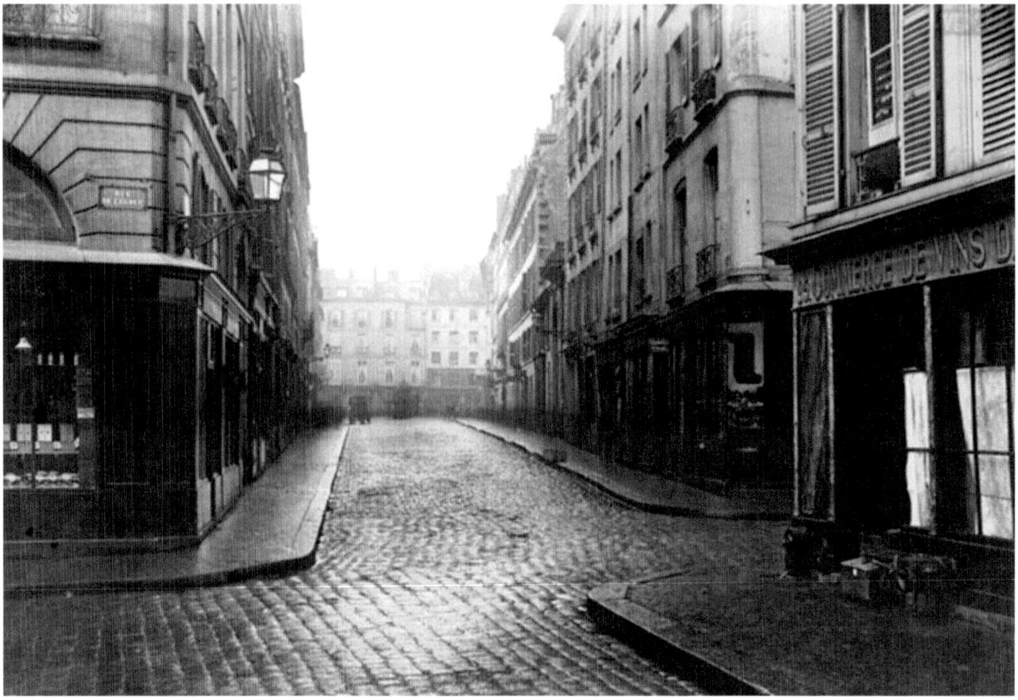

Figure 5.2 Charles Marville, *Rue Taranne, from Saint Benoît Intersection, Paris VIe*, c. 1866, print from digitally restored negative (original albumen print), 14 1/4 x 9 3/4 in. Bibliothèque Historique de la Ville de Paris. Credit: Laurent Gloaguen, *Vergue.*

Old Paris: Pedestrian-level Viewpoints; Haptic Effects; Spatial Compression or Expansion

In the Old Paris series, Marville positioned his street-level camera to capture the pedestrian's viewpoint, inviting the viewer to enter the pictorial space. Marville enhanced this immersive effect by using low-angle shots, positioning the camera below the viewer's sight line and pointing it upwards. This low-angle shot also brings into sharp relief the cobblestones and the axial gutters, whose gleaming line emphasizes the course of the street, as seen in *Rue Traversine* (1866) (Figure 4.6). From this perspective, we can also observe how Marville produced vivid haptic effects. Using a precisionist lens to capture tactile surfaces forces our close attention to the sharply defined details of the smooth or decayed textures of building facades, broken rooflines, deep recesses, and slippery, wet cobblestones.

The low-angle shot in the Old Paris series also affects the viewer's sense of enclosure. Marville directs the viewer's gaze to where the orthogonal lines converge, as seen in the vertical photographs, such as *Rue des Marmousets, from Rue St. Landry*. However, the distant view is abruptly blocked off by a structure or crossway oblique to the street's axis, creating spatial compression and labyrinthine confinement. As in *Rue du Bon Puits from Rue Traversine*, these effects are further exaggerated by the low-angle shots of the tall buildings that line the narrow streets and restrict direct sunlight on the pavement and exposure to the sky (Figure 4.5).

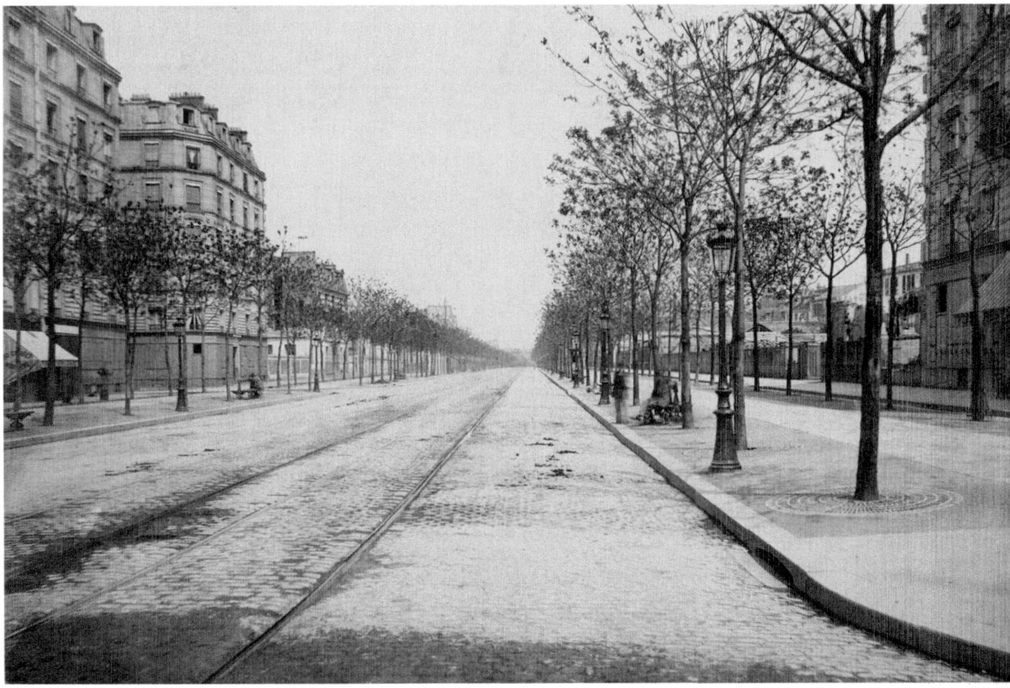

Figure 5.3 Charles Marville, *Boulevard St. Marcel, from Avenue des Gobelins*, 1876–7, albumen print, 10 1/2 x 13 7/8 in. State Library of Victoria, Melbourne. Image: Wikimedia Commons.

New Paris: Pedestrian-level Viewpoints; Haptic Effects; Spatial Compression or Expansion

In the New Paris series, Marville also sets his camera at the street level, but with different effects. As seen in *Boulevard Saint-Germain. Paris VIIe* (c. 1877), in this series, he aims the lens directly at the horizon line to focus on the orthogonals' vanishing point. (Figure 4.9) He also combines the horizontal frame with the large format negative, composing wide-angle views that span the major features of Haussmann's boulevards. He thus diminishes haptic effects in favor of uniformly lit and smooth street surfaces and building facades. In other views, such as *Boulevard St. Marcel, from Avenue des Gobelins* and *Boulevard Sébastopol*, Marville manipulated the wide-angle effect of the large format negative to enhance the breadth of newly constructed intersections and roundabouts and to capture their unencumbered circulation of light and air (Figures 5.3 and 5.4).

The Empirical Study

Both my art historical discussion and the present-day curatorial views recognize that Marville's photographs represented a movement from documenting Old Paris to promoting the modernity and reconstruction of the new city while providing a sense of reflective nostalgia and optimism for the future. However, for reasons discussed (later in this

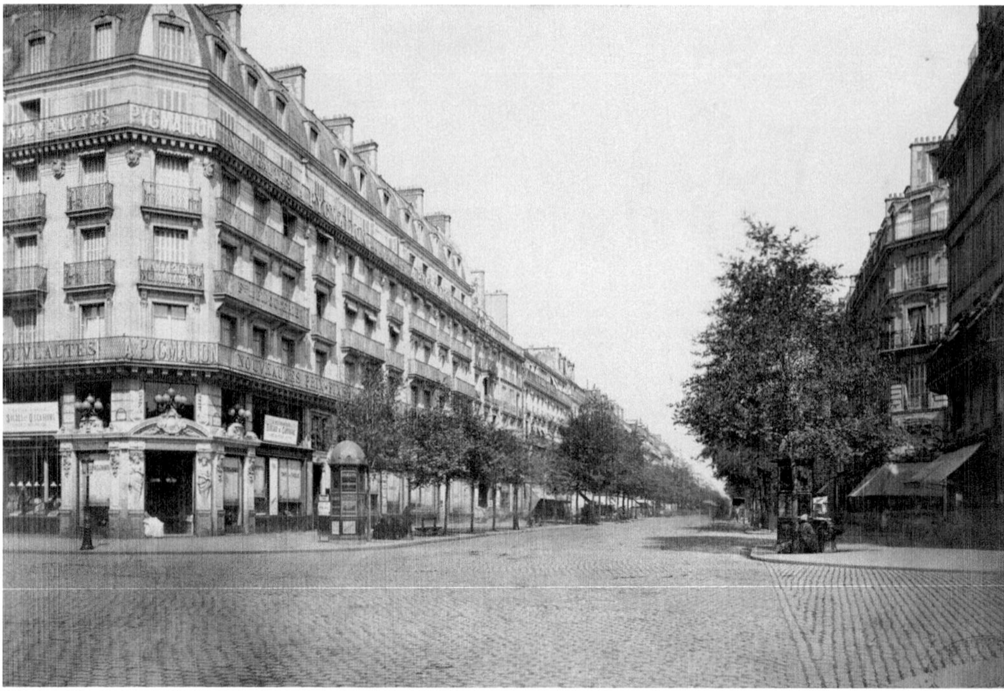

Figure 5.4 Charles Marville, *Boulevard Sébastopol*, 1876–7, albumen print, 9 3/4 x 14 1/2 in. State Library of Victoria, Melbourne. Image: Wikimedia Commons.

chapter), curatorial discussions favor Old Paris over New Paris photographs. As our formal analyses suggest, Old Paris's surface textures may pique the viewer's desire to explore the space and touch the surfaces more than New Paris's minimalist uniformity. In the Old Paris series, Marville uses a mobile viewpoint and sense of enclosure that ostensibly invites the viewer to imagine walking through these photographs. Whether describing the Old Paris or New Paris series, curatorial and art historical analyses suggest that the presumed embodied and affective responses to these works of art might be considered "universal."[22]

However, as my psychologist collaborators interrogated, is this presumed embodiment universally "true"? No matter how vexing this question was for me, it was pivotal to my original question for this study: can we describe and reconstruct the lived experience of the nineteenth-century spectator's actual traversal through the changing urban terrain? Furthermore, can we hypothesize how a nineteenth-century viewer might respond to the embodied and haptic aspects of Marville's photographs as described in curatorial writings?

The Empirical Experiment

The following brief description of our empirical study addresses the above implications and offers our attempt to bridge art criticism and historical discussion with empirical assessments. To investigate if the photographic images trigger distinct associative effects in present-day individuals, we invited 41 people to participate in an online empirical study.[23] We chose 18 photographs in each of three categories (total 56)—Old Paris series, divided into vertical and horizontal formats, and New Paris series, horizontal format—and randomly showed them to our participants. Then, we compared the viewers' responses to images of old Paris and new Paris.

With each photograph, we asked participants to rate, on a 7-point scale (1 = not at all, 7 = very much), several dimensions addressing their emotional, haptic, and embodied responses and associations. The terms comprising the survey questions were chosen from the above curatorial/critical review noted explicitly in the arguments for Old and New Paris. Our research question was: would an untrained viewer have similar responses?

First, we asked the participants about their subjective, felt experiences of nostalgia, optimism, sublime, and fear while viewing the photographs with the following four questions:

> How *nostalgic* does the image make you feel?
> How *optimistic* does the image make you feel?
> How *sublime* does the image make you feel?
> How *fearful* would you feel in this space?

Second, and following a similar aim, we assessed the haptic and embodied experience of the space captured in the photographs by asking the participants nine questions:

> How much do you feel you are *inside* the space?
> How much do you want to *explore* the space?
> How much would you like to *touch* the place?
> How easily do you think you could *walk* in this space?
> How much would you want to *meet* others in the space?
> How *open* is the space?
> How *smooth/rough* do you think the space would feel?
> How *slow/fast* is the space?
> How *rhythmic* is the space?

Importantly, we expected participants to be unfamiliar with the provenance of these photographs or the specific art historical aspects of Marville or Haussmann. For most participants, the photographs were merely black-and-white images of a (presumably French-speaking) city.

The Results[24]

The summary below focuses on the general emotional, haptic, and embodied responses and associations garnered from our study.[25] As expected, we found pronounced differences between responses to the Old Paris and New Paris series. The data provide clear evidence that different photographic techniques created distinct reactions in our participants. However, participants' responses challenged our expectation that they would support the art historical-curatorial view. Contrary to our expectations, viewers reacted to the random presentation of the New Paris photographs with strong haptic and embodied responses that evoked a desire to touch, meet, and explore, and a sense of fastness, smoothness, openness, and ease of walking through the new Paris space.[26] Their ratings concurrently reflected a sense of nostalgia, optimism, and sublime for the New Paris photographs.

Interestingly, regarding horizontal versus vertical Old Paris photographs, the ratings on the four emotive scales showed slightly stronger agreement with feeling these ways for the horizontal formats. Furthermore, as our historical reconstruction suggested, New Paris evokes feelings of optimism and sublime in the viewer. However, regarding nostalgia, we found patterns opposite to expectations. Contrary to a curator's fearful response to the New Paris photographs (discussed below), our participants' responses showed that fear was rated higher for the Old Paris photographs than for the New Paris photographs.

On the one hand, these results suggest that Marville's aesthetic decisions translated into a strong embodied, emotional, and haptic response. In this light, Marville succeeded under nineteenth-century critical standards. On the other hand, for my art historical project, the outcome that favored New Paris over Old Paris photographs posed a potential dilemma. How was it that present-day viewers did not feel what my curatorial colleagues and I felt? Do these results undermine the historical legitimacy/integrity of my interpretations? To resolve this dilemma, I posited the curatorial gaze as distinct from that of the general viewer-participants in our survey, a position to which my collaborators agreed. Indeed, the curatorial gaze also conforms to nineteenth-century critical and aesthetic standards. Thus, because of our findings, we distinguish between informed viewers, who are literate or trained in the sources, references, and historical meanings consistent with the curatorial gaze, and the uninformed (untrained) viewers who participated in our study.

As discussed in Chapter 4, Marville's Old Paris series also corroborated literary writers' and pictorial artists' nostalgic portrayals of the ancient city. This cultural discourse on old Paris is foundational to the curatorial gaze and the phenomenological effects it carries. In conclusion, I would like to re-examine the curatorial gaze as a matrix for substantiating nineteenth-century viewers' embodied responses to Marville's photographs and, in turn, their lived experience of Haussmannization.

Reflections on the Curatorial Gaze: Evidence of the Embodied Aesthetic Experience

As we discussed above, Marville developed a documentary style that combines both compositional harmony and precision of details and, at the same time, aims to instil a mobile, embodied gaze within the photographic fabric. My formal analyses of Marville's Old and New Paris series are informed by the curatorial gaze, which facilitates a mobile embodied gaze in the descriptions of Marville's Old Paris photographs and imagined walks through the old Paris streets.[27] The curatorial gaze is pivotal for providing concrete evidence of the embodied aesthetic experience. Coincidently, these viewers register an affective response that manifests as nostalgia. Given this evidentiary support, it is useful to examine exemplary curatorial texts that describe such imagined walks through the old Paris streets.

Writing in 1994, curator-archivist Marie de Thézy acknowledges the insalubrious conditions seen in the Old Paris photographs (as per the 1878 exhibition) but also admits that modern-day viewers "forget the smell, the nauseating odor of these very dirty streets." What remains, she argues, is the image itself: "*The empty streets, fixed by an early morning hour, appear to invite the spectator on some peaceful and solitary promenade*"[28] (emphasis added). Thézy also reveals the associative effects that Marville's photographs trigger: "Marville's art transfigures the condemned streets where the spectator is transported into an endless dream of nostalgic wandering."[29]

Sarah Kennel, the curator of the 2013 Marville retrospective organized by the Metropolitan Museum of Art and the National Gallery of Art, demonstrates in the exhibition catalogue how Marville's artistic devices provide maximum visual data that trigger the viewer's sense of mobility and spatiotemporal immersion.[30] Here, she describes the *Passage Saint-Benoit* (1865–8):[31] "The vertical path in the paving stones and the inclusion of the wall on the right edge of the frame accentuate the picture's extraordinary spatial depth, *inviting the viewer to imagine crossing the enclosure*" (emphasis added). Immersed in this space, Kennel observes how the "remarkable clarity of details *draws our lingering gaze*." "These riveting details," she concludes, "*slow us down and ask that we pay attention to this place before we leave it*"[32] (emphasis added).

While the city commissioners intended the 1878 exhibition of Marville's Old and New Paris series to celebrate the city's modernity and condemn old Paris, coincidently, curatorial writers have generally ignored the New Paris series in their formal analyses. Thézy attributed this critical lacuna to the aesthetic boredom compelled by Haussmann's strict geometry and perspectival precision.[33] However, we can assume that the contemporary critical reception of the New Paris series was less about the quality of Marville's photographs than the subject of his commission—the uniform buildings along the new boulevards. In this series, photography's documentary function exceeded its aesthetic function. Nineteenth-century critical rhetoric underlies this eclipse. When the photographer Henri de la Blanchère defined the merits of "artistic photographs," he disqualified "the photographs of historic buildings" from this designation.[34] This stricture seems to have applied equally to modern buildings and their urban setting.

Unknowingly, perhaps, present-day critics have adapted Blanchère's restraints to describe Marville's photographs of Haussmann's Paris. Curatorial viewers have emphasized the coincidence of the geometric precision and harmony of the photographs and the subject they portray.[35] Yet for these viewers, the lack of details in the New Paris series compels aesthetic boredom. Kennel thus echoes her curatorial precursors in her description of Marville's *Avenue d'Iéna* (c. 1877),[36] which stands for all the New Paris views included in our empirical study. As she puts it, the avenue has been "emptied of intricate visual detail, leaving us to contemplate, nothing but the endless, barren trees and the myriad traces left by carriage wheels on the macadam."[37]

Aesthetic boredom notwithstanding, Kennel responds to the emptiness with an embodied affect of fear: "It is as if we were witnessing the aftermath of some terrifying catastrophic event that caused the entire city to flee."[38] Whether voicing boredom or fear, contemporary writers experience a cognitive stasis. They cannot imagine entering or walking through the New Paris photographs, perhaps resulting from an adverse feeling of the sublime.[39] That Marville's contemporaries also experienced the sublime in the face of Haussmann's boulevards, lined by uniformly designed building facades and evenly spaced trees, pavers, and sidewalk slabs, can be traced to Victorien Sardou's 1866 play *Maison Neuve*. The playwright described the newly opened boulevard Malesherbes as if standing in the photographer's fixed position as that "eternal sidewalk going on and on forever."[40]

As an art historian, I must ask: is the curatorial bias a matter of associative functions (nostalgia) or fine art functions (formal attributes)? To answer this question is to accept a merging of these functions that results from a perfect match between Marville's artistic choices and technical skills and the subject of his commission. As curator Françoise Reynaud asserts:

> The extraordinary talent Marville displayed in picturing the Paris streets that were 'to be demolished' was probably more attuned to a historical subject. One wonders, though: is it romantic nostalgia that makes us prefer Marville's Old Paris photographs, or are some subjects more artistic than others?[41]

As our formal analysis shows, Marville's artistic choices and technical skills in portraying new Paris also perfectly match his subject, a fit that earned him the title of "Haussmann's man."[42]

Curatorial inattention to Marville's New Paris series can also be explained by the curators' historical moment. For these viewers, the photographer's compelling portrayal of urban austerity induced their negative association with Haussmann's legacy in the mid-to late-twentieth-century urban renewal schemes. As curator Maria Morris Hambourg surmises:

> We come by [our] nostalgia naturally. Having devastated parts of our older cities through extravagant demolition, we have turned to more conservative measures. Denigrating what now seem to be false prophets of modernism (from Haussmann to Robert Moses), we polish up our past and sympathize with historical figures whose sentiments, we think, resemble our own.[43]

Given this legacy, the contrast between old and new Paris is sustained in the *informed* viewer's bias for Marville's Old Paris series.

Conclusion: Minding the Gap between Neuroscience and Art History Research

Our combined art historical and empirical research suggested that laypeople (uninformed/untrained viewers) and curatorial audiences had different haptic, embodied, and affective reactions to Marville's techniques, thus highlighting the photograph's power to influence the viewer. Given the disparity between these two groups of viewers, our experiment sheds light on a critical discourse regarding the gap between neuroscience and humanities/art historical research.

This critical discourse concerns scientists more often than art historians. For these investigators the gap pertains to art historians' disregard of psychological and neuroaesthetic studies of art appreciation. Conversely, most art historians reject the scientific studies because of the latter's disregard for contextualization. As neuroscientist Anjan Chatterjee admits:

> Understanding the context in which the work was produced, the intent of the artists, and the historical and cultural conversation in which it was engendered enhances one's appreciation. Neuroscience methods do not easily grasp these complex aspects of the richly textured meaning of art.[44]

Furthermore, sometimes scientists' and art historians' assumptions and goals are misaligned. Psychologists Nicolas J. Bullot and Rolf Reber address these problems in the research article, "The Artful Mind Meets Art History: Toward a Psycho-Historical Framework for the Science of Art Appreciation Study."[45]

Bullot and Reber's study reviews the scientific theories of art appreciation that attempt to bridge the science-humanities gap, laying the foundation for their modeling of a "psycho-historical framework for the science of art appreciation." Their model comprises "three modes of appreciation: *basic exposure* to an artwork, the *artistic design stance*, and *artistic understanding*." While Bullot and Reber claim that the three modes are not successive, they assert that the third mode, *artistic understanding*, is the "natural" response of the art history specialist, involving "an attitude whereby appreciators develop their sensitivity to art-historical contexts by means of inquiries into the making, authorship, and functions of artworks."[46] However, they propose that the layperson can acquire the third mode by progressing through the second mode, the *artistic design stance*, "a requisite for artistic understanding." This three-modal approach can be adapted to resolve the discrepancies in our empirical study of viewers' valuation and interpretation of Marville's photographs. In fact, my collaborators suggested a similar approach for further study, wherein laypeople would be provided with contextual information prior to viewing the Old and New Paris photographs.[47] Bullot and Reber's *artistic understanding* mode also supports my formulation of the "curatorial gaze" (which I modeled independently of the authors).[48]

These concurrences between our collaborative, transdisciplinary study and Bullot and Reber's model for a "psycho-historical framework for the science of art appreciation" marks our study as a pioneering contribution to closing the gap between scientific and art historical research. Our study also shows how collaborative dialogues can repair the messy chasms that may arise between quantitative data of experimental research and contextual interpretation of the historical archive.

Notes

1 Professor, Department of Psychology, University of Vienna.
2 Doctoral Candidate (2022–23), Department of Psychology, University of Vienna.
3 Weingarden, Pelowski, and Kühnapfel, "Embodying the Photographic Image: Charles Marville's Old Paris and New Paris Series," in *Art as Experience of the Living Body*, 261–92.
4 We use upper-case letters for Old Paris and New Paris to designate Marville's series, as opposed to the general historical references to old and new Paris.
5 Except for the commercially published volume, *Marville: Paris* by Marie de Thézy (Paris: Hazan, 1994), all monographic studies of Marville are published as exhibition catalogues. Curators and contributing authors of exhibition catalogues are tasked with interpreting art historical research to a general, non-specialist audience. Thézy's publication falls within this category. Thézy was chief conservator/curator at the Bibliothèque Historique de la Ville de Paris (BHVP; Historical Library of the City of Paris), in charge of photographic collections. *Marville: Paris* comprises a *catalogue raisonné* of Marville's photographs of Paris before, during, and after Haussmannization, which Thézy amassed primarily from the BHVP collection and secondarily from national archival collections; the volume constitutes the first such publication available to the public.
6 In our study, we adapted the concept of "the gaze" from cultural and visual studies. For an art historical and visual culture approach to the gaze, see Margaret Olin, "Gaze," in *Critical Terms for Art History*, ed. Robert S. Nelson and Richard Shiff (Chicago: University of Chicago Press, 1996): 208–19. As Olin asserts, the gaze does not only apply to communication between people: "The possibilities of the gaze thus extend to ornament on objects of the applied arts or architecture, to landscapes, and even to abstract art. Certain constructions of the gaze can be associated with artistic styles" (217).

7 See Iser, "The Reading Process: A Phenomenological Approach" and Fish, *Is There a Text in This Class?: The Authority of Interpretive Communities*. They argue that any individual interpretation of a text is created in an interpretive community of minds of participants who share a specific reading and interpretation strategy.

8 I have also adapted the informed/uninformed viewer model from the literary reader-response theory.

9 For a history of nineteenth-century photographic technologies, see Rosenblum, *A World History of Photography*, 192–9.

10 As discussed in Chapter 4, photography had a third documentary function during Haussmannization—to record the demolition and rebuilding process. While we acknowledge that Marville's photographs of demolition sites were central to the history of urban renewal and Haussmann's campaign, we chose to exclude them in our empirical aesthetic study.

11 As Paul-Louis Roubert has noted and as I discuss in Chapter 4, Baudelaire's rhetoric about photography's "true duty" as a "record-keeper" did not occur in a vacuum. Photography's documentary function was established at the outset of its invention (1839), as were the prohibition of photography as a fine art and the censure of painting's dependency on photography. These primary texts are reprinted in Baudelaire, "Le Public Moderne et la Photographie," https://journals.openedition.org/etudesphotographiques/185#ftn25.

12 In "The Queen of the Faculties," the section that follows "The Modern Public and Photography" in "Salon of 1859," Baudelaire counters photography's "deputizing" the fine arts with the sanctioning of "the right of deputizing [the fine arts] that imagination has, in virtue of its divine origin [...]," in *Selected Writings*, 300.

13 Baudelaire, "The Salon of 1859: The Modern Public and Photography," in *Selected Writing*, 297–8.

14 Francis Wey, "De L'influence de L'héliographie sur les Beaux-arts," *La Lumière*, no. 1 (February 8, 1851): 2; qtd. and trans. in Karen Hellman, "Real/Ideal: Photography in Mid-Nineteenth-Century France," in Karen Hellman, ed., *Real/Ideal: Photography in Mid-Nineteenth-Century France* (Los Angeles: J. Paul Getty Museum, 2016): 8.

15 Charles Nègre, "Héliographie sur Papier Ciré et à Sec," *La Lumière*, no. 14 (May 11, 1851): 56; qtd. and trans. in Hellman, "Real/Ideal," 7.

16 "Rapport présenté par M. E. Durieu, au nom de la Commission chargée de l'examen de l'Exposition ouverte dans les salons de la Société française de photographie, du 1 août au 15 novembre 1855," *Bulletin de la Société Française de Photographie* 2 (January 1856): 63; qtd. and trans. in Paul-Louis Roubert, "A Bourgeois Ambition: Photography and the Salon des Beaux-Arts during the Second Empire," in Hellman, *Real/Ideal*, 32.

17 Notably, later historians focused on the Old Paris series rather than the New Paris series. Given this bias, our visual analyses of Marville's photographs were weighted by features of the Old Paris photographs.

18 Thézy, *Marville: Paris*, 29; Kennel, "Charles Marville: Hidden in Plain Sight," in Kennel, *Charles Marville*, 29, 39. It should be noted that neither Thézy nor Kennel, nor other sources researched for this study, provide primary sources for contemporaneous commentary that conveys positive-negative associations regarding old Paris/new Paris for this side-by-side display.

19 Unless otherwise noted, all New Paris photographs are dated 1876–77.

20 Thézy, *Marville: Paris*, 31.

21 Françoise Reynaud adds an immersive description of Marville's method of positioning his camera so that "the signs located to the left or right of an intersection, which gave the names of the streets, squares, and other thoroughfares," making it "generally easy to identify the position of the camera simply by consulting a Paris map of the period," "Marville and Old Paris," in Kennel, *Charles Marville*, 199. To this, I add that for Marville's nineteenth-century viewers, this topographic specificity might elicit their embodied, spatial memory of the place.

22 Here, we use "universal" as a psychological concept defined as "the condition of existing everywhere, often in a similar or identical form. In psychology, universality is more specifically: 1. the tendency to assume that one's personal qualities and characteristics, including attitudes and values, are common in the general social group or culture;" "APA Dictionary of Psychology," 2023, https://dictionary.apa.org/universality. We argue against this concept, specifically that some of the curatorial claims seem to hold the implication that everyone would respond to the scene or aspects of the scene similarly. We hold that "universal" responses are not "true," and thus, one

should be careful when making such claims. For a psychologist's challenge to this "absolutist" concept, see Walter J. Lonner, "Universals," *The Encyclopedia of Cross-Cultural Psychology*, published online September 3, 2013, 1308–11, https://doi.org/10.1002/9781118339893.wbeccp553.

23 Participants were recruited from the courses and student database of Florida State University and the University of Vienna via personal contacts. Participants were eligible to participate in the study if they were older than 18 and could speak English. Twenty-three participants took part for course credit; the other participants were not compensated for participation.

24 For the diagrammed quantitative results, see https://www.researchgate.net/publication/372499063_Embodying_the_photographic_image_Charles_Marville's_Old_Paris_and_New_Paris_series#fullTextFileContent.

25 Lastly, in our survey, we asked participants how old they thought the image was (in date years) and how much time they would like to spend in the space (in minutes). Participants estimated that the Old Paris photographs were indeed older than the New: New Paris photographs from 1877, which were estimated to have been taken between 1849 and 1985 (M = 1909). Old Paris photographs (taken between 1864 and 1869) were estimated to have been taken between 1828 and 1976 (M = 1903, horizontal) and (M = 1902, vertical). When asked how much time they would spend in each space in minutes, participants wanted to spend almost twice as much time in New Paris scenes (mean 34.72 min) than in Old (17.46 min).

26 The one question that did not show an overall difference among the three photograph categories was how much viewers felt *in the space*. For us, this was rather interesting and contrary to our assumption that viewers might routinely 'enter' a photographic space and, thus, report a feeling of being inside the scene of a photograph or other representative media. Following the participant reports, they did not feel they were 'inside' any photograph, regardless of their format. The photographs, one might surmise, while inviting a conceptual, affective, or embodied response, kept the viewer outside the image.

We also surmised that the the differences between our participants responses and curatorial/art historical descriptions could be connected to broader patterns of understanding and response to technologies, including photographs, which could differ between now and the 1980s and 1990s. With the rise of interactive, dynamic media and screens, it may be that we are finding a changing response—especially at the haptic level—in our contemporary audience, suggesting a more significant gap between now and two or three decades ago than that between then and the nineteenth century.

27 The curators' imagined "walk through" Marville's photographs is inflected with Baudelaire's *flânerie*. Kennel ascribes *flâneusement* to Marville (Kennel, *Charles Marville*, 15). Here, I avoid using the "flâneur" trope as overdetermined and instead connect the excursive modes of representation and writing with the picturesque, a practice that Baudelaire inherited from earlier Romantic traditions.

28 Thézy, *Marville: Paris*, 36.

29 Thézy, *Marville: Paris*, 36.

30 Kennel, *Charles Marville*, 29.

31 Marville's *Passage Saint-Benoît* was not included in our survey; for the image, see "Passage Saint-Benoît, 6ème Arrondissement, Paris," Paris Musées, https://www.parismuseescollections.paris.fr/fr/musee-carnavalet/oeuvres/passage-saint-benoit-6eme-arrondissement-paris#infos-principales.

32 Kennel, *Charles Marville*, 31.

33 Thézy, *Marville: Paris*, 34.

34 Henri de la Blanchère, "Études Photographiques," *La Lumière* 6 (January 31, 1857): 19; qtd. and trans. in Roubert, "A Bourgeois Ambition," in *Real/Ideal*, 31.

35 In addition to Thézy, *Marville: Paris*, see Kennel, *Charles Marville*, 38.

36 Marville's *Avenue d'Iéna* was not included in our survey; for image see "Charles Marville, Avenue d'Iéna, ca. 1853–70," Wikipedia Commons, 2010, https://commons.wikimedia.org/wiki/File:Charles_Marville,_Avenue_d%27I%C3%A9na,_ca._1853%E2%80%9370.jpg.

37 Kennel, *Charles Marville*, 38.

38 Kennel, *Charles Marville*, 38.

39 This understanding of the sublime follows a largely Burkean definition, considering the sublime to be a mental configuration incurred when the viewer perceives immeasurable magnitude, infinite expanse, or indeterminate repetition; see Edmund Burke, "Section XIII 'The Effects of

SUCCESSION in Visual Objects Explained,'" in *A Philosophical Enquiry into the Origin of Our Ideas of the Sublime and Beautiful* [orig. 2nd ed., 1759], ed., intro, and notes by Adam Phillips (Oxford/New York: Oxford University Press, 1990): 128. Viewed from a distance, these vistas compelled feelings of awe or fear.

40 Victorien Sardou, *Maison Neuve: Comédie en 5 Actes* (Paris: Théâtre du Vaudeville, 1866); qtd. in Kennel, Charles Marville, 39.

41 Reynaud, "Marville and Old Paris," in Kennel, *Charles Marville*, 205.

42 Maria Morris Hambourg, "Charles Marville's Old Paris," in Jacqueline Chambord, *Charles Marville: Photographs of Paris at the Time of the Second Empire on Loan from the Musée Carnavalet, Paris* (New York: French Institute/Alliance Française, 1981): 10.

43 Hambourg, "Charles Marville's Old Paris," 7. Thézy references Haussmann's legacy as a "champion of urban vandalism" (*Marville: Paris*, 6).

44 Anjan Chatterjee, "Neuroaesthetics: Researchers Unravel the Biology of Beauty and Art," *The Scientist Magazine* (May 1, 2014): [6/7].

45 Nicolas J. Bullot and Rolf Reber, "The Artful Mind Meets Art History: Toward a Psycho-Historical Framework for the Science of Art Appreciation," *Behavioral and Brain Sciences* 36, no. 2 (March 18, 2013): 123–37.

46 Bullot and Reber, "The Artful Mind Meets Art History," 123.

47 Weingarden et al., "Embodying the Photographic Image," 288–9.

48 I have also proposed that further experiments could employ eye-tracking glasses to record each viewer's gaze. An eye-tracking experiment might lessen the gap between the two groups, revealing the viewers' real-time engagement with and focal points within Marville's Old Paris series and New Paris series. In this case, the beholder's "lingering gaze," or rather a penetration into the pictorial space, might raise the question of whether and how such viewing patterns relate to the emotional, embodied, and haptic experience of Marville's photographs. See Weingarden et al., "Embodying the Photographic Image," 289.

6 Manet's Erotic and Ironic Gaze

Photography, Pornography, and Censorship

In this chapter, I establish how Édouard Manet's scandalous paintings, *Le Déjeuner sur L'herbe* (*Luncheon on the Grass*) (1863) and *Olympia* (1863), belong to a discursive space on legal censorship of pornographic photography and literature during the 1850s and 1860s (Figures 6.1 and 6.2).[1] Pornography, like prostitution, was a commonplace commodity of the newly built Paris boulevards. Such public displays of moral rupture resonate with the material rupture of urban spaces. I explore how Manet parodied traditional masterpieces within this cultural context by referencing censored erotic photographs. The hostile, critical discussions of his paintings also reiterated the vocabulary used in the 1857 censorship trials of Gustave Flaubert's novel *Madame Bovary* (1856) and in Charles Baudelaire's erotic poems in *Les Fleurs du Mal* (1857). When viewed together, social practices, literary texts, and visual media blended to form a cultural discourse on pornography's contamination and containment.[2]

To illustrate the dynamics of cross-contaminations between word and image and between fine art and popular culture, I turn to Manet's Salon paintings, *Le Déjeuner sur L'herbe* (Salon des Refusés, 1863) and *Olympia* (Salon of 1865).[3] These paintings, with their slippage between legal (academic) and illegal (pornographic) photography of female nudity, are pivotal within the context of the censorship laws formulated during the 1850s.[4] I extend this visual dynamic to the literary by examining Manet's paintings and pornographic photography within the context of the 1857 censorship trials of *Madame Bovary* and *Les Fleurs du Mal*. The transcripts of the trials are replete with anxiety about literary obscenities and the contamination of the public senses and minds. The prosecutorial discourse not only shaped the critical reception of *Le Déjeuner sur L'herbe* and *Olympia* but also contaminated the reception of Manet's *Nana* (1877), even after the laws had been revoked in 1870 (Figure 6.9).

Salient events in the history of moral censorship in nineteenth-century France frame my investigation. *Madame Bovary* and *Les Fleurs du Mal* were prosecuted under the May 17, 1819 law, which aimed to suppress the exhibition, distribution, or sale of any printed matter that constituted an offense to public and religious decency and morals ("un outrage a la morale publique et religieuse et aux bonnes moeurs").[5] In 1835 the 1819 law was extended to graphic matter—drawings, prints, and paintings.[6] Then, in 1852, the 1835 law was revised to include photography, an addition that later contaminated the response to Manet's Salon paintings.[7]

DOI: 10.4324/9780429442315-7

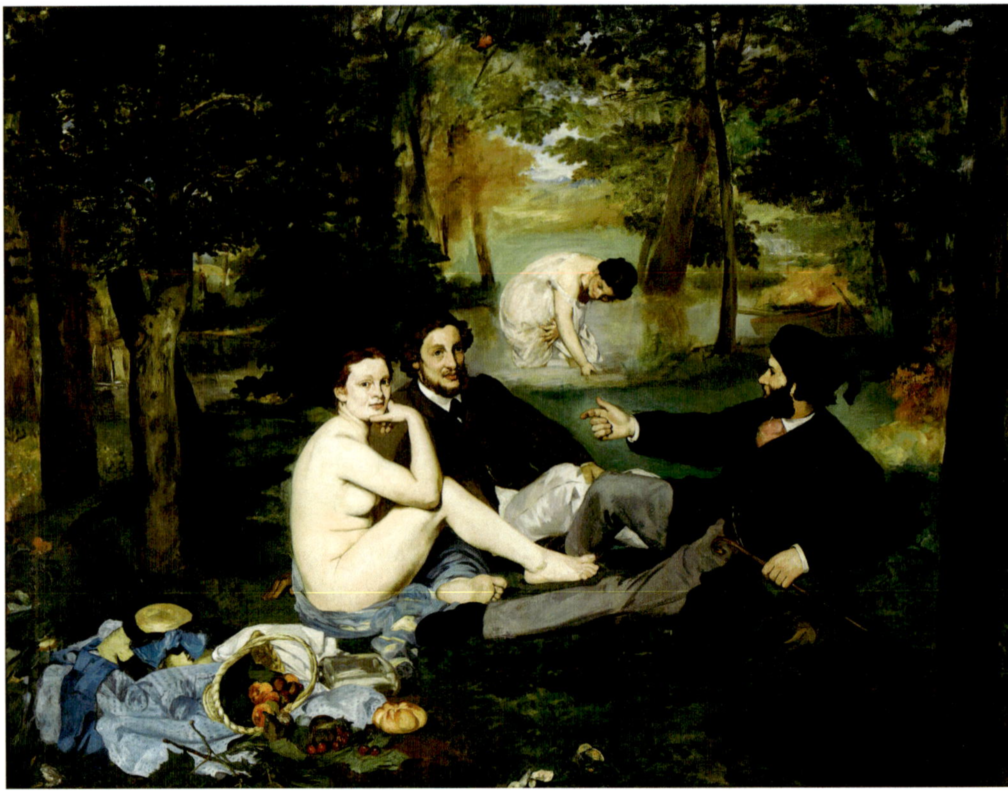

Figure 6.1 Édouard Manet, *Le Déjeuner sur L'herbe* (*Luncheon on the Grass*), 1863, oil on canvas, 82 x 104 in. Musée d'Orsay, Paris, France. © RMN-Grand Palais/Art Resource, NY. Photo: Benoît Touchard/Mathieu Rabeau.

Manet's Photographic Contamination of Salon Painting

As it is well known, Manet modernized museum masterpieces in *Le Déjeuner sur L'herbe* by referencing Raphael's river gods in *The Judgment of Paris*, by way of Marcantonio Raimondi's engraving (1510–20), and in *Olympia* by referencing Titian's *Venus of Urbino* (1538).[8] Less known is his way of modernizing the classical nude by referencing photographic re-presentations of the female paradigm.[9] As Elizabeth Anne McCauley has demonstrated, this new genre, labeled "*académies*" or "*études aprés nature*" (studies after nature), emerged with the development of commercial photography during the early 1850s (Figure 6.3).[10] *Académies* were legitimate photographic studies of nude or partially clothed models. Within a few years, many photographers turned their technical devices to the more profitable commodity of pornographic photography. My archival research at the Bibliothèque nationale de France (BnF) further shows that Manet not only used licit photographs (*académies*) but more flagrantly used illicit photographs of erotic nudes.

In what follows, I examine how and why Manet incorporated photographic images and their social and aesthetic discourses into *Le Déjeuner sur L'herbe* and *Olympia* in 1863. My argument builds on scholarship treating the nineteenth-century genre of the photographic nude, its erotic/pornographic derivatives, and the censorship it spawned.[11]

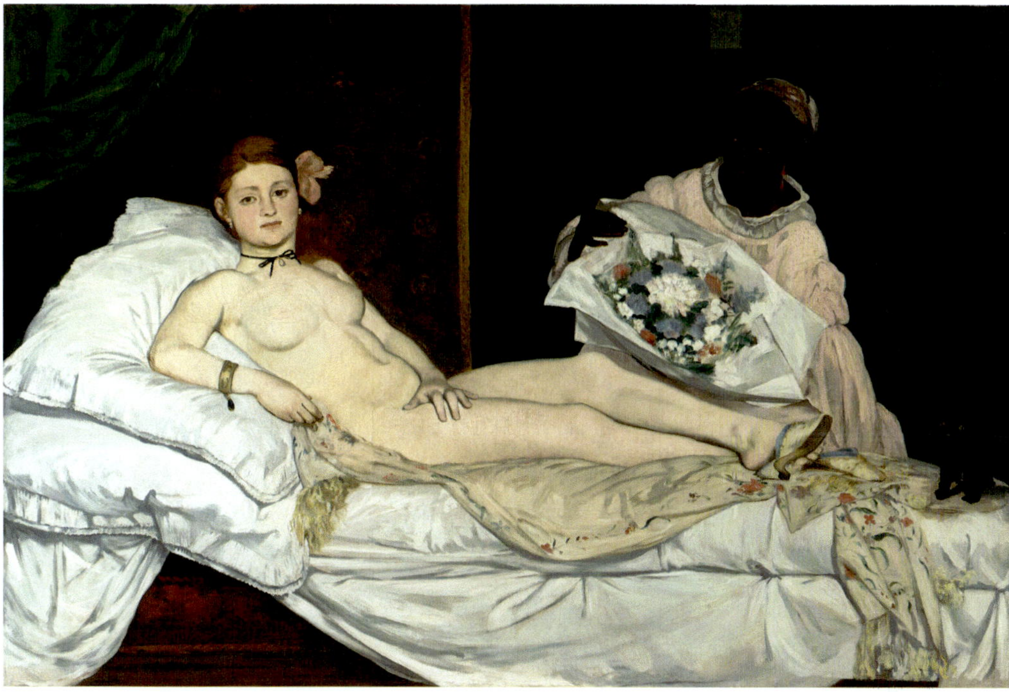

Figure 6.2 Édouard Manet, *Olympia*, 1863, oil on canvas, 40 3/4 x 74 5/6 in. Musée d'Orsay, Paris, France © RMN-Grand Palais/Art Resource, NY. Photo: Patrice Schmidt.

Viewed in this context, we shall see that Manet exposed his photographic references at a highly volatile moment in the medium's history and that he did so to define himself as a Baudelairean painter of modern life. Manet thereby mastered the slippage between fine art and erotic photography, manifested in the Venus paradigm. To locate Manet within the history of the photographic genre, I have adopted André Rouillé's rubrics of the aesthetic, erotic, and pornographic Venus.[12] While these categories are not easy to maintain, they help to show how the aesthetic Venus immediately morphed into the erotic Venus and the erotic into the pornographic.

Initially, *académies* could be categorized as aesthetic Venuses. By the end of the 1840s, a decade after the birth of photography, the female nude became an aesthetic object of study; photographers, often trained as academic painters, would arrange the female model in classical poses and under studio light that replicated chiaroscuro effects in painting. In the early 1850s, commercial photographers began marketing these photographic nudes to painters, eliminating the cost of hiring live models and providing a storehouse of Venus references, ranging from ancient to modern prototypes.[13] However, because most viewers regarded the photograph as a mimetic image of the thing itself, the model posing for aesthetic purposes represented no one other than herself, a real woman inscribed with a specific social identity.

For nineteenth-century viewers, the female studio model was quickly identified as a prostitute. Her social standing was even more vividly marked by the realism that photography exposed. These models are not the idealized product of artistic creation but embody the brute reality of folded and blemished flesh, body hair, and filthy feet. As recorded in

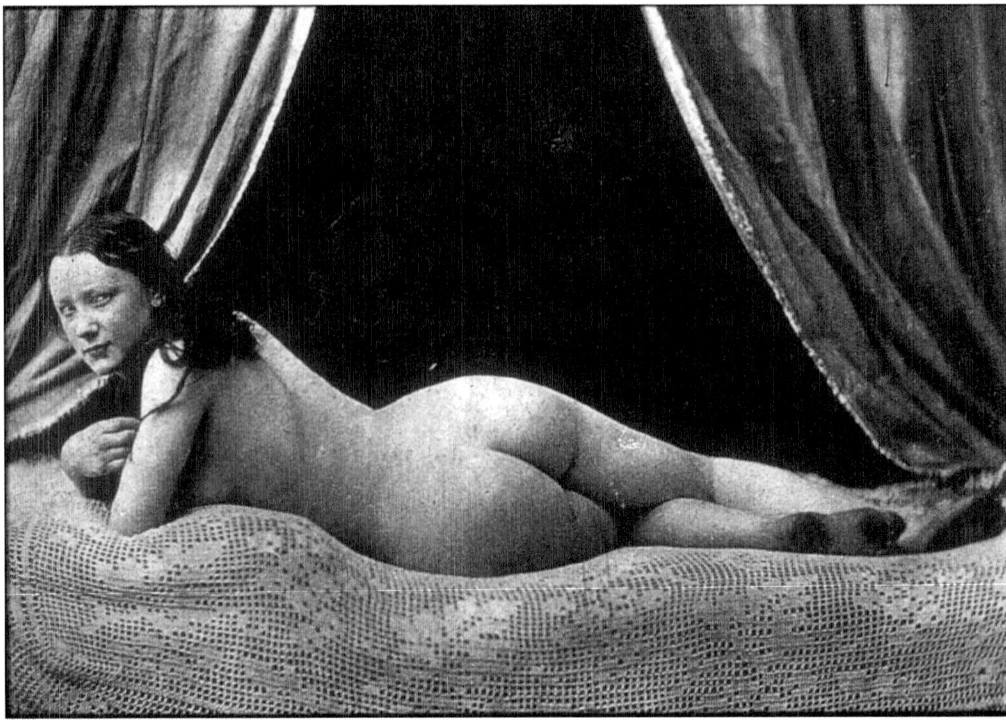

Figure 6.3 Jacques Antoine Moulin (Quinet Frères, eds), *Étude D'après Nature* (Study After Nature), c. 1854, Stamped: "Autorisée sans exposition à l'étalage" (Permitted except for showcase display), unknown dimensions. Bibliothèque nationale de France, Paris.

the labeling of Jacques Antoine Moulin's *Étude D'après Nature*, legally designated "restricted use" ("Autorisée sans exposition à l'étalage" [Permitted except for showcase display]),[14] this tangible slippage between the artistic and the commonplace blurred the cognitive difference between the aesthetic and erotic, a slippage government censors were quick to arrest.

As McCauley has shown, "the 1850s witnessed the height of increased censorship and public outrage [against] the marketing of photographic nudes." Government seizures of photographic nudes began in 1851. In 1852, copyright and censorship laws were extended to photography, "requiring that copies of photographic prints be deposited at the Ministry of Interior or the Prefecture of Police." Here, it was determined whether photographs could be offered for public sale or be restricted to classroom or studio use. In 1855, the police established a separate register to record arrests of models, photographers, hand-tinters, street peddlers, and merchants of erotic/pornographic images for offending public morals. Police records also show how these illicit photographs were sold and displayed everywhere: cafés, street corners, dance halls, and brothels, in the backs of print shops, and in established photographers' studios. Although "public condemnation became so strong that it forced the nude photographic market underground," that market thrived, as indicated by "a surge in arrests that began in 1859 and continued in the early 1860s."[15]

Photography and art critics likewise decried the prostitute's familiar personage in the photographic medium and paintings derived from these photographs. For example, in 1855, a writer for the *Revue Photographique* criticized a photographer of legitimate artists' nudes (Moulin) for also marketing "'*académies de la rue*,' or 'street académies.'"[16] In 1863, the conservative art critic Maxime Du Camp criticized Alexandre Cabanel's *Birth of Venus* (1863) for merely depicting an *académie*—an actual paid model—and nothing more: "C'est Venus! Non point, c'est un modèle, et rien de plus." (It's Venus! No, it's a model, and nothing more.) (Figure 6.4) In his words, she is not the "chaste" Venus born of the waves, as embodied by "Titian's Venus, Correggio's Danaë, Raphael's Galatea." "They are goddesses," Du Camp proclaimed, "they have none of the provocations of women." On the contrary:

> [Cabanel's] Venus is not born, she is waking up. [...] she seems to solicit the admiration of the spectator and say to him: See how beautiful I am! look, I am here so you can contemplate me at your ease; the sea is a pretext, my name a come-on. I am a woman, nothing more, but nothing less[.][17]

If viewers like Du Camp did not distinguish between painted and photographed Venuses in the official Salon, then Manet's Venuses must have enraged the moral guardians to the extreme. In his paintings, everything is laid bare; no effort is made to veil his actual model and her social status. Manet replaced allegorical props with the mundane picnic and bedroom props of a modern prostitute-model. On this point, Émile Zola was as brazen as Olympia herself, stating that "she is a girl of sixteen, doubtless some model whom Édouard Manet has quietly copied just as she is."[18] Analyzing the critical responses to

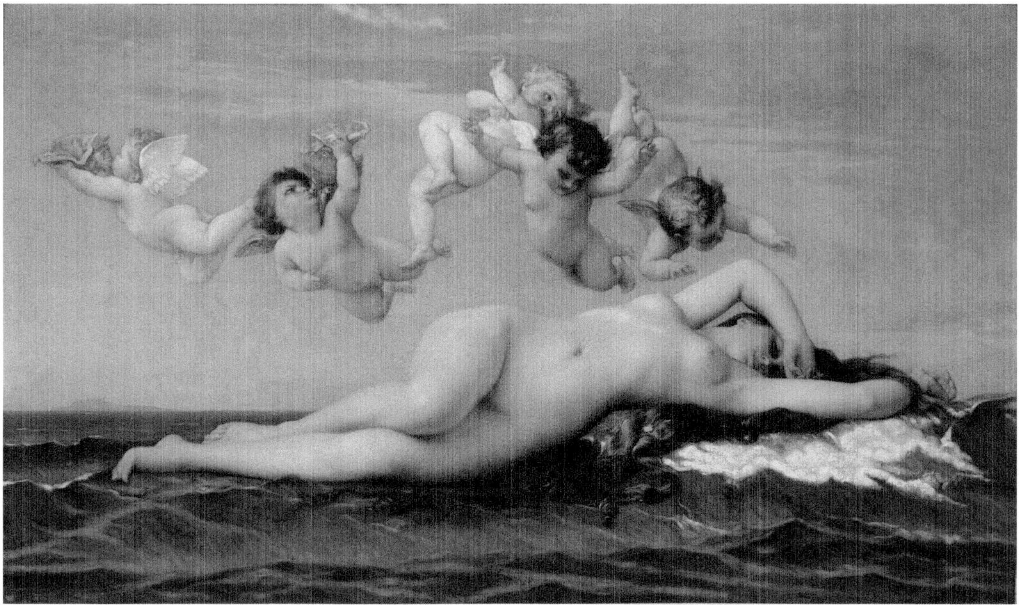

Figure 6.4 Alexandre Cabanel, *Naissance de Vénus* (*Birth of Venus*), 1875 [artist's replica; original 1863], oil on canvas, 41 1/8 x 71 7/8 in. Metropolitan Museum of Art, New York. Gift of John Wolfe, 1893.

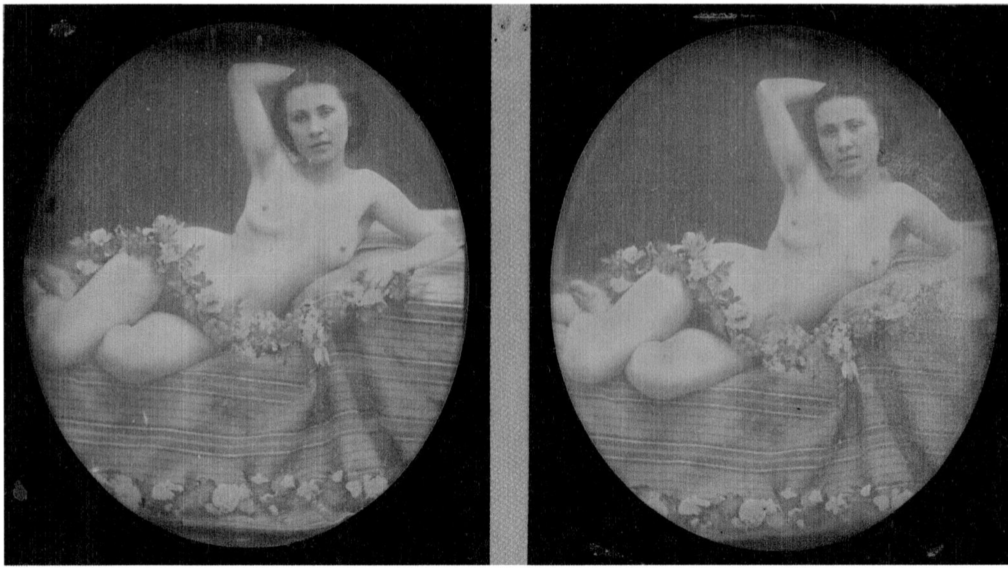

Figure 6.5 Attributed to Auguste Belloc, *Jeune Femme Allongée sur un Sofa* (Young Woman Lying on a Sofa), colored stereoscopic daguerreotype, 1850–2, unknown dimensions. Musée Niépce, Chalon-sur-Saône, France.

Olympia, T. J. Clark observed that, while a few critics identified her as a prostitute, most others skirted the issue by discussing "[the] violence done to the body, its physical uncleanliness, and [its] general air of death and decomposition."[19] To this, I would add the visual evidence of Manet's contamination of the aesthetic Venus with a popular, prurient commodity. Given the public's widespread access to and ensuing scandal over erotic photographs, critics would have immediately recognized *Olympia* as the equivalent—or surrogate—of a beguiling erotic photograph, now enlarged to the life-size proportions of a full-scale Salon painting (Figure 6.5).[20]

Olympia shares more than socio-economic status and tawdry studio attire with her photographed counterparts. What is most provocative about Manet's Venus is the directness of her gaze. Yet, during the decade before Manet's painting, photographers had already inverted the male gaze by having the model look head-on into the camera aperture. What I have called the "female gaze"[21] now dominates the exchange between the viewer and the viewed. The female gaze here functions as both subversion and titillation of the male gaze and should be seen as central to the gestures and glances that the censors determined as "explicit sexual invitation."[22] In this context, the female gaze distinguishes the legitimate *académie* from the illicit one. It also distinguishes Manet's Venus from Cabanel's, where Venus's gaze toward the viewer is veiled by half-closed, drowsy eyelids and shadows cast by her hand and arm. Manet thus signaled his modernity in two photographic ways: by replacing the model's indirect or sideways glance with a direct female gaze and by painting a copy of an erotic photograph rather than an *académie*.

Embodying Contamination Through the Auratic Female Gaze

Walter Benjamin's idea of the auratic gaze, in terms of the subjective reciprocity between viewer and image viewed, offers a paradigm for assessing the "reality" of the photographic nude and its sensual presence in Manet's paintings.[23] In "The Work of Art in the Age of Its Technological Reproducibility" (1936), Benjamin celebrated photography as a revolutionary invention for democratizing the fine arts. Benjamin here introduced the concept of a work's "aura," which he associated with "the whole sphere of authenticity" comprised by the artwork's physical, unique presence, historical tradition, and the context in which it is viewed. These attributes, he asserted, are not reproducible for a work of art, so its reproduction "devalue[s] the here and now of the artwork." As Benjamin stated:

> One might focus these aspects [of authenticity] of the artwork in the concept of the aura, and go on to say: what withers in the age of the technological reproducibility of the work of art is the latter's aura. [...] By replicating the work many times over, it substitutes a mass existence for a unique existence. And in permitting the reproduction to reach the recipient in his or her own situation, it actualizes that which is reproduced.[24]

While this devaluation of the aura also applies to a natural object and its reproduction—in a photograph, we can see but not smell a rose—it does not apply to experiencing the natural object's authenticity or uniqueness. According to Benjamin, "in the work of art this process [of technological reproduction] touches on a highly sensitive core, more vulnerable than that of any natural object. That core is its authenticity."[25] A photograph of a rose is still a rose. Viewed in this light, Manet's unidealized, painted re-presentation of the photographic nude is analogous with the natural object's auratic authenticity and the viewer's actualization "of that which is reproduced."

While this analogy is an inversion of Benjamin's critique of an artwork's authenticity, his argument provides a means for reasserting the power of the aura when the photographed object, the female nude, returns the viewer's gaze. In his 1939 essay, "Some Motifs in Baudelaire," Benjamin elaborated on the continuum between the gaze and the aura. What is important here is that Benjamin introduces the "gaze" as that which mediates the artist's expressive intentions to the viewer. As such, the "gaze" is a function of the "aura," wherein the latter is "attach[ed] to the object of a perception," and the former is a cognitive experience contingent on the object's unique presence. The gaze thus provides a guide for determining photography's role in the modern "decline of the aura." Looking at this decline from a historical perspective, Benjamin discerns the aura's power in early figurative photography:

> What was inevitably felt to be inhuman [...] in the daguerreotypy [*sic*] was the (prolonged) looking into the camera, since the camera records our likeness without returning our gaze. Inherent in the gaze, however, is the expectation that it will be returned by that on which it is bestowed. Where this expectation is met, *there is an experience of the aura in all its fullness*.[26]
>
> (emphasis added)

In this context, photography sustains the continuum between the aura and the gaze because, as Charles Haxthausen has observed, each "exerts the semblance of human subjectivity"[27] in the photographed object. The opposite effect of the aura occurs—it declines—when the artwork is the object of photographic reproduction.

For the nineteenth-century viewer, the female gaze, as photographed and painted, embodies human subjectivity for both the female model and the (white) male observer.

In the latter case, this subjectivity extends to the photographer. Benjamin misjudged the relation between the sitter and camera and, as a result, was blind to the photographer's voyeuristic gaze, always present either in front of, beside, or mirrored in the apparatus (Figures 6.6, 6.7, and 6.8). As such, the model was already poised to return the photographer-qua-viewer's gaze. This auratic gaze, in my view, is what frequently blurs the boundary between the erotic and the aesthetic function of the *académies*. That is, the erotic model's direct frontal gaze ensnares the viewer's gaze into active complicity with her nakedness. In contrast, an *académie*'s averted gaze enables the viewer's passive contemplation of her nudity.

Stereoscopy and the Corporeal Female Gaze

The female gaze is even more pronounced, as is its erotic function, in another invention that made its public debut in the early 1850s—stereoscopic photography.[28] During 1853–4, it became a thriving industry, bringing home or to fairgrounds, real-life scenes of distant lands, rural pastimes, or farcical tableaux. But it was most effective in capitalizing on the voyeuristic experience of viewing a three-dimensional figure through an apparatus held by a single viewer. With the technology of stereoscopic photography, the nude became even more real, endowing her corporeal presence with the "aura" of authenticity that Benjamin attributed to the natural object.[29] Thus, the stereoscopic image increases the auratic-qua-erotic value of the female nude not only because of her three-dimensionality but also because she dominates the viewer's field of vision. In this format, *académies* could no longer pass as aesthetic Venuses, and stereoscopic Venuses pushed the erotic to the pornographic extreme.[30] The technical novelty of a popular pastime and its notorious underside set the stage for Manet's contrivance of public spectacle into high art.

By viewing Manet's *Olympia* and *Le Déjeuner sur L'herbe* alongside stereoscopic photographs, features that have been considered technical aberrations can be better understood as deliberate copying of stereoscopic effects. These effects include harsh frontal lighting on surfaces parallel to the camera lens—which diminishes tonal contrasts and shadows of the flesh. Conversely, an appendage—an arm or leg—projecting at an oblique angle to the lens obtains shaded and highlighted effects. As a result, those figures, limbs, or objects closest to the camera lens are seen as the most relief-like when viewed through binocular stereoscopic lenses, while those in the distance look flatter. Similar effects also caused critics to see Manet's nudes as poorly modeled and flat, and those who saw Olympia's more fully modeled left hand as an obscene gesture may have been right.[31] However, we can also consider how the hand and crossed feet project to the picture plane at an angle, simulating the stereoscopic effect of projecting anatomical parts in high relief.[32]

Verbal-Visual Contamination: Attacks on Realism

We can now examine the highly volatile moment in which Manet exposed his photographic references in *Le Déjeuner sur L'herbe* and *Olympia*. Both paintings are inscribed within a pre-existing cultural discourse on obscenity and censorship. This discourse was engendered not only by the photographic nude during the 1850s and 1860s but more aggressively by Realist painting and literature. Reviewing the 1857 obscenity trials of Flaubert's *Madame Bovary* and Baudelaire's *Les Fleurs du Mal*, we can see how word and

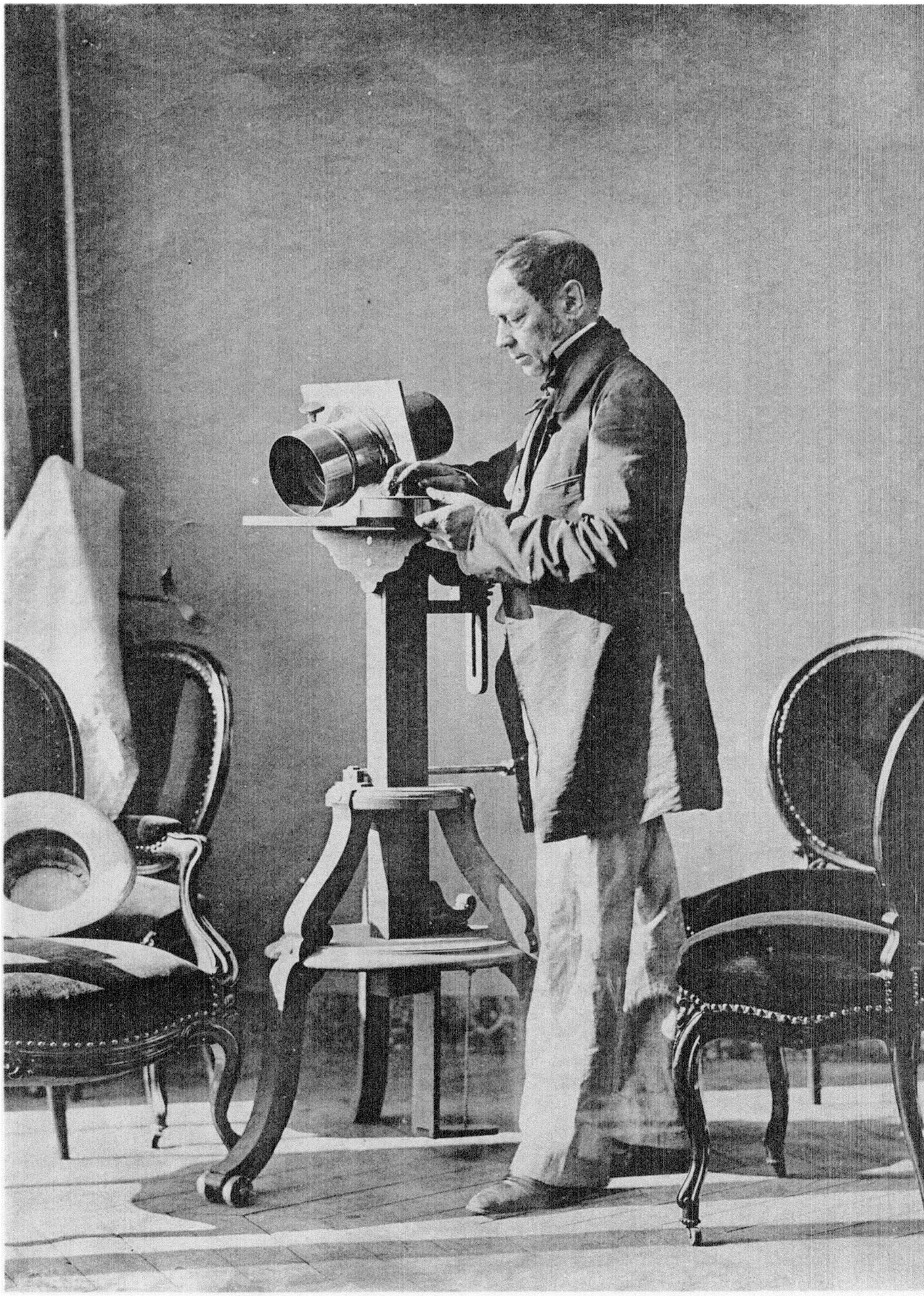

Figure 6.6 Hippolyte Bayard, *Self-portrait With Camera*, 1850, collotype, 7 5/8 × 5 3/8 in. Courtesy of the Eastman Museum, Rochester, NY.

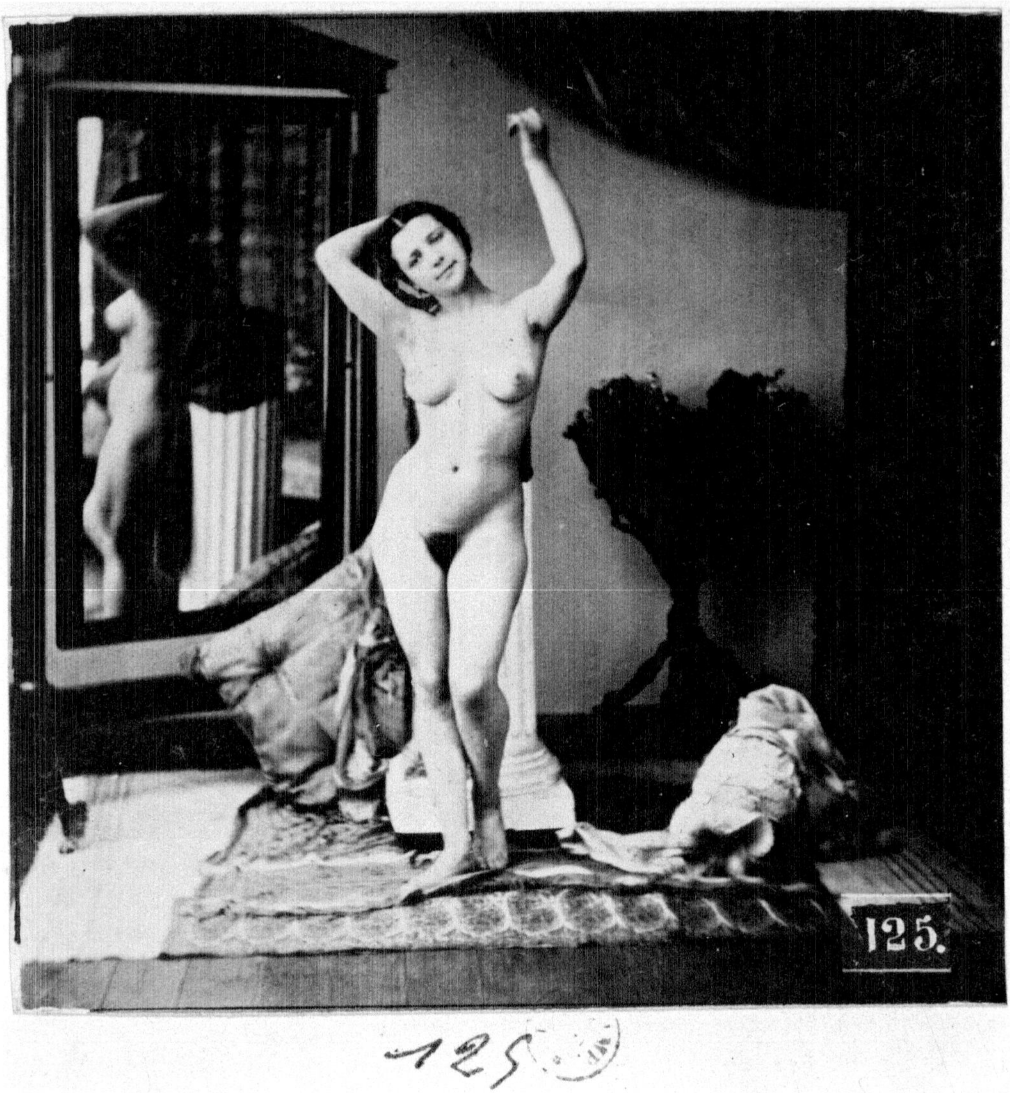

Figure 6.7 Louis-Camille d'Olivier. Untitled (composition with nude), 1856, salted paper print, 2 3/4 x 2 5/8 in. (overall). Courtesy of the Eastman Museum, Rochester, NY.

image intersect in the pejorative language used for critical and legal discussions. The rubric "Realism" stands at the center of this lexicon, signifying a literary style, a pictorial iconography, and a mimetic portrayal of contemporary types exposed with dirty bodies, indecent poses, and corrupt morals. The real contagion of Realism is made explicit in Count de Montalembert's (1810–70) speech in August 1857 before the five academies of the Institut de France. As Claude Pichois states, "Montalembert attacked those modern young men who were devoid of any 'passion for elevated things'" and corrupted by crude empiricism. Montalembert asserted that "under the name of Realism, a word less barbaric yet than the thing itself, this fatal influence is already contaminating literature, art, and even philosophy."[33]

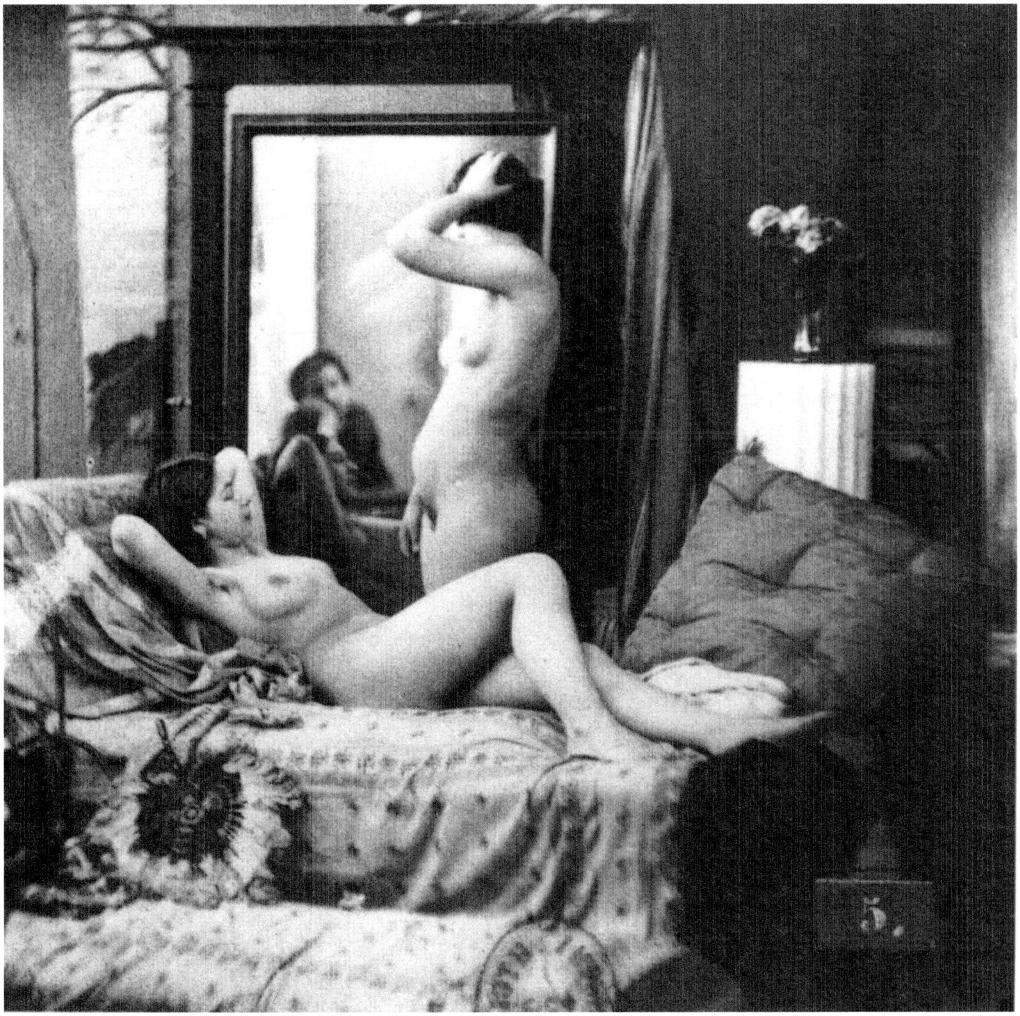

Figure 6.8 Louis-Camille d'Olivier. Untitled (self-portrait with two nudes), c. 1856, half of a stereoscopic view, albumen paper print mounted on cardboard, 3 3/8 x 6 5/8 in. Bibliothèque nationale de France, Paris.

Ernest Pinard, the prosecuting attorney in Flaubert's and Baudelaire's trials, set the stage for the state's attacks on Realism.[34] Thus, his arguments provide a lexicon for an originary discourse on the obscene and register the social anxiety of contaminating obscenities. Furthermore, this discourse includes the frequent use of pictorial analogies to describe the writers' methods of representation and the mediums against which writing should be condemned.[35] In these legal proceedings, the literary had already become contaminated by the pictorial.

Just two years earlier, Gustave Courbet (1819–77) coined the term "Realism" in his "Realist Manifesto," a pamphlet he distributed at his Realist Pavilion, housing a one-person exhibition set alongside the official exhibition at the 1855 Universal Exposition. In the manifesto, Courbet proclaimed his "goal… to translate the customs, the ideas, the appearance of my time, according to my estimation," and, he would add, without idealization.[36]

Here, Courbet showed his painting *The Artist's Studio* (1854–5; Musée d'Orsay, Paris), which had been rejected by the official 1855 exhibition.[37] This painting, and others' works depicting non-idealized nudes, were ridiculed as contaminations of the academic practices fostered by the official Paris salons and juries. Courbet's *The Bathers* (1853; Musée Fabre, Montpellier), with a fleshy nude as its focus, was condemned as especially subversive when first exhibited at the Salon of 1853.[38] Théophile Gautier (1811–72) pronounced Courbet's offenses against the Academy with vitriolic prose. Given the nude's "monstrous rump," the critic compared her to the Hottentot Venus.[39] (As we have seen, in 1863 Maxime Du Camp invoked the same fear of contamination when comparing the "chaste," classical Venuses with the *académies de la rue* Venuses.) Gautier then attacked Courbet's realistic depiction of a female body as a contamination of classical beauty:

> What did the painter have in mind in exposing this surprising anatomy? Did he want to fling down the gauntlet before the beautiful forms of Antiquity and make his own protest against the deceitful white marbles of Paros and Penteli? Was it through hatred for the Venus de Milo that he depicted this squalid body emerging from the dark water? Did he decide to challenge her immortal body with his own version of the rear view of a woman? Does this bather represent his ideal of beauty, or was he merely content to copy an obese model, her fat unequally distributed, as she posed naked on the studio platform?

Gautier continued his rant by recounting the painful experience of viewing the naked truth of female corpulence and supposed deformity:

> We do not deny that these strange forms, this puffy flesh with its folds, rolls, indentations and bulges, are depicted with the most rigorous truthfulness, but why inflict this distressing sight on us? Is not a girl or a young woman with a pure and elegant figure, with smooth and fresh skin, just as natural as an enormous matron, padded with adipose tissue and deformed by accumulations of lard? Is truth to be found only in the Ugly?[40]

While Gautier advanced the Academy's battle between Idealists and Realists, he also invoked *The Bathers' succès de scandale* in the public sphere.[41] According to Pierre-Joseph Proudhon, Empress Eugénie was repulsed by the painting, comparing the nude to the French-bred draft horse Percheron, known for its strong rump.[42] Prosper Mérimée compared the bather to butchered meat.[43] Eugène Delacroix responded by lamenting the painter's immoral thinking as signified by the nude's corpulent form: "[T]he vulgarity of forms is nothing, it is the vulgarity and uselessness of thought that are abominable[,]"[44] a condemnation that resonated in the Flaubert and the Baudelaire obscenity trials.

Cross-Contaminations on Trial

Madame Bovary Trial: January 30–February 7, 1857

Throughout the Flaubert trial, Pinard attempted to forestall the mimetic contagion that migrated from painted Realism to literature. In his preliminaries, Pinard found it difficult to explain the novel's "offenses to public morals and to religion" without reading *Madame Bovary* in its entirety.[45] Instead, he first summarized the story without criticism. He then "cite[d] incriminating texts, and finally answer[ed] the objections that may arise against the general method of indictment."[46] Pinard cunningly introduced the citations by retitling the novel, from *Madame Bovary: Moeurs de Province* (*Madame Bovary: A Tale of Provincial Life*) to *Madame Bovary: Histoire des Adultères d'une Femme de Province* (*Madame Bovary: Story of the Adulteries of a Provincial Woman*).[47] The citations were extensive and specifically culled from "four scenes, or rather four tableaux."[48] As this strategy suggests, Pinard used pictorial analogies throughout his argument to reinforce the novel's offensiveness.[49] The new title thus framed the four tableaux within the ready-made doctrines of public offense. Of these framings, Emma's "fall with Rodolphe" and then "with Leon" addressed the moral offenses. Pinard declared, "The offense against public morals lies in the lascivious pictures which I have brought before your eyes; the offense against religious morals consists in mingling voluptuous images with sacred things."[50] Ironically, Pinard justified his indictments with his lascivious picture-making:

> I bring to your attention two things here, gentlemen, an admirable picture, the product of a talented hand, but an execrable painting from a moral point of view. Yes, M. Flaubert knows how to embellish his paintings with all the resources of art, but without the discretion of art. With him there is no gauze, no veils, it is nature in all her nudity, in all her crudity![51]

Pinard's concluding statement in the Flaubert case reifies the verbal-visual contamination provoked by Courbet's intervention:

> This code of morals stigmatizes realistic literature, not because it paints the passions: hatred, vengeance, love—the world sees but the surface and art should paint them—but not paint them without bridle, without limits. Art without rules is not art. It is like a woman who discards all clothing.[52]

Echoing critics of Courbet's *Bathers*, Pinard argued that *Madame Bovary* was too close to reality, lacking moral edification or aesthetic refinement. At the same time, Jules Sénard (1800–85), Flaubert's attorney, argued in favor of "a fidelity wholly Daguerreian in the reproduction of the type of things, and in the intimate nature of the thought of the human heart."[53] Using the name of Louis Daguerre (the French inventor of photography) as an adjective, Senard framed his defense with photography's mimeticism. Senard thereby defended Flaubert's "unbridled" realism as a sign of the author's morality and religiosity, leaving it to the reader to see moral vice and then avoid it. Accordingly, Senard turned Pinard's censure of offensive tableaux against him, accusing the prosecutor of perverting the text. Addressing the tribunal, Senard declared:

> [M. Gustave Flaubert] affirms before you that the thought in his book, from the first line to the last, is a moral thought; and that, if it were not perverted (and you have seen during the last hour how great a talent one may have for perverting a thought) it would be (and will become again presently) for you, as it has been already for the readers of the book, an eminently moral and religious thought capable of being translated into these words: the excitation of virtue through the horror of vice.[54]

Later in the trial, Senard repeated his accusation against Pinard's claim of textual and moral perversity.[55] To prove that Flaubert's beneficent moral thought is transmitted through the text to the reader, Senard read a passage that elicits "an excitation of virtue through a horror of vice, as the author himself calls it, and which the reader, no longer perplexed, cannot fail to see, *unless influenced by ill-will*"[56] (emphasis added). Here, again, Senard pointed his finger back to Pinard.

This finger-pointing was successful. Senard convinced the judges that obscenity lies in the eyes of the reader/beholder. In acquitting Flaubert of obscenity charges, the judges deemed that while "the incriminated passages, viewed abstractly and isolated," remained "reprehensible," they were "few in number as compared with the extent of the work." They further found that "these passages, either in the ideas they expose, or in the situations they represent, bring out as a whole the characters which the author wished to paint." Therefore, the judges determined that it "does not appear that [Flaubert's] book has been written like certain other books, with the sole aim of giving satisfaction to the sensual passions, to a spirit of license and debauch[.]"[57]

Spinning the Web of Contamination: Baudelaire's Response to the Flaubert Trial

Baudelaire's review of *Madame Bovary*, published on October 18, 1857, attests to the threat of Flaubert's novel not only in the person of Pinard but in the social imaginary. Here, Baudelaire follows Senard's defense, but with a twist. Flaubert chose the Realist style for the "commonplace" subject of adultery, not so much to transform social mores but to awaken the "desiccated souls" of his "nineteenth-century readers."[58] Accordingly, Flaubert "imprison[s] the warmest, the most burning feelings within the most trivial adventure. The most solemn, the most fateful words… fall from the lips of the most vacuous." However, what Baudelaire most extols about Flaubert's realism is how the author "divest[s] himself of his sex and become[s] a woman. The result is a miracle." Since Flaubert could not completely rid himself of his masculinity, Emma becomes a "strange androgyne [who] has retained all the seductions of a virile soul in a charming feminine body."[59] Emma's "exclusive love of ['seduction'] and domination" are among her virile qualities at which Baudelaire marvels. For our purposes, Baudelaire's essay sheds light on the social imaginary of the obscene, the fear of its contamination of public morality, and the novel's contamination of hegemonic masculinity.

Fleurs du Mal Trial–August 20, 1857

Baudelaire's trial was held in August 1857, one month after *Fleurs du Mal* was published.[60] Not surprisingly, Pinard rehearsed his accusations against Flaubert in the Baudelaire trial. He accused the poet of intentionally breaking "artificial rules" of traditional art in order "to paint everything, to bare everything [*mettre à nu*]."[61] Although Pinard had not used the artistic term Realism,[62] he railed against mimetic realism as an "unhealthy fever which causes [one] to paint everything, to represent everything, to say everything, as though the crime of offending public morality was abrogated, and as though this morality did not exist."[63] Accordingly, Pinard assailed Baudelaire's method as a "search of human nature in its most private recesses; he will use the most vigorous and impressive tones to render it, he will especially exaggerate its most repugnant sides; he will magnify it profusely, so as to create an impression, a feeling."[64] While Pinard conceded that "the judge is not a literary critic," he argued that "the judiciary has a discretionary authority to recognize whether [public] morality is offended, if the limit has been crossed. The judge is a sentinel who must not let [moral offenders] across the border.

That is his mission."[65] Continuing the gate-keeping metaphor, Pinard invoked his guardianship: "In the current trial, should the prosecution sound the alarm?" Accordingly, the prosecutor repeated his strategy from the Flaubert trial, vowing to "cite those parts in this collection which we cannot allow to pass without protesting."[66] He, therefore, read to the trial judges passages from the condemned poems to elucidate "unhealthy impressions" from images presented to the reader, leading to "the [acquired] taste for lascivious frivolities."[67] Pinard convinced the judges to charge Baudelaire with the offense against public morality and decency in this trial, which lasted only one day.[68] The guilty verdict reiterated Pinard's pictorial metaphors and moral abhorrence.

> Whereas the error of the poet, with the object he wished to attain, and in the course which he followed, whatever effort of style he may have made, whatever may be the blame which precedes or follows his paintings, cannot destroy the harmful effect that the pictures present to the reader, and which, in the incriminated plays, necessarily lead to the excitation of the senses by a gross realism, and offensive to modesty.[69]

Pinard's picture-making litigation was not only shaped by reading Flaubert's and Baudelaire's texts. As one of the "desiccated souls" of the nineteenth-century readership, Pinard was equally contaminated by the lewd and lascivious photographs displayed in his court. Flaubert and Baudelaire were tried in the same court for prostitutes and other "delinquents."[70] Those considered delinquents included pornographic photographers, printers, models, and vendors whose production and commerce, since 1854–5, had been forced underground due to public condemnation.[71] As earlier noted, in 1855 the Paris police established a separate register to record these offenders of outrage to public morals, following the 1852 extension of the censorship laws to photography. As the State's prosecutor, Pinard would have presided over this legislation at the time of Flaubert's and of Baudelaire's trials when the matter of what constituted illicit photography came to the tribunal's attention.

The litigation of literary texts not only signified the public's fear of, and contamination by, pornography; it also signified some photographers' fear of cross-contamination within their métier. In September 1857, several ostensibly legitimate cameramen, who felt unjustly persecuted for trying to sell nudes that they considered artistic, addressed a petition to the emperor and the Ministry of the Interior.[72] Their petition worked otherwise. As McCauley asserts, "rather than letting up on arrests, the Parisian police... became more active than ever," resulting in "a surge in arrests" between 1859 and the early 1860s.[73] The photographers' fears of guilt by association were justified. In 1862, a journalist for *Le Figaro* reported that all photographers who owned a shop would display dozens of portraits of bare-fleshed actresses in the window. According to this report, "The passer-by stops, upset and troubled—in front of these exposed women."[74] In the same year, a self-styled former photographer advised on attracting mostly male passers-by: "In a decent frame, there must be mostly female portraits, as most people wandering about are going to be men. Female portraits must always be pretty... Make sure to add one or two actresses in costume, and wearing as little as possible."[75] For the *Figaro* reporter, these displays were an incitement to drain "the nation's strength[:] Each Parisian on average wastes a quarter of an hour a day looking at photographs, which by the end of the year amounts to 90 hours!"[76]

The relevance of pornographic photography to Baudelaire's trial and the condemned poems can be clarified by viewing both the event and the poetry in tandem with the critical reception of Manet's 1863 and 1865 Salon paintings and his appropriation of stereoscopic photography therein. Although *Le Déjeuner sur L'herbe* and *Olympia* were not officially censored, they extended the discourse on censorship in which the attacks on literary and artistic Realism converge in the public imaginary on the obscene.

Viewing Stereographic Pornography with Manet–Baudelaire–Pinard

The Parisian prostitute and her representation were at the center of this convergence. As we have seen, photography and art critics decried the prostitute's familiar personage in the photographic medium and paintings derived from these photographs. However, the moniker *académies de la rue* may also infer the photographic display of "exposed" or scantily dressed women along Parisian boulevards, a sight as familiar as actual prostitutes. These photographs contaminated the "passer-by" and the portrait-sitters displayed among them. As one critic asked in 1861, what Parisian promenader "has not seen the noble faces of virgins and empresses exposed pell-mell beside nude courtesans?" Echoing both Courbet's critics and Pinard's attacks on Realism, this writer attests that whether such "cynical photographs boldly showing insolent details" are displayed in showcases or back rooms, they are merely "*realist* studies [that] leave nothing to be desired"[77] (emphasis added). Art critics voiced the same rhetoric of repulsion, disgust, and public offense in reviewing Manet's *Le Déjeuner sur L'herbe* and *Olympia*.

In describing the seated nude in *Le Déjeuner sur L'herbe*, critics agreed that she was "the most vulgar of nude women."[78] One critic observed, "[A] commonplace woman of the demimonde, as naked as can be, shamelessly lolls between two dandies dressed to the teeth." However, the critic's consternation was not only directed to the nude—her nudity contaminated her male counterparts. He thus berated the young men for their fake performance of manliness, signified by their dalliance with the naked woman-qua-prostitute: "These [dandies] look like schoolboys on a holiday, perpetrating an outrage *to play the man*"[79] (original emphasis in the French text). No doubt, the inference of illicit sex compelled one critic to call the subject "fairly scabrous."[80]

When *Olympia* was exhibited in the official Salon two years later, the painting became the discursive site for even more explicit obscenity charges.[81] While critics identified Olympia as a prostitute,[82] they further invoked her moral depravity by associating Manet's abbreviated modeling, indicated by loosely painted black contours, with actual physical dirtiness (as Courbet's nude bather had been).[83] Rehearsing Pinard's charges against the Realists' exaggerated vulgarity, critics labeled Manet a Realist. As one critic asked, "[W]hy do the Realists choose unclean women as their models, and having done so, reproduce even the filth which clings to their contours?"[84] Most expressive of the correlation between filth and lasciviousness was the critic from *Le Monde Illustré*: "The *auguste jeune fille* is a courtesan, with dirty hands and wrinkled feet;... her outlines are drawn in charcoal and her greenish, bloodshot eyes appear to be provoking the public, protected all the while by a hideous Negress."[85]

Despite these detractors, in the Realist critics' eyes, the same salacious attributes signaled Manet's elevated moral stature (as was the case for Flaubert and Baudelaire). We have seen that Zola hailed *Olympia* as an *académie de la rue*. As he stated, *Olympia* is nothing but:

> some model whom Manet has... copied just as she was. And everyone exclaimed that this nude was indecent.... [H]ere in the flesh is a girl whom the artist has put on canvas

in her youthful, slightly tarnished nakedness. When other artists correct nature by painting Venus, they lie.

For Zola, Manet's morality—his refusal to lie—stands above that of his detractors:

Manet asked himself why he should lie. Why not tell the truth? He has introduced us to Olympia, a girl of our own times, whom we have met in the streets pulling a thin shawl of faded wool over her narrow shoulders.[86]

Zola's Realist defense of Manet reverses the pejorative rhetoric lashed upon Olympia and her photographic counterpart. While Zola engaged in the infamous scandal to deride it, Baudelaire engaged in flaunting it. All the attributes that make Olympia "real" and salacious—her nudity, gaze, pose, and setting—are equally found in Baudelaire's censored poem, "Les Bijoux" ("The Jewels").[87] As we will see, the stereoscopic nude contaminates both the poem and its censor, shaping a discourse that would saturate Manet's *Le Déjeuner sur L'herbe* and *Olympia*.

Pinard foregrounded "The Jewels" as the first among the six censored *Fleurs du Mal* poems he cited. He condemned the poem as "constitut[ing] lascivious painting, offending public morality."[88] Furthermore, during the trial, when Pinard re-titled the poem, "Naked Woman, Trying on Poses Before Her Fascinated Lover,"[89] he (inadvertently) provided Manet with a visual cue and an iconographic image to compare the painted Olympia with the poem. Indeed, the new title invokes Olympia's frontal gaze and the seductive image her lover gazes upon. In "The Jewels," the narrator responds to the sight of his lover, nude but for her jewelry, who seductively poses on the bed with her gaze fixed directly on him. Little commentary is necessary to see the visual analogies between the painted image and the poem's picture in lines 1–16.

Knowing my heart, my dearest one was nude,
Her resonating jewelry all she wore,
Which rich array gave her the attitude
Of darling in the harem of a Moor. /
When dancing, ringing out its mockeries,
This radiating world of gold and stones
Ravishes me to lovers' ecstasies
Over the interplay of lights and tones. /
Allowing love, she lay seductively
And from the high divan smiled in her ease
...
A tiger tamed, her eyes were fixed on mine,
With absent air she posed in novel ways,
Whose candor and lubricity combined
Made charming all her metamorphoses.[90]

As noted earlier, the directness of the model's gaze and its erotic function were especially pronounced in stereoscopic photographs from the early 1850s.

Ironically, Baudelaire was among the cultural critics of pornographic stereoscopy. In 1859, a year of increased arrests for pornographic photography, Baudelaire decried the profusion of stereoscopy as a popular pastime. In his notorious diatribe against photography, written in his Salon of 1859 review (discussed in Chapter 4), he suggested an analogy between the public's obsession with photography, especially pornographic stereography, and masturbation.[91] Shortly after the invention of stereography, Baudelaire reports:

> [A] thousand pairs of greedy eyes were glued to the peepholes of the stereoscope, as though they were the skylights of the infinite. The love of obscenity, which is no less deep-rooted in the natural heart of man than the love of himself, could not let slip such a glorious opportunity for self-satisfaction.[92]

Notwithstanding Baudelaire's (apparent) distress, he was subject to, and he exploited, the psychic and visual impressions of stereoscopic pornography.

By considering Baudelaire's "The Jewels" and Manet's *Olympia* alongside stereoscopic images, the features that Pinard considered most obscene evoke the salient anatomical effects of stereoscopic photography. Again, these effects include harsh frontal lighting that brings bent or raised arms and legs, buttocks, or breasts into strong relief and bold projection. These anatomical parts also captured Pinard's attention in "The Jewels."

During the Baudelaire trial, Pinard cited lines 17–28 from "The Jewels" as those that "constitute lascivious painting." For our purpose, the first six lines are noteworthy regarding a photographic analogy with *Olympia*:

> Her shoulders and her arms, her legs, her thighs,
> Polished with oil, undulant like a swan,
> Passed by my tranquil and clairvoyant eyes;
> Then belly, breasts, those clusters on my vine
> Came on, tempting me more than devils could
> To break the peace my soul claimed as its own.

These lines are provocative for cataloging the figure's seductive body parts and how they pictorially function in the poem. In fact, we might mistake Baudelaire's "picture" for a stereoscopic nude. Like the protruding limbs, hips, and breasts that project from actual stereoscopic photographs, "the interplay of lights and tones" upon the poem's figure brings the lover's sexuality into stark relief. As readers, we are lured into the poem by the figure's nudity and witness the lovers' intimate moment. Likewise, as viewers of the painting, we are thrown into the privacy of a prostitute's boudoir, colluding with her waiting client's sexual desire.

By invoking the pornographic stereoscopic gaze, Manet not only contaminated the Salon and its classical ideals but also fomented the cross-contaminations of the Realist debate. In doing so, he forced the viewers of *Le Déjeuner sur L'herbe* and *Olympia* to reflect upon their predisposition for perceiving Realism as obscene. This visual coding also compels the reader/viewer's voyeuristic gaze, making everyone complicit with the painter and photographer, as well as the censor and litigator, in the construction of the obscene. In his painting *Nana* (1877), Manet pushed this cross-contamination to the extreme (Figure 6.9).

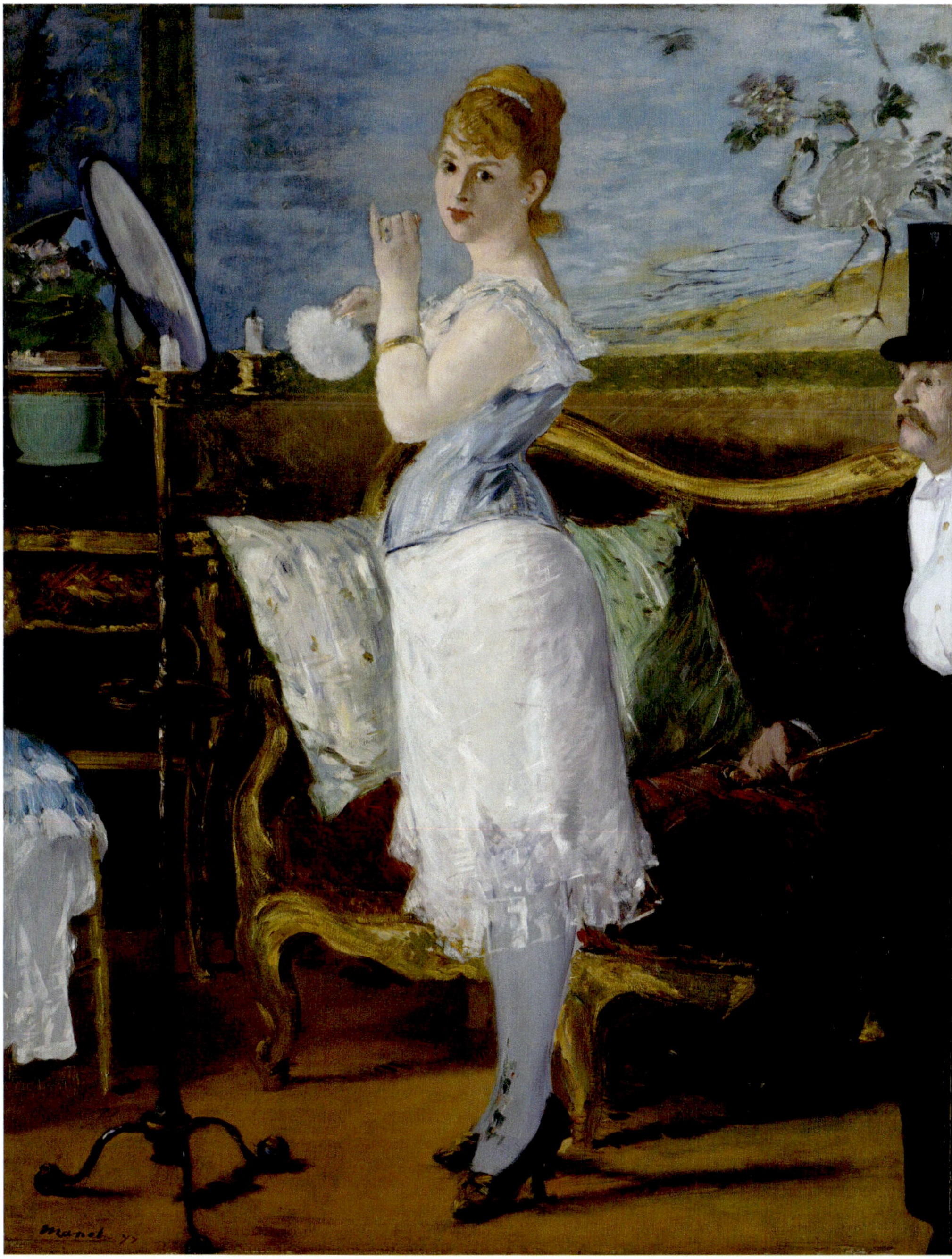

Figure 6.9 Édouard Manet, *Nana*, 1877, oil on canvas, 45 1/4 x 60 5/8 in. Hamburger Kunsthalle, Hamburg. Photo: bpk/Elke Walford.

Enduring Cross-Contaminations: Manet's *Nana* in the Public Imaginary

In the painting *Nana*, a young woman stands before the mirror applying makeup but turns her head to pose before an implied but absent viewer.[93] Her pose and props recall contemporary photographic conventions and the photographer-qua-viewer's voyeuristic gaze. As seen in a photograph by Louis-Camille d'Olivier, the model often posed in front of a full-length "psyche" mirror, reflecting multiple views of her figure, and stared directly at the photographer (Figure 6.7). In Manet's painting, the viewer's gaze replaces the photographer's gaze. But Nana's gaze is doubly reflected. Her brazen self-image is mirrored in the viewer's gaze and her companion's gaze. In each case, our/his gaze reproduces Nana's self-image as it reflects back to her, creating a narcissistic circuit of self-absorption and domination. In this respect, Nana reifies Emma Bovary as a "strange androgyne [who] has retained all the seductions of a virile soul in a charming feminine body" (Baudelaire).

Although the 1852 censorship laws were repealed in 1870, the public reception of *Nana* marks the enduring efficacy of the legal, literary, and visual cross-contaminations in constructing obscenity in the social imaginary. J.-K. Huysmans' 1877 review of Manet's *Nana* mirrors this excess.[94] In his review, Huysmans conflated the painting and Zola's novel *L'Assommoir* (1877), in which the courtesan Nana is first featured.[95] When Huysmans entered the fray, the painting enjoyed a *succès de scandale*. Rejected by the official Salon because of its lewdness, *Nana* was prominently exhibited in a storefront window on a busy Parisian boulevard. In this context, *Nana* elicited the same frenzied responses as the bare-fleshed female portraits in the photographers' shop windows. Huysmans states, "From morning to night crowds gather before this canvas, and it draws indignant screams and laughter from a mob stupefied by the canvases that [the Academic painters] smear and exhibit at the salons each year." For Huysmans and the gawkers, the association between Manet and Zola was immediate: "Here is the subject of the painting: Nana, the Nana of *L'Assommoir*." Huysmans lauded Manet for presenting "the quintessential *fille*."[96] Following eroticized descriptions of Nana's undergarments and anatomy, Huysmans pronounced:

> Manet was right to present us, in his *Nana*, with one of the perfect samples of the type of prostitute that his friend and dear master, Emile Zola, is about to depict for us in one of his next novels. Manet has made her appear as such as she will be forced to be, with her complicated, knowing vice, her extravagance, and her salacious luxury.

In Huysmans' description of the painting, extra-textual references exceed the image, conflating an extant and a future novel, *Nana* (1880), the popular imaginary of prostitution, and the memory of censored obscenities. Indeed, Huysmans was admittedly seduced: "If I were not afraid of hurting the prudishness of readers, I would say that M. Manet's picture feels like an unmade bed, it feels, in a word, like what he wanted to represent—the show-off and hussy." Subsequently, in Zola's novel, Nana also stands half-dressed before the mirror, similarly seducing and corrupting her lover. These visual-textual cross-contaminations revert the seductive gaze to the viewer/reader to locate the perception of obscenity in the eyes of the beholder. At this level of nineteenth-century spectatorship, the efficacy of pictorial realism exceeds the effectiveness of legal restrictions. Whether sustained in text or image, the visual construct of obscenity evades suppression in the public imagination and fortifies the cultivation of pornographic consumption.

Manet's Ironic Baudelairean Gaze: Disjunction and Desire in Modern Life

In conclusion, by reviewing Manet's engagement with stereography and its ocular properties, we can further align his contamination of the idealized Venus with the ruptures and fragmentation that define Baudelairean modernity. Manet first experimented with stereoscopic effects in *Surprised Nymph* (1861; Museo Nacional de Bellas Artes, Buenos Aires).[97] As in a stereoscopic view, the figure is pushed to the frontal plane, and the flatly painted landscape intensifies its relief effect. This backdrop replicates the photographer's actual studio prop. This compositional disjunction closely corresponds with the spatial disruptions of the stereoscopic view that Jonathan Crary describes as "a derangement of the conventional optical cues. [...] If perspective implied a homogeneous and potentially metric space, the stereoscope discloses a fundamentally disunified and aggregate field of disjunct elements."[98]

An even more complex juxtaposition of three-dimensional figuration and flattened backdrop provides the basis for re-viewing *Le Déjeuner sur L'herbe* and its affinities with stereoscopic imagery. These affinities are general regarding a rural scene typical of the medium and specific regarding compositional disparities (Figure 6.10). In the latter case, the foreground figures are uniformly lit and surrounded by a spatial vacuum lacking orthogonal and atmospheric depth. Broad bands of greens and ochres and vertical swathes of black-brown tree trunks appear flat but tonally modulated by the open sky at the apex of the figurative pyramid. Given the figures' scale and illumination, they appear disproportionate to the landscape setting. What is most problematic is the contradiction between the female figure in the foreground, which appears cutout or flat, and the background figure, which is fully modeled and projects from a sun-drenched landscape

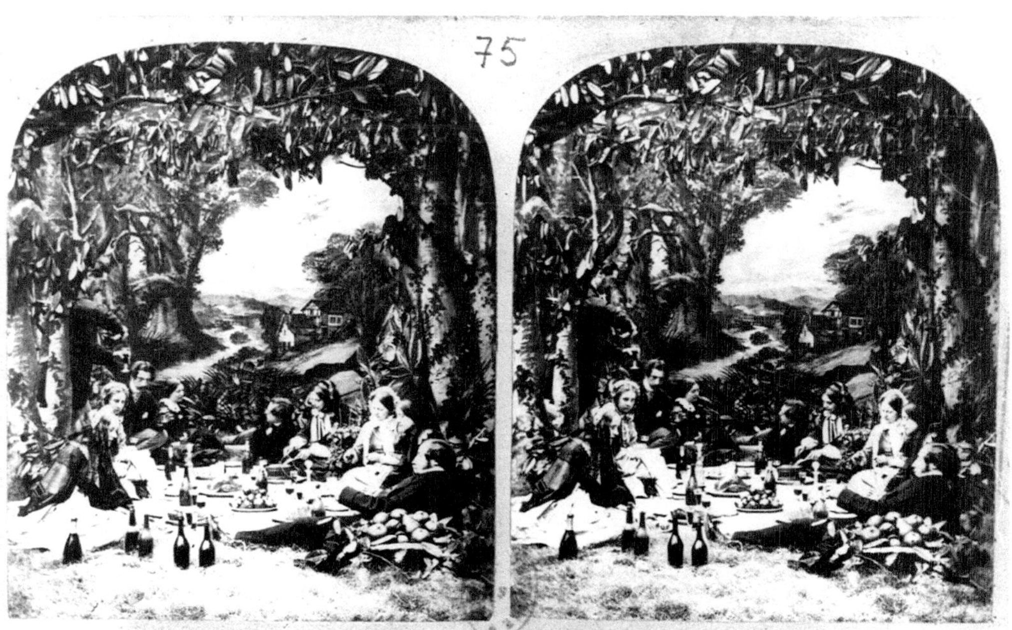

Figure 6.10 Charles Gaudin, Untitled (figures in landscape), stereoscopic photograph, 1850s, unknown dimensions. Bibliothèque nationale de France, Paris.

background. This discrepancy can also be explained as translations of stereoscopic views, which force our reading of the composition as a "fundamentally disunified and aggregate field of disjunct elements."

I suggest that Manet combined two stereoscopic effects in a single painting, exaggerating the effects of "vertiginous uncertainty about the distance separating forms."[99] As in the figure of Olympia, the high-toned surfaces and minimal modeling of the seated nude in *Le Déjeuner sur L'herbe* replicate the raking light of stereoscopic nudes that occupy the picture foreground (Figures 6.5 and 6.11). While this lighting flattens the torso and limbs parallel to the camera lenses, it renders relief-like projections (toward the camera) of her elbow, facial features, and left-hand fingers. (The same effects are found in the representation of her male companions, especially in the pointing hand of the right-side figure and the left-side figure's right hand placed behind the nude's buttocks.) This is one cutout effect of stereoscopic photography; however, the other renders foreground objects in high relief, as seen in the background figure. Viewed separately from the foreground, she appears in her own "disturbing[ly] palpable [...] airless space." Again, the model is posed in front of a planar landscape backdrop, and there is no logical spatial distance between the two. Further evidence of Manet's borrowings from stereoscopic photography is that this half-draped figure, framed by trees and crouching in a pond, clearly refers to a series of photographs titled *Baigneuse à L'étang* (Bather at a Pond) attributed to Billon,[100] a photographer who specialized in both *académies* and pornographic nudes (Figure 6.12). Finally, when the distant figure is viewed within the entire landscape composition, another technological indicator comes to the fore: by rendering the most distant figure in high relief and at dead-center, Manet overtly supplanted orthogonal perspective with stereoscopic projection.

Manet's references to stereoscopic images destabilize the painting's compositional unity and the viewer's gaze. Indeed, our attention keeps shifting from background to foreground with no transition. In this respect, Manet was re-presenting the actual optical experience of viewing stereoscopic images. As the inventors made known, two similar (but not exact) images fuse into one because of the convergence of the optic axes; this

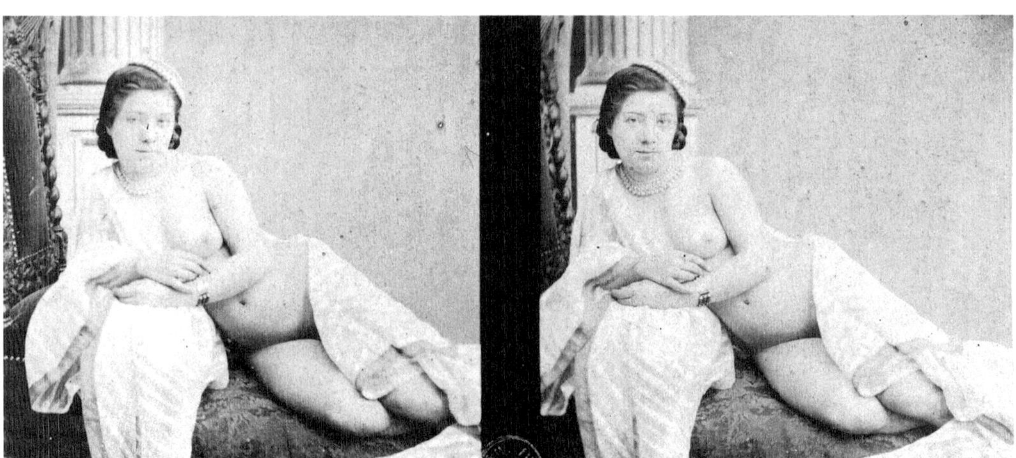

Figure 6.11 Félix Jacques Antoine Moulin, Untitled (nude reclining on a sofa), stereoscopic photograph, c. 1855, unknown dimensions. Bibliothèque nationale de France, Paris.

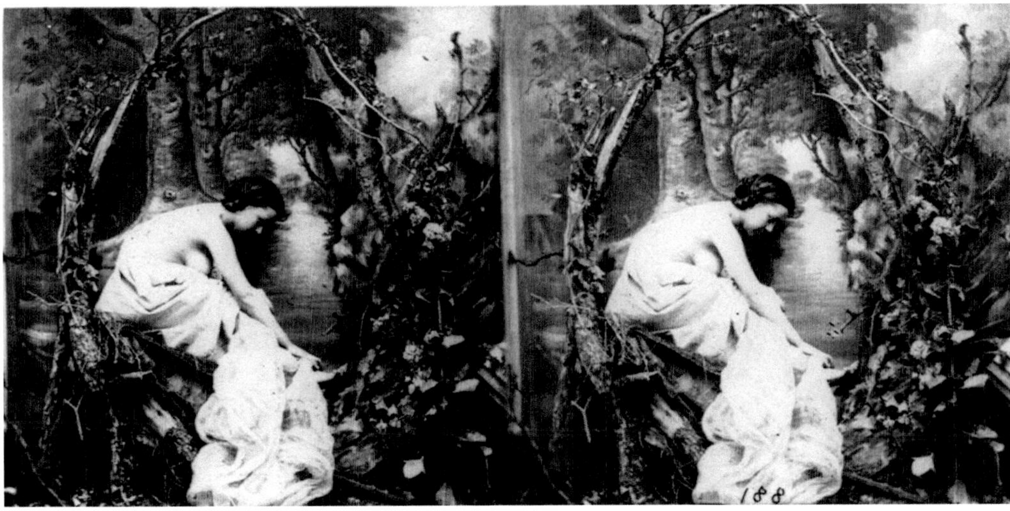

Figure 6.12 Billon (Alfred-François Cordier, called Billon-Daguerre), *Baigneuse à L'étang* (Bather at a Pond), stereoscopic photograph, 1861, unknown dimensions. Bibliothèque nationale de France, Paris.

convergence, however, takes place over time, "uniting in rapid succession, similar points of the two pictures."[101] Crary further states, "[N]o matter how 'vivid' the quality of illusion, [...] the composite, synthetic nature of the stereoscopic image could never be fully effaced."[102] Thus, the stereoscopic experience can be characterized as a doubling of opposite effects: disjunctive, divergent, unifying, and convergent.

This sustained doubleness distinguishes Manet's use of photography from his contemporaries and marks his originality as a Baudelairean painter of modern life. As discussed in Chapter 1, aesthetic dualism and ironic self-reflection issue from Baudelaire's modernist credo formulated in "The Painter of Modern Life." Baudelaire defined modernity as a constant state of flux between "the transient, the fleeting, the contingent" on the one hand and "the eternal and immutable" on the other.[103] The artist can achieve an aesthetic dualism by rendering the familiar unfamiliar—"strange," "mysterious," and "ever new"—and by "elud[ing] the rules and analyses of the school."[104] As demonstrated in *Le Déjeuner sur L'herbe* and *Olympia*, for Manet, ironic parody became the means to achieve this Baudelairean effect. Thus, we can re-view these Salon masterpieces—a term then applied to them—through the lens of ironic parody as a subversive yet restorative artistic device. This approach follows Linda Hutcheon's definition of modern parody as inherently double-coded since it inscribes both "continuity" with and "critical distance" from the old (targeted) text in the new text.[105] Parody's dualism, in turn, induces an ironic attitude or situation for the viewer/reader and fosters critical self-reflection. This ironic gaze thereby advances an ongoing process of artistic renewal. Most compelling for establishing Manet's Baudelairean modernity is how the artist doubled parody's ironic effect by re-presenting old masterpieces through the lens of modern photography.

By inscribing both continuity with and critical distance from Titian's *Venus of Urbino* and Raphael's *River Gods* in his new masterpieces, Manet effectively subverted the viewer's aesthetic expectations and, thereby, promoted critical self-reflection. Manet's subversive strategies also engage the photographic medium for ironic effects. In viewing

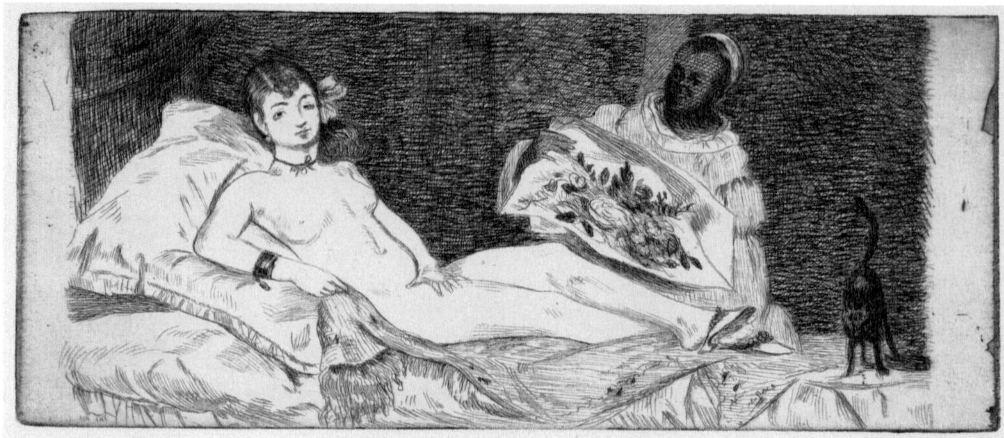

Figure 6.13 Édouard Manet. *Olympia*, 1857, etching, second state of six (published plate), 3 3/8 x 8 1/8 in. (plate). Metropolitan Museum of Art, New York. The Elisha Whittelsey Collection, The Elisha Whittelsey Fund, 1983.

Le Déjeuner sur L'herbe and *Olympia*, we slip and slide between pictorial harmony and fragmentation, an effect replicating the perceptual slippage while gazing at a nude through a stereoscope. The (photographic) directness of the models' gaze compels attraction and distraction and complicates our thoughts and senses. Initially, we are locked into erotic complicity with the directness of their gaze; subsequently, we are distracted from their sensual provocations by the lush brushstrokes and pigments that their presence offers.

Manet's ironic parody signals how technological reproduction obtains an aura of originality, a modernist paradox that Manet inscribed in a portrait of Émile Zola (1868) (Figure 9.4). In that work, the artist painted his own etching of *Olympia* (1867), who here turns her seductive gaze to the sitter (Figure 6.13). (Manet's etching was reproduced in Zola's 1867 pamphlet on Manet, whose blue cover appears in the painting's still-life.)[106] Zola's gaze, in turn, had sanctioned Manet's modernity. In his review of *Olympia* and *Le Déjeuner sur L'herbe* in 1867, the critic recognized a Baudelairean dialectic between the eternal and the transitory that marked artistic originality. This modernity extended the rupture of the Parisian cityscape to the rupture of the canonic female body and persona.

Notes

1 Earlier versions of this chapter were published in Lauren S. Weingarden, "Cross-contaminated Bodies: Constructing Obscenity in the Public Imaginary," in *Pornography: Interdisciplinary Perspectives*, ed. Frank Jacobs (Bern, CH: Peter Lang, 2019): 85–108; and "The Photographic Subversion: Benjamin, Manet and Art(istic) Reproduction." *Aletria: Revista de Estudos de Literatura* 14, no. 2 (December 31, 2006): 224–45 [online], https://doi.org/10.17851/2317-2096.14.2.224-245; also published in *InterCulture* (extinct online journal), 51 (Spring 2008): 1–24.

2 Obscenity and pornography solicit slippery definitions. This study is premised on the idea that interpretations of obscenity and pornography are in the eye of the beholder/adjudicator. Interpretative slipperiness is also a matter of linguistic translation. The original texts analyzed here are English translations of French. In the French texts, the word "pornography" is absent. Instead, "obscenity" is used to denote both that which is erotic (sexually titillating) but not illegal and that which intentionally elicits sexual excitation and, therefore, is deemed illegal.

Thus, pornography is an interpretive rubric culled from our contemporary English lexicon. Pornography here refers specifically to images that denote an explicit intent of sexual arousal, as explicit as the sexual acts or body parts exposed, and an explicit intent to sell these images to stimulate the consumer's sexual arousal. See Maxence Rodemacq, "L'industrie de L'obscénité à Paris (1855–1930)," *Romantisme* 167, no. 1 (2015): 13–6. In my interpretive discussion, obscenity and pornography are largely interchangeable, even if they are not so in the cited French translations. But here's the catch: even though the word pornography is absent in legal and critical texts, the descriptive words used to characterize obscene images or texts are often more sexually nuanced than the generic label the speakers avoid.

3 *Le Déjeuner sur L'herbe* was rejected from the official 1863 Salon but exhibited in the Salon des Refusés. The rejected artists, two-thirds of those who entered the juried competition, protested and were granted the alternative Salon, uniquely held in that year. *Olympia* was accepted for and exhibited at the 1865 Salon.

4 For a detailed history of these photographic practices, see Elizabeth Anne McCauley, "Braquehais and the Photographic Nude," in *Industrial Madness: Commercial Photography in Paris, 1848–1871* (New Haven/London: Yale University Press, 1994): 149–94.

5 Nicolas Harrison, *Circles of Censorship: Censorship and its Metaphors in French History, Literature and Theory* (Oxford: Clarendon Press, 1995): 52; George D. McKee, "Jurisprudence: Crimes of the Image," Chronology to accompany online resource *Image of France 1795–1880*, published online September 21, 2014, web.archive.org, https://web.archive.org/web/20140921044520/https://sites.google.com/site/imoffrdocs/home/jurisprudence. All subsequent references to legislation are from this chronology and from George D. McKee, "Legislation," Chronology to accompany online resource *Image of France 1795–1880*, published online October 7, 2020, web.archive.org, https://web.archive.org/web/20201007192519/https://sites.google.com/site/imoffrdocs/home/legislation.

6 Article 20, of September 9, 1835.

7 On February 17, 1852, Article 22 repeated Article 20 of September 9, 1835. McCauley includes photography in this decree, but McKee does not; see McCauley, *Industrial Madness*, 155.

8 Nineteenth-century viewers engaged Titian's *Venus of Urbino* and Goya's *Nude Maja*, two of Manet's "masterpiece" models, in the discourse on obscenity. I would argue that Manet deliberately invoked their discursive meanings in *Olympia*. For Titian's painting, see Clark, *The Painting of the Modern Life*, 86, 93. Salon of 1865 critic Alfred Sensier (pseud. Jean Ravenel) bemoaned Manet's obvious borrowing from Goya for *Olympia* and hinted at the tainted reputation of Goya's *Nude Maja*; see Clark, *The Painting of the Modern Life*, 139–40, 143. As Janis A. Tomlinson explains, Goya's *Maja*s were subject to official censorship. In 1814, the *Clothed Maja* and *Nude Maja*, formerly in the collections of Manuel Godoy, were sequestered and seized by the Inquisition as among "the most provocative and obscene paintings" which tried to "corrupt our customs;" "Goya and the Censors," in *Suspended License: Censorship and the Visual Arts*, ed. Elizabeth C. Childs (Seattle: University of Washington Press, 1997): 142. That the two *Maja* paintings were well-known within Manet's literary-artistic circle is evidenced by Baudelaire, who had seen copies of them at an art dealer's gallery in 1859. Baudelaire also wrote to photographer Félix Nadar that Théophile Gautier had seen the copies in Spain and "begged" Nadar to photograph and, if possible, to buy them; Juliet Wilson-Bareau, "Manet and Spain," in Gary Tinterow and Geneviève Lacambre, *Manet/Velázquez: The French Taste for Spanish Painting* (New York/New Haven: Metropolitan Museum of Art/Yale University Press, 2003): 228. According to Charles Yriarte, in the mid-1860s, the *Nude Maja*, along with the *Clothed Maja* housed at the Royal Academy of San Fernando, were still hidden from public view "sur prétexte d'une pruderie déplacée" ("under the pretext of a displaced prudery"); *Goya* (Paris: Plon, 1867): 89.

9 See Weingarden, "The Photographic Subversion," 224–45.

10 McCauley, *Industrial Madness*, 97.

11 McCauley, *Industrial Madness*, 158, 160. Other histories of erotic photography have expanded upon McCauley's findings; see Sylvie Aubenas and Philippe Comar, *Obscénités: Photographies Interdites d'Auguste Belloc* (Paris: Bibliothèque nationale de France, 2001). See also Serge Nazarieff, *Early Erotic Photography* (Köln: Taschen, 1993).

12 André Rouillé and Bernard Marbot, *Le Corps et Son Image: Photographies du Dix-neuvième Siècle* (Paris: Contrejour, 1986): 49.

13 McCauley, *Industrial Madness*, 157–8. As Rodemacq shows, in the 1850s, new photographic techniques and formats, including the carte-de-visite, also contributed to the growth of pornography. In 1860, stanhopes or stanoples were introduced in the form of microscopic photographs mounted on small magnifying glasses. The very small size reduced the costs of production and purchase ("L'industrie de L'obscénité à Paris," 19).

14 This restrictive designation is marked on the original photograph.

15 McCauley, *Industrial Madness*, 156–7. As records indicate, illicit erotic and pornographic objects, from photographs and stereographs to carved pipes, were widely available, which suggests that to identify *Olympia*'s kinship with such items would be either to state the obvious or to admit collusion with the underground market. I have confirmed this directly during my review of the police register at the Paris headquarters of the Prefecture de Police. As Rodemacq argues, during the Second Empire, the "industry of obscenity" manifests in the diverse media in which obscene goods were produced together with the vast quantities of their production. In fact, "obscenity was adapted to nearly thirty different visual forms in 1868" ("L'industrie de L'obscénité à Paris," 16).

16 Anon., "Biographie: M. Moulin," *Revue Photographique* I (Dec. 1855): 26–7; cited in McCauley, *Industrial Madness*, 156.

17 Maxime Du Camp, *Les Beaux-arts à L'exposition Universelle et Aux Salons de 1863, 1864, 1865, 1866 et 1867* (Paris: Vve J. Renouard, 1867), 11, 29, 31–5; partially trans. and qtd. in McCauley, *Industrial Madness*, 163. As McCauley notes here, in Du Camp's 1863 salon review, "Cabanel's *Birth of Venus* and [Paul-Jacques-Aimé] Baudry's *Pearl and the Wave* were criticized as *académies* surrounded by allegorizing but unbelievable props."

18 Émile Zola, "A New Style of Painting: Édouard Manet," qtd. in and trans. Linda Nochlin, *Realism and Tradition in Art, 1848–1900; Sources and Documents*, ed. Linda Nochlin (Englewood Cliffs, NJ: Prentice-Hall, 1966): 77.

19 Clark, *The Painting of the Modern Life*, 96. Clark discusses the critical reception, in the form of suppression, of Olympia as a sexualized "naked" contemporary prostitute and her recognizable working-class social status, as opposed to the frequently discussed contemporary mythological, sexualized nudes, such as Cabanel's *Birth of Venus*. Phylis A. Floyd, however, argues that Olympia was a high-class courtesan, perhaps Napoléon III's mistress ("The Puzzle of Olympia," *Nineteenth-Century Art Worldwide* 3, no. 1 (2004), https://www.19thc-artworldwide.org/spring04/285-the-puzzle-of-olympia).

20 Hand-painted erotic photographs are an even closer mnemonic match for *Olympia*; a color image of Belloc's, *Jeune Femme Allongée sur un Sofa* (Fig. 6.5) can be accessed at Google Arts and Culture https://artsandculture.google.com/asset/jeune-femme-nue-allong%C3%A9e-sur-un-sofa-anonyme/nQENdCa2NAj_vA); see also Félix Jacques Antoine Moulin, *Nude*, 1851–1853, hand-painted stereoscopic daguerreotype, National Gallery of Art, Washington, DC (https://www.nga.gov/collection/art-object-page.212275.html).

21 I first used the phrase "female gaze" in a conference paper, "The Photographic Subversion: Benjamin, Manet, and Art(istic) Reproduction," presented at the 2004 Association of Art Historians Annual Conference, Leeds, England, and was subsequently published under the same title (see note 1). Temma Balducci also employs the phrase "female gaze" in *Gender, Space, and the Gaze in Post-Haussmann Visual Culture: Beyond the Flâneur* (London/New York: Routledge), 2017. In Chapter 2, Balducci uses the phrase to deconstruct the "male gaze," which has dominated Baudelairean studies of the flâneur and rendered female looking powerless. Balducci examines Manet's paintings, among others, to reconstruct the female acts of looking in the (bourgeois) social and (urban) public spheres during the late Second Empire and the Third Republic, "thereby allowing a space to theorize women's looking at a time when they lacked such power" (67). While Balducci's argument effectively deconstructs both the male gaze and its correlated bourgeois social structure and lends agency to the female "flâneuse," she remains engaged with the critical discourse she critiques. My use of the term female gaze renders the model-qua-prostitute an active agent, endowed with erotic power that ensnares the male viewer and renders him complicit in acts of obscenities.

22 McCauley points out that the model's pose determined whether the photograph was obscene. She also discusses how "explicit sexual invitation," registered in gestures and glances, transformed *académies* into erotic photography (*Industrial Madness*, 168, 170, 172).

23 See Weingarden, "The Photographic Subversion." In his essay, "The Work of Art in the Age of Its Technological Reproducibility," Benjamin's main concern is ideological. For that reason, he sanctioned the mechanical reproduction of art, as opposed to the artwork's "aura of authenticity," as a mandate for social change and the de-centralization of cultural authority. See Walter Benjamin, "The Work of Art in the Age of Its Technological Reproducibility: Second Version," in *Walter Benjamin: Selected Writings. Volume 3: 1935–1938*, eds. Howard Eiland and Michael W. Jennings, trans. Edmund Jephcott and Harry Zorn (Cambridge, MA: Belknap, 2002): 101–33.

24 Benjamin, "Work of Art," 100–1.

25 Benjamin, "Work of Art," 103.

26 Walter Benjamin, "Some Motifs in Baudelaire" (1939), in *Walter Benjamin: Selected Writings. Volume 4: 1938–1940*, eds. Howard Eiland and Michael W. Jennings, trans. Edmund Jephcott (Cambridge, MA: Belknap, 2006): 338.

27 Charles W. Haxthausen, "Reproduction/Repetition: Walter Benjamin/Carl Einstein," *October* 107 (Winter 2004): 53.

28 For the history and techniques of stereoscopic photography, see Helmut Gernsheim, *The Rise of Photography 1850–1880: The Age of Collodion* (London/New York: Thames and Hudson, 1988): 61–9; Jonathan Crary, *Techniques of the Observer: On Vision and Modernity in the Nineteenth Century* (Cambridge: MIT Press, 1999): 116–36; and Denis Pellerin, *La Photographie Stéréoscopique Sous le Second Empire* (Paris: Bibliothèque nationale de France, 1995).

29 Benjamin, "The Work of Art," 104–5. At the same time, these photographic images of natural objects function like the "technological reproduction" of artworks. Benjamin claimed that "by permitting the reproduction to reach the recipient in his or her own situation, it actualizes that which is reproduced." From this position, the stereoscope also responds to Benjamin's tenet that "the desire of the present-day masses to 'get closer' to things, [...] to get hold of an object at close range, in an image, or better, in a facsimile, a reproduction." Crary also quotes Benjamin's essay to note that the stereoscope was a realization of this "present-day" need to "get closer to things" (*Techniques of the Observer*, 127).

30 For the history of stereoscopic photography's appropriation of erotic/pornographic *académies* and the censorship imposed on the trade, see Pellerin, *La Photographie Stéréoscopique*, 30–42. Crary also notes "the very close association of the stereoscope with pornography" (*Techniques of the Observer*, 127). Several erotic/pornographic stereoscopic images that closely resemble Olympia's pose and gaze are reproduced in Aubenas and Comar, *Obscénités*, and Nazarieff, *Early Erotic Photography*.

31 See Clark's discussion of the critics' complaints about the technical errors in the painting, including the flexed, strongly modeled hand, in *The Painting of Modern Life*, 133–9, and Floyd's discussion in "The Puzzle of Olympia." McCauley identifies Braquehais's "depiction of pubic hair" and the model's clutched hand in his *Academic Study—No. 6* as "an allusion to masturbation," comparable to Olympia's (*Industrial Madness*, 173–4, fig. 66).

32 See Crary's description of the stereoscope's elusive "reality effect," of which three-dimensionality is a part, in *Techniques of the Observer*, 124–5.

33 Qtd. in Pichois with Ziegler, *Baudelaire*, 233.

34 As Harrison observes, "the crime, [in fact], seems to have been realism" as a mode of representation (*Circles of Censorship*, 48, 51).

35 Given the repetition of pictorial analogies, I have translated Pinard's use of "*peindre*" and "*peinture*" as "to paint" and "painting," respectively, as opposed to "portray" and "portrayal" as used by literary historians. Pinard's analogies between literary and pictorial realism in both indictments are explicit and pejorative. In the Flaubert trial, he used variations of "*peindre*" and "*peinture*" 21 times; in the briefer Baudelaire trial, he used these words five times. Pinard also frequently uses the word "tableau(x)" when describing scenes from the novel and poems. In the Flaubert trial, Pinard used the word "tableau(x)" 13 times; in the briefer Baudelaire trial, he used the word three times.

36 Gustave Courbet, "Realist Manifesto (1855)," in Nochlin, *Realism and Tradition*, 33–4. Originally published as Gustave Courbet, *Exhibition et vente de 40 tableaux et 4 dessins de l'œuvre de M. Gustave Courbet, avenue Montaigne* (7 Champs-Elysées, Paris, 1855), n.p.

37 The Exposition Universelle of 1855 was an international exhibition held on the Champs-Élysées in Paris from May 15 to November 15, 1855.

38 Contemporary caricatures accent the nude's deformities. See, for example, Bertall, "Le Salon Depeint et Dessine par Bertall"/"La Terrible Savoyarde, par Courbet," *Le Journal pour Rire* 91 (June 25, 1853): 1; and Cham, "Prémiere Promenade à L'exposition, par Cham"/"Une Baigneuse, par M. Courbet," *Le Charivari* 22, no. 149 (May 29, 1853): [3]; the full caption: "Femme de 45 ans sur le point de se laver pour la première fois de sa vie, dans l'espoir d'apporter un soulagement à ses varices."

39 "Hottentot Venus" was the pejorative moniker ascribed to Sarah Baartman (1790–1815). Baartman was one of two Khoikhoi women who were exhibited as freak show attractions in nineteenth-century Europe due to their large buttocks. As such, Baartman was at the center of a Eurocentric racist discourse and "the object of the 'ethnopornographic' gaze;" Andrew P. Lyons, "The Two Lives of Sara Baartman: Gender, 'Race,' Politics and the Historiography of Mis/Representation," *Anthropologica* 60, no. 1 (May 2018): 337–46.

40 Théophile Gautier, "Salon de 1853: Onzième Article," *La Presse* 17 (July 21, 1853): [2]; qtd. in Dominique de Font-Réaulx, et al., *Gustave Courbet* (New York/Ostfildern: Metropolitan Museum of Art/Hatje Cantz, 2008), 445–6.

41 Gautier, "Salon de 1853," 445.

42 As Pierre-Joseph Proudhon reported, "The Empress had just seen Rosa Bonheur's *The Horse Market*, and had been informed of the draft horses' breed, the French Percheron, a workhorse known for its strong rump. When she reached Courbet's *Bathers*, The Empress could not restrain a cry of surprise: 'Is she also a Percheron?'." *Du Principe de L'art et de sa Destination Sociale* (Paris: Garnier Frères, 1865), 214–5.

43 Georges Riat, *Gustave Courbet, peintre* (Paris: H. Floury, 1906): 105.

44 Eugène Delacroix, *Journal*, vol. 2, ed. Paul Flat and René Piot (Paris: Plon, 1893), 158–160.

45 Ernest Pinard, "The Public vs. M. Gustave Flaubert: Speech of the Prosecuting Attorney, M. Ernest Pinard," in Gustave Flaubert, *The Complete Works of Gustave Flaubert*, vol. 2 (New York/London: M. W. Dunne, 1904): Appendix: 3–4; and Ernest Pinard, "Procès Intenté à L'auteur: Réquisitoire," in Gustave Flaubert, *Oeuvres Complètes de Gustave Flaubert: Madame Bovary*, t. 2 (Paris: L. Conard, 1910): 558.

46 Pinard, "Speech Prosecuting Attorney," 3; "Réquisitoire," 559.

47 Pinard, "Speech Prosecuting Attorney," 10; "Réquisitoire," 562.

48 Pinard, "Speech Prosecuting Attorney," 10; "Réquisitoire," 562.

49 Pinard also used the metaphor of portrait paintings when he described the novel's individual characters.

"Without doubt, it is the husband who begins and who terminates the book; but the most serious portrait of the work, the one that illumines the other paintings, is that of Madame Bovary" ("Speech Prosecuting Attorney," 5; "Réquisitoire," 559). The defense attorney, Jules Senard, and the ruling judges continued these pictorial analogies and metaphors as descriptors of incriminating passages in the novel.

50 Pinard, "Speech Prosecuting Attorney," 10; "Réquisitoire," 562.

51 Pinard, "Speech Prosecuting Attorney," 26; "Réquisitoire," 571.

52 Pinard, "Speech Prosecuting Attorney," 36; "Réquisitoire," 577.

53 Jules Senard, "The Public vs. M. Gustave Flaubert: Plea for the Defense, by M. Senard" in Flaubert, *Complete Works*, vol. 2, Appendix: 47; "Procès Intenté à L'auteur: Plaidoirie," in Flaubert, *Oeuvres Complètes*, t. 2, 580. According to Michèle Hannoosh, Senard followed a traditional mode of textual interpretation; "Reading the Trial of the *Fleurs du Mal*," *Modern Language Review* 106, no. 2 (2011): 377–8.

54 Senard, "Plea for the Defense," 37; "Plaidoirie," 579.

55 As Dominick LaCapra has argued, Pinard's choice of passages and his vehemence reflect the attorney's own psychological fragility: "Pinard may have been 'poisoned' or 'contaminated' by the 'painting of passion' in the novel and, himself facing temptation, may have set out in a self-preservative quest to find a legal antidote to Flaubert's writing;" *Madame Bovary on Trial* (Ithaca: Cornell University Press/Phaidon, 1982), 27. As per my conclusion to this chapter and the thesis of the following chapter, I agree with LaCapra that Pinard's response signifies the social imaginary and affective responses to pornography.

56 Senard, "Plea for the Defense," 90; "Plaidoirie," 579.

57 "Appendix: The Decision," Flaubert, *Complete Works*, 123–4; "Jugement," in Flaubert, *Oeuvres Complètes*, 630.

58 Charles Baudelaire, "*Madame Bovary* par Gustave Flaubert," *L'Artiste* (October 18, 1857); "*Madame Bovary* by Gustave Flaubert," in *Selected Writings*, 248. Charvet translates "*ces vieilles âmes*" as "desiccated souls" to convey Baudelaire's idea that the nineteenth-century reader is blasé, jaded, world-weary, and passionless or stuck up (afraid to experience the intensity of the world or go beyond propriety). As translated by Aimée Boutin, email correspondence, May 19, 2017.

59 Baudelaire, "*Madame Bovary* par Gustave Flaubert," *Selected Writings*, 249–50.

60 As Pichois recounts, a report was issued for the Sûreté Publique of the Ministry of Interior, on July 11, 1857, declaring that *Les Fleurs du Mal* constituted "an act of defiance in contempt of the laws which safeguard religion and morality." Based on the report's recommendation that the matter be referred to the Public Prosecutor [Pinard], the Minister of Interior accordingly "apprised the Public Prosecutor of his official opinion. The book was a clear outrage to 'public morality;'" Pichois with Ziegler, *Baudelaire*, 224–5; original texts included.

61 Pinard, "Le Procès des 'Fleurs du mal': Réquisitoire d'Ernest Pinard [Baudelaire]," in Charles Baudelaire, *Oeuvres Complètes. Les Fleurs du Mal*, t. 1, ed. Claude Pichois with Jean Ziegler (Paris: Gallimard, 1975): 1206. See Harrison's summary of Pinard's indictment against Baudelaire compared with his indictment against Flaubert, in *Circles of Censorship*, 49.

62 Pichois with Ziegler, *Baudelaire*, 233.

63 Pinard, "Réquisitoire [Baudelaire]," 1209.

64 Pinard, "Réquisitoire [Baudelaire]," 1206. Pinard's sensitivity to the sensory titillation of Baudelaire's poems was unique to this trial, marked by his later repetition of the accusation: "the author strives to force every situation as though he held the challenge to give senses to those who do not feel any more" (1208).

65 Pinard, "Réquisitoire [Baudelaire]," 1206.

66 Pinard, "Réquisitoire [Baudelaire]," 1206.

67 Pinard, "Réquisitoire [Baudelaire]," 1208.

68 Hannoosh explains why Baudelaire was convicted and Flaubert was not: "The court, in its ruling, had acknowledged the morality of Flaubert's objective, while nevertheless complaining about the coarseness of the representation. The judges had stated, hypothetically, that the representation of immorality for the sole purposes of character portrayal or local colour could constitute an offence, but they had concluded that the moral context of the book ultimately outweighed this. In Baudelaire's case, the force of the 'immoral' representation was seen to prevail not only over the 'moral' parts, but also over the moral which the work as a whole actually contained;" "Reading the trial of the *Fleurs du Mal*," 384. Pichois attributes Pinard's successful prosecution to Count de Montalembert's attack on Realism in front of the assembled Academies on August 17, 1857. Presented three days before the Baudelaire trial, Montalembert's attack marked the official status of Realism as a "synonym of immorality" and of a Realist as "someone who should be treated as a public enemy;" Pichois with Ziegler, *Baudelaire*, 234.

69 "Le Procès [Jugement]," in Baudelaire, *Oeuvres Complètes*, t. 1 (1975), 1181–2.

70 Dominick LaCapra, "Memory, Law, and Literature: The Cases of Flaubert and Baudelaire," in *History, Memory, and the Law*, ed. Austin Sarat and Thomas R. Kearns (Ann Arbor: University of Michigan Press, 1999): 104.

71 McCauley, *Industrial Madness*, 157–8.

72 McCauley, *Industrial Madness*, 157–8.

73 McCauley, *Industrial Madness*, 158.

74 Aurélien Scholl, "Les Coulisses," *Le Figaro* 9, no. 726 (January 30, 1862): 5–6; cited in Marie Robert, "The Photographer, the Model and Body Trafficking," in *Splendours & Miseries: Pictures of Prostitution in France, 1850–1910* (Paris: Musée d'Orsay/Flammarion, 2015): 75. Scholl opens his article with a report that there is talk of establishing a Ministry of Photography. But given the extravagant hold on the public's attention, he recommends that "just as we have inspectors of the Beaux-arts, we should also have inspectors of the 'Vilains-arts'" (5).

75 Durandard, "Petit Manuel de Philosophie Photographique," *Le Monde Illustré* 6, no. 272 (June 28, 1862): 408–9; qtd. in Robert, "The Photographer," 74.

76 Scholl, "Les Coulisses," 5; qtd. in Robert, "The Photographer," 73.

77 Gabriel Pélin, *Les Laideurs du Beau Paris* (Paris: L'écrivain et Toubon, 1861) : 13, 14, 16; qtd. in McCauley, *Industrial Madness*, 158.

78 A. Paul, "Salon de 1863: Les Refusés," *Le Siècle* 28, no. 103009 (July 19, 1863): [2]; qtd. in Alan Krell, "Manet's *Déjeuner sur L'herbe* in the *Salon des Refusés*: A Re-appraisal," *Art Bulletin* 65, no. 2 (June 1983): 317.

79 Louis Etienne, *Le Jury et ses Exposants: Salon des Refusés* (Paris: E. Dentu, 1863): 30; qtd. in George Heard Hamilton, *Manet and His Critics* (New Haven/London: Yale University Press, 1987): 45.

80 Didier de Monchaux, "Salon de 1863–Les Refusés," *La Patrie* 23 (May 21, 1863): 2; qtd. in Elizabeth Anne McCauley, "Sex and the Salon: Defining Art and Immorality," in *Manet's "Le Déjeuner sur L'herbe,"* ed. Paul Tucker (New York: Cambridge University Press, 1998): 41.

81 Because of the uproar over the painting's indecencies, it had to be relocated from a lower position to one high above a doorway. See Félix Jahyer, *Salon de 1865: Étude sur les Beaux-arts* (Paris: Dentu, 1865): 282; qtd. in Clark, *The Painting of Modern Life*, 85.

82 As Clark observes, "There was nothing very remarkable in [the critics' identifying Olympia as a prostitute]; it had become an established critical tactic in the 1860s to detect the contemporary, even the bourgeois, *courtisane* beneath the skin of a Venus or Phryne." (*The Painting of Modern Life*, 86).

83 Critics identified Olympia as "a coal lady from Batignolles," "whose modest outlines had never been outraged by water, that banal liquid." Another critic admonished, "The tone of her flesh is dirty, the modeling nonexistent." (Clark, *The Painting of Modern Life*, 87, 96).

84 Louis de Laincel, *Promenade aux Champs-Élysées: L'art de la Démocratie, Causes de Décadence, le Salon de 1865, ...* (Paris: E. Dentu, 1865): 13; qtd. in Clark, *The Painting of Modern Life*, 96.

85 Ego, "Courrier de Paris," *Le Monde Illustré* 9, no. 422 (May 13, 1865): 291; qtd. in Clark, 86. Here the oft-cited phrase "auguste jeune fille" refers to Zacharie Astruc's poem that was published in the first edition of the exhibition catalogue (and then removed from later editions). See Cachin et al., *Manet*, 181. While her contemporaries and later scholars have identified Victorine Meurent as posing for Olympia, Olympia's "Negress" maid's identity as a well-known artist's model, Laure, has only recently been examined. For an in-depth discussion of Laure's personal history and her place in Parisian modernity and its Caribbean population, see Denise Murrell, "*Olympia* in Context: Manet, the Impressionists, and Black Paris," in Denise Murrell, *Posing Modernity: The Black Model from Manet and Matisse to Today* (New Haven/London: Yale University Press, 2018): 7–26. See also Darcy Grimaldo Grigsby, "Still Thinking about Olympia's Maid," *The Art Bulletin* 97, no. 4 (October 2, 2015): 430–51.

86 Émile Zola, "Edouard Manet: Etude Biographique et Critique," *La Revue du XIXe Siècle* (January 1, 1867); rpt. *Éd. Manet: Étude Biographique et Critique* (Paris: E. Dentu, 1867): 36; trans. in Nochlin, *Realism and Tradition*, 77. As noted above, Victorine Meurent (1844–1927), a professional model and painter in her own right, posed for *Olympia*; she also modeled for *Le Déjeuner sur L'herbe*; "Self-Portrait, Victorine Meurent (French, 1844–1927)," Boston Museum of Fine Arts, http://collections.mfa.org/objects/696863/selfportrait. See also Eunice Lipton, *Alias Olympia: A Woman's Search for Manet's Notorious Model & Her Own Desire* (New York: Charles Scribner's & Sons, 1992).

87 Hamilton delves into the iconographic relationship between "The Jewels" and *Olympia* (*Manet and His Critics*, 78–79). Significantly, I have independently arrived at similar conclusions, extending the text-image correlations to the discourse on obscenity, the censorship trial of *Les Fleurs du Mal*, and stereoscopic photography.

88 Pinard, "Réquisitoire [Baudelaire]," 1206. The report from the Sûreté Publique [Public Security] of the Ministry of Interior cited "The Jewels" as "offering in every line the most licentious images rendered with the utmost brutality;" qtd. in Pichois with Ziegler, *Baudelaire*, 224.

89 Pinard, "Réquisitoire [Baudelaire]," 1208.

90 Baudelaire, "Les Bijoux," in *Fleurs du Mal* [1st ed., 1857], in trans. "The Jewels" in *The Flowers of Evil*, trans. James McGowan (Oxford: Oxford University Press, 1993; reis. 2008), 47–8.

91 Baudelaire's analogy was on target. As Marie Robert explains: "A kind of social welfare for men's sexuality, much like prostitution, pornography came into being at the same time as photography. From the 1840s, photographers and their models produced a vast repertoire of poses, attitudes, and 'obscene' couplings. These two industries—that of sex and of the image—fed off one another throughout the century;" Robert, "The Photographer," 72.

92 Baudelaire, "The Salon of 1859: The Modern Public and Photography," in *Selected Writings*, 295–6. See also Elissa Marder, "A Thousand Hungry Eyes: Baudelaire's Stereoscope," *Yale French Studies*, no. 139, Photography and the Body in Nineteenth-Century France (2021): 85–99. Marder argues for a stereoscopic (double) viewing/reading of Baudelaire's essay, comprising "the form of a parodic theological-historical account" (89).

93 Hollis Clayson also comments on Nana's gaze and the flirtatiousness directed to the viewer outside the picture frame, but she does not associate this aspect with photography; see Clayson, *Painted Love: Prostitution in French Art of the Impressionist Era*, 69.

94 J[oris].-K[arl]. Huysmans, "La *Nana* de Manet," *L'Artiste de Bruxelles* (May 13, 1877): 148–9; rpt. http://www.huysmans.org/artcriticism/nana.htm. All subsequent quotations are from this online text.

95 *L'Assommoir* was published in installments, from April to June 1876, first in *Le Bien Public*, and then in *La Republique des Lettres*, from June 1876 to January 1877; it was published as a single volume in February 1877; Robert Lethbridge, "Introduction," in Émile Zola, *L'Assommoir*, trans. Margaret Mauldon (Oxford: Oxford University Press, 1998): xiv. The novel's title is a period term referring to a "drinking den," as some translators render it, where the working class could drown their sorrows.

96 *Fille* signifies a prostitute, shortened from "fille de joie" (lady of the night); see http://www.wordreference.com/fren/fille%20de%20joie.

97 As I have argued in "The Photographic Subversion" (9).

98 Crary, *Techniques of the Observer*, 125.

99 This and the following quotations are from Crary, *Techniques of the Observer*, 125.

100 Billon is the pseudonym of Alfred François Cordier, dit Billon-Daguerre; Pellerin, *La Photographie Stéréoscopique*, 104.

101 Crary describes this temporal convergence, quoting Sir David Brewster's *The Stereoscope: Its History, Theory and Construction* (London, 1856), in *Techniques of the Observer*, 120, 122.

102 Crary, *Techniques of the Observer*, 133. See also his discussion of the stereoscope's "principle of disparity," which endured even in the apparatus's later forms (125–6, 128–9).

103 Baudelaire, "The Painter of Modern Life," *Selected Writings*, 392, 403.

104 Baudelaire, "The Painter of Modern Life," *Selected Writings*, 395, 402, and Baudelaire, "The Exposition Universelle 1855," in Baudelaire, *Art in Paris, 1845–1862*, 124.

105 Linda Hutcheon discusses modern parody as double-coded in *A Theory of Parody: The Teachings of Twentieth-Century Art Forms*.

106 The pamphlet, *Éd. Manet* (Paris, 1867; see note 86), is depicted with its blue cover on Zola's desk; see Cachin, *Manet, 1832–1883*, 280–4. She writes (184): "It is hard to tell whether this [the image of *Olympia*] is a lost print or an enlarged photograph of the painting." More recently, Anne Higonnet argues for the photograph of the painting ("Manet and the Multiple," *Grey Room* 48 [July 2012]: 109). Higonnet's argument is convincing in that Manet was collapsing diverse reproductive technologies available in the mid-nineteenth century. However, an equally strong argument favors the etching since Manet made several print versions after *Olympia* (a woodblock and two etchings) and because one of the etchings was made for the Zola portrait, in particular, the 4th state of the "small plate." For a discussion of these etchings, see *Manet, 1832–1883*, 186–9. Furthermore, Manet may have been referencing Baudelaire's critical acclaim of Manet as a leader among the etchers emerging in contemporary art; see Baudelaire, "Peintres et Aquafortistes" (1862), trans. "Painters and Etchers," *Art in Paris 1845–1862*, 217–20. Perhaps this is another example of Manet's skill in representing historical transience and cognitive slippage.

7 Brain Mapping Censorship and Desire in Second Empire Paris

Introduction: On the Conflation of Prostitutes and Pornography during Haussmannization

This empirical chapter explores the ways in which the conflation of prostitution and pornography during Haussmannization constituted a discursive network for understanding the critical reception of Manet's paintings *Le Déjeuner sur L'herbe* and *Olympia*. An examination of this network reveals the critical rhetoric as a manifestation of the embodied, experiential effects of obscenity and as a public response to feelings of arousal, disgust, and anger. As previously discussed, the increased visibility of prostitutes and pornographic photographs was a product of Haussmannization itself. Their display on city streets mediated the public's reception of *Le Déjeuner sur L'herbe* and *Olympia*, as did the model's apparent identity as a prostitute in the latter.[1] Similarly, crowds who gathered to see *Olympia* were also seen as susceptible to her corruption:

> The mob, as at the Morgue, crowds around the spicy Olympia[.] Art sunk so low doesn't even deserve reproach. 'Do not speak of them; observe and pass on,' Virgil says to Dante while crossing one of the abysses of hell. But Manet's character belong[s] rather to Scarron's hell than to Dante's.[2]

This critical reception was inflected by the conflation of prostitution and pornography in the public imaginary and the discourses of condemnation and contamination it spawned.

Two additional examples of period commentary serve to recall the rhetoric of fear and repulsion with which critics attacked pornography and prostitution and their widespread urban presence. In 1862, Aurélien Scholl complained about the proliferation of pornographic photography in the form of scantily clad or nude photographs of actresses, in shop fronts of photographers' studios. Scholl's complaint not only portends the nation's supposed moral corruption but, given the hours Parisians spent gazing at these shop displays, pornographic photography would also lead to its economic degradation.[3] Writing in 1870 in *La Prostitution à Paris et à Londres 1789–1870*, C. J. Lecour voices similar concerns about the omnipresence of Parisian prostitutes along the boulevards, in popular and commercial establishments, and aboard trains and carriages:

> Uncontrolled ['nonregistered'] prostitutes… are everywhere, in the cafés-concerts, the theatres, the balls. One meets them in public establishments, railway stations, and even railway carriages. They are there on all the promenades, in front of most of the

DOI: 10.4324/9780429442315-8

cafés. Late into the night they circulate in great numbers on all the finest boulevards, to the great scandal of the public[.][4]

The rhetoric that Lecour and Scholl share further reveals the analogy between pornographic photography and prostitution in the public imaginary.

Methods and Models

In this section, I examine the vitriolic language against obscenity and the hostile critical reception of Manet's *Le Déjeuner sur L'herbe* and *Olympia* through the lens of neurobiology and cognitive linguistics. I do so to argue that the critical and legal hostility was not only institutionally motivated but, crucially, triggered by embodied visceral responses to pornographic texts and images. My interpretation contrasts with that of T. J. Clark, who, in assessing the negative criticism surrounding Manet's *Olympia*, argued that the intoned anger was feigned: "Certainly the critics' descriptions belong to a shifty, knowing, hypocritical game of make-believe: make-believe anger, make-believe morality, counterfeit concern for art."[5] Notwithstanding the value of Clark's compendium of critical texts, I disagree with his assessment and argue that the critics' anger was real and embodied. My counterargument entails viewing the hostile rhetoric as consonant with both the psychophysiology of arousal and the civic litigation against visual and literary obscenity. As we have seen in the previous chapter, government censors used the same pejorative language to restrict the sale and public display of the pornographic photography that Manet adapted to his art. As neural research on internet pornography addiction has established, "neurobiological mechanisms underlie neuropsychological operations."[6] Together with other neurological studies, this research registers the spectrum of arousal and disgust that accompanies pornographic viewing. In turn, cognitive linguistics, modeled on George Lakoff's experientialist epistemology, registers the neuropsychological operations manifest in the pejorative language that functioned to mask male sexual arousal.

In what follows, I explore the embodied and social discourse on obscenity from two interrelated perspectives. I first use phenomenological and neurobiological studies that record the affective, physiological, and psychological responses to pornography. From this perspective, I model a spectrum of arousal and disgust and a consonant lexicon of "dirt" that signifies these embodied, physiological responses. For this, I adapt Lakoff's modeling of epistemological metaphors as embodied affects. From the second perspective, I demonstrate how arousal and disgust further manifest in the psychobiological spectrum of anger and, in turn, the social sphere of condemnation and censorship. In linking the subjective with the social sphere, I extend Lakoff's metonymic modeling of anger as a signifier of embodied, physiological states of containment and explosion. Here, embodied anger (triggered by arousal and disgust) motivates social anger and functions as a mode of self-regulation in the name of obscenity and censorship. Mobilizing these neurobiological and cognitive linguistic perspectives allows us to view the critical responses to Manet's paintings in a new light.

Phenomenology of Arousal and Disgust: Attraction/Repulsion Spectrum

Central to my investigation of nineteenth-century constructs of obscenity is the attraction/repulsion spectrum that pornography and its censorship elicit. Philosopher

Matthew Kieran's "On Obscenity: The Thrill and Repulsion of the Morally Prohibited" provides a phenomenological model for arguing that "cognitive-affective responses [to] sexual desire" compel moral judgments of obscenity.[7] Accordingly, Kieran identifies the conditions under which an object is judged obscene, eliciting the attraction/repulsion spectrum. First, "the object of judgment must naturally elicit or commendingly solicit cognitive-affective responses that [override] internalized moral prohibitions." Second, "the object must do so in a way that gives rise to or is judged to warrant feelings of repulsion and yet is nonetheless found... to merit attraction or compulsion."[8] "Thrill and repulsion" thus arise from visceral responses more akin to feeling than thought. Yet, those visceral feelings migrate to the cognitive faculties, wherein judgments are formed. Notably, neurobiologists call this migration "bottom-up" processing.

Information studies scholar Patrick Keilty also applies a phenomenological, bottom-up analysis of the attraction/repulsion spectrum to online pornography, but he extends his analysis to the cognitive faculties. Borrowing from cinematic models, Keilty describes how "browsing pornography has a carnal and conscious relation to pornographic images, conjoined by the always reflexive and reflective exchange of sensation that defines the auto-erotic goals of this particular browsing process."[9] Here, I contend that online pornography historically aligns with the nineteenth-century erotic photography that Manet appropriated. Both mediums depended on new technologies of communication and reproduction, which, in turn, spawned an epidemic of pornographic consumption. As noted in the previous chapter, Baudelaire decried how the public—both male and female and even schoolchildren—became obsessed with stereographic pornography; "it was the rage of society." He thus recounts that "thousands of pairs of greedy eyes were glued to the peepholes of the stereoscope," providing humankind's "love of obscenity... a glorious opportunity for its own satisfaction." Even "society" women abandoned their "modesty" to satisfy their "love."[10]

Keilty's focus on the always reflexive and reflective exchange of sensation helps to support the "carnal, fleshy"[11] foundations of the attraction/repulsion dynamic that compelled Baudelaire's outrage and, more importantly, nineteenth-century critical and legislative attacks on objects judged obscene. In mapping the carnal and conscious relation to pornographic images, Keilty describes how "the lived body's capacity to sense itself, when considered in the context of browsing pornography online, coupled with [a] sensing [of] one's own sensuality, [is] a form of self-touching," or masturbation.[12] However, "the reflexive and reflective exchange" between the embodied self and the erotic other "can be broken, not only by conscious reflectivity but also by reflexive feelings of shame, disgust, or the panic of being discovered by someone outside the scene intruding upon the moment (entering the picture, if you will)."[13]

Turning back to our nineteenth-century rhetorical examples, it is no wonder the critics and litigators were angry and hostile in judging erotic art and photography. Following Keilty's research, what transpires in the process of viewing pornography, whether online or as photographs, suggests that viewers cannot help themselves from embodying and entering the erotic acts or poses that unfold before them, and willingly or not, engaging in "spontaneous eroticism," or "self-eroticism." These adjudicators had already embodied "the reflexive and reflective exchange" between the self and the erotic other when viewing photographic images in the men-only backrooms of cafés and photography studios or private domestic quarters.

Brain Anatomy of the Attraction/Repulsion Spectrum

I now want to demonstrate how brain anatomy may function in the attraction/repulsion spectrum and the contingent rationale for judgment of the obscene. Specifically, I am interested in whether arousal and disgust may activate common brain areas and, if so, how they may deviate (Figure 7.1). Here, we can begin with a 2005 study led by cognitive psychologist Rudolf Stark, which shows that the same areas of the brain process erotic and disgust-inducing pictures.[14] These areas function at basic levels for visual processing (occipital lobes) and in three structures within the temporal lobes for memory (hippocampus), processing of body sensation (thalamus), and processing of affective information (amygdala). Stark's team also found that erotic pictures are further processed by the ventral striatum (basal ganglia), which activates reward systems.[15] Other studies show that the amygdala plays a central role "in the processing of erotic stimuli." Specifically, the amygdala channels the involuntary responses to the more controlled brain centers: "structures associated with emotional attention (e.g., occipital, orbitofrontal, and parietal cortices) and reward processing (e.g., the ventral striatum)."[16]

When I shared these observations with cognitive psychologist Matthew Pelowski, he pointed out that these anatomical studies were limited; they register basic brain functions but not valence. What is valence? Valence processes attractiveness (goodness), aversiveness (badness), and approach-avoidance.[17] While valence registers emotion, it also leads to cognitive appraisal, judgment, and choice. For my purposes, valence became crucial for modeling the attraction/repulsion spectrum of nineteenth-century viewers of pornography. I envisioned a neural split or separation from the shared brain areas, with the reward circuit activated in one stream and appraisal of disgust in another. When I proposed this hypothesis to Pelowski, he directed me to the relevant brain anatomy.

Given these directives, I searched for studies that connected the brain anatomy of arousal and disgust with valence (appraisal). My first step was to locate where and when valence takes place. Minming Zhang and colleagues' study of male homosexual versus male heterosexual neural responses to erotic versus disgusting visual stimuli provided an answer.[18] Zhang's team confirmed Stark's and others'[19] location of arousal in the occipital (vision) and hippocampus (memory) areas. Still, they also found more diverse areas for

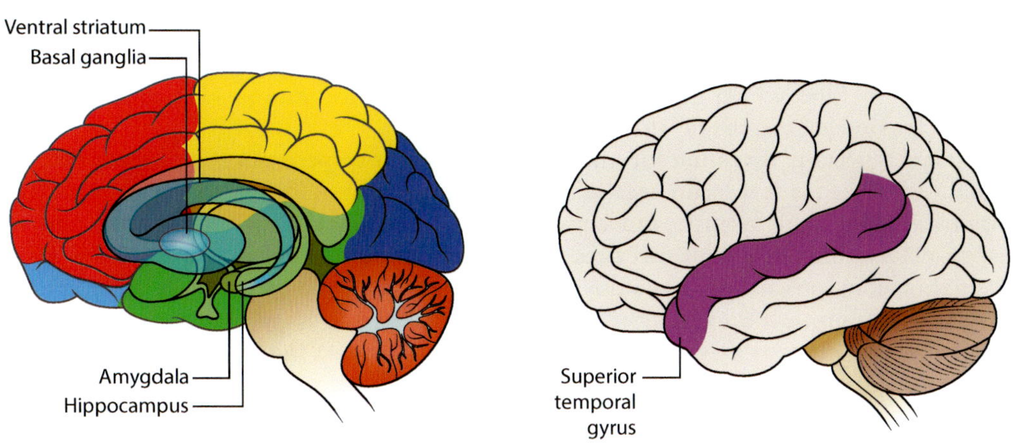

Figure 7.1 Composite brain anatomy. Illustration by Susan Tyler.

disgust-induced activation by erotic images. Significantly, some areas that disgust activates are also in arousal and reward systems. Equally important, several regions are more active in processing emotional valence and affect.

Regarding these shared and distinct areas, Zhang and colleagues' study showed that the activations related to sexually disgusting stimuli (in both groups) included the thalamus.[20] The thalamus is part of the limbic system, which is responsible for combining higher mental functions and primitive emotions into one system (up-bottom or bottom-up processing).[21] The frontal lobe was also activated. It is associated with (among other processes) reward, attention, and motivation, while it limits and selects sensory information arriving from the thalamus to the forebrain.[22] For my purposes, since the frontal lobe is associated with both attention and reward, it may be responsible for the onset of ambiguity in registering attraction while simultaneously processing disgust. The right cerebellum, at the back of the brain just above the brainstem, was also activated and may cause ambiguation between pleasure and disgust. While the cerebellum is essential in motor control, it may also regulate fear and pleasure responses.[23] This dual function is confirmed in Zhang and colleagues' study. As they state, "the cerebellum was found to be typical for mood induction elicited by sexual stimuli in both groups." At the same time, "the cerebellum[,] also known to engage in emotion processing,... was considered to participate in neural circuits related to aversive stimuli." The activation of the left superior temporal gyrus (STG) is especially relevant for mediating the physiological sensations to the cognitive faculties and into the social sphere (bottom-up processing). While it is involved in auditory processing, including language, it also has been implicated as a critical structure in social cognition.[24] Taken together, these brain areas and their contingencies with language are essential for translating the arousal/disgust responses into the social sphere and, therefore, were likely activated in the pejorative discourse on obscenity in nineteenth-century visual culture.

Zhang and colleagues' study is also helpful in teasing out what is specific to heterosexual men's viewing of disgusting erotic imagery. Most important for my modeling of embodied responses is the activation of the autonomic nervous system through the insular cortex, located deep within the brain.[25] The autonomic nervous system is a control system that acts unconsciously and regulates bodily functions such as heart rate, digestion, respiratory rate, pupillary response, urination, and sexual arousal.[26] Likewise, aversive stimuli also increased activation of the left cuneus, a primary visual processing region in the occipital lobe, which Zhang's team noted "might reflect more vivid mental imagery induced by viewing disgusting stimuli[.]"[27]

What is especially revealing for my attraction/repulsion spectrum is that these heterosexual responses mark a brain area where involuntary physiological functions, affective responses, emotional disgust, and cognitive self-reflection coincide. This spectrum of responses also indicates a neural process whereby external images induce more vivid internal imaging processes. Given this data, I propose that the men who labeled images obscene likely produced even more vivid mental portrayals than met their seeing eye. Moreover, this internal/external dynamic speaks to the role that embodied feelings of disgust play in activating social cognition. Thus, I can further posit that nineteenth-century critics and adjudicators made their embodied, emotional, and neural experience of arousal and disgust social through their executive powers activated on the right hemisphere of their brain—that is, through language.

A Neurobiological Model of Language

Neuropsychologist Stefan Koelsch and his team have focused on the dynamic neurobiological connection between affect and symbolic language coding. Koelsch and colleagues formulate "a neurobiological theory of emotions" that includes, among other processes, "emotions which are uniquely human (such as complex moral emotions) and considers the role of language for emotions." Based on this theory, Koelsch's team "propose[s] that four classes of emotions originate from four neuroanatomically distinct cerebral systems," what they call "the quartet theory of human emotions.... [E]ach of these systems generates a specific class of affects (e.g., ascending activation, pain/pleasure, attachment-related affects, and moral affects)."[28] The affect systems interact with each other and emotional effector systems (bodily organs that become active in response to stimulation), including language. Koelsch and colleagues explain, "These effector systems include motor systems (which produce actions, action tendencies, and motoric expression of emotion), peripheral physiological arousal, and attentional and memory systems. The activity of affect systems and effector systems is synthesized into an emotion percept (pre-verbal subjective feeling), which can be transformed (or reconfigured) into a symbolic code such as language" [bottom-up processing]. It follows that language functions as part of a cognitive appraisal (qua-valence) system: "conscious cognitive appraisal (involving rational thought, logic, and usually language) can regulate, modulate, and partly initiate, activity of affect systems and effector systems"[29] [up-bottom processing].

For my purposes, Koelsch and colleagues' identification of the affective-effector function of language specifically supports my adaptation of Lakoff's experiential epistemology to the embodied, experiential effects of obscenity and the rhetoric of dirt, anger, and censorship. Koelsch and colleagues' functional model of language is particularly useful for understanding how language is deployed to control internal affects and emotions in the public sphere:

> Language has two main functions for human emotions: (a) With regard to subjective feelings, language is an important means of expressing as well as communicating emotions (and, therefore, also to elicit emotions in other individuals). (b) With regard to conscious appraisal, language is an important means of regulating emotions.[30]

From Neurophysiological Processing to Linguistic Processing of Arousal and Disgust

To demonstrate how linguistic codes function in tandem with physiological and neural processing, I have adapted George Lakoff's experientialist model of metaphors, whereby linguistic expressions signify embodied physiological states. Lakoff's experientialist view of epistemology is especially germane for analyzing metaphors of embodied arousal and disgust. In my analysis, I first focus on "dirt" as a primary metaphor for (hidden) physiological and affective responses to obscenity and perceived social contamination (bottom-up processing). This focus is justified by the repetition of "dirt" in descriptions of visual images, particularly of Manet's *Olympia* and related photographic images of studio models. In the second analysis, I focus on "obscenity" as a metaphor informing the litigation language. Obscenity (as a metaphor), in turn, generates linguistic codes that shroud lust and anger in imagery deemed repulsive and immoral (up-bottom processing).

Foundational to my use of Lakoff's experientialist model of metaphors is that "basic-level physical experience" informs our contact with the world and that language is the cognitive expression of that experience. He thus describes his model of metaphors as bridging "language and thought on one hand and the world on the other."[31] Equally important to my affective/cognitive spectrum of obscenity is what Lakoff calls "image schema."[32] An image schema is a cognitive counterpart to and precondition for metaphor; it mediates thought and language within the subject [person]-world continuum.

Obscenity: Metaphors of Contamination and Social Self-Control

To show that "the metaphor is natural in that it is motivated by the structure of our experience," Lakoff charts a three-part system of metaphor's mediating function: "each metaphor has a source domain, a target domain [abstract concept], and a source-to-target mapping."[33]

Adapting Lakoff's model to the dirt metaphor, one can identify a source domain [AROUSAL], a target domain [DIRT], and a source-to-target mapping [LUST]. Here, the source-to-target mapping of AROUSAL→DIRT is motivated by our physical sensation of sexual desire and the ensuing affect of shame.

Because DIRT is also a material bodily condition, it can also function as a source domain, with OBSCENITY as the target domain (metaphor).[34] As with AROUSAL→DIRT mapping, the DIRT→OBSCENITY mapping substantiates and reinforces the violation of social norms. In this case, the source-to-target mapping (DIRT→OBSCENITY) is motivated by the embodied experience of feeling UNCLEAN and needing to get rid of dirt.

As we have seen, phenomenological and neurophysiological models support the DIRT→OBSCENE mapping of a complex psychological-social linguistic system in the context of nineteenth-century criticism and censorship, as captured, for example, in Bertall's caricature of *Olympia* (1865) (Figure 7.2). The caption makes Olympia-qua-prostitute's dirt explicit:

> The cat's tail, or the coal-peddler of Batignolles. / Everyone admires this beautiful coal-woman whose modest contours have never been exposed to so commonplace a liquid as water. Let us admit it boldly: the coal-woman, her paper-wrapped bouquet, M. Manet, and his cat are the lions of the Salon of 1865. Three cheers for M. Zacharie Astruc.[35]

The caricature's metaphoric function is motivated by the social body's need to control and eliminate dirt. In this regard, OBSCENITY is correlated with and may be substituted by CONTAMINATION, as further signified by Randon's caricature of *Le Déjeuner sur L'herbe* (1867) (Figure 7.3). The caption of Randon's image reads: "Le Déjeuner sur l'herbe, or the Triumph of virtue. The virtue of these gentlemen, let us say it between us, must not cost them much; they even seem to say, like the Auvergnat: 'we are very quiet here.'" Here, the women's moral filth is transferred to the otherwise virtuous "gentlemen," who pay cheaply for their lustful desires. Their diminished virtue is reinforced by the reference to "l'Auvergne," a moniker that demotes the urban bourgeois sophisticates to country bumpkins.[36] CONTAMINATION resonates with the transference or projection from the subjective, physiological source domain (lust/dirt) to the social sphere in art criticism and visual caricature.

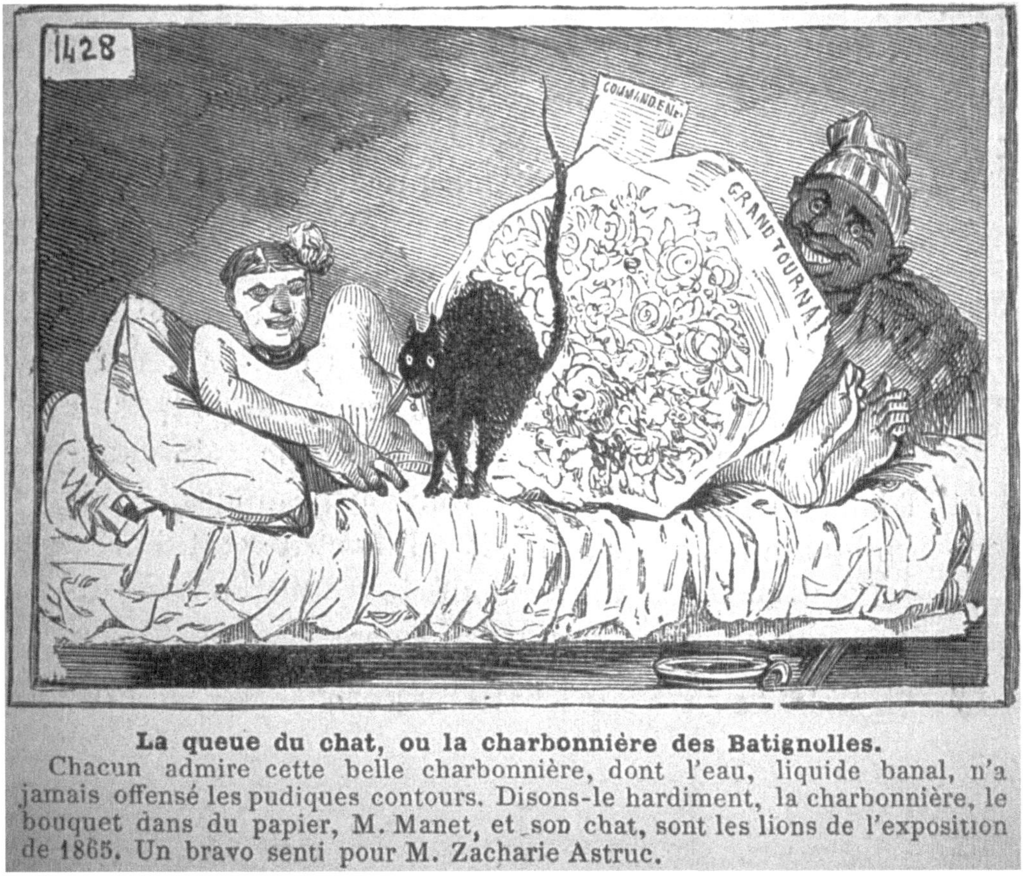

Figure 7.2 Bertall, "La Queue du Chat, Ou la Charbonniére des Batignolles" (The Cat's Tail, or The Coalwoman of Batignolles). In *L'Illustration*, vol. 45, no. 1162 (June 3, 1865): 341. Courtesy of Northwestern University Libraries. CC 4.0.

Nineteenth-Century Psychology of Sexual Arousal

Nineteenth-century psychophysiology further supports linguistic modeling of the convergence of sexual arousal and social contamination in the public sphere. Dr. Jacques-Joseph Moreau's study on sexual pathologies, published in *La Psychologie Morbide* (1859), bears witness to the medical claims that regarded the physiological effects of untamed sexual arousal as both a moral perversion and a cause for madness. While Moreau's focus is on hereditary madness, his conflation of subjective pathologies and their social manifestations resonates in the discourses on obscenity and the censorship of visual and literary works judged obscene. In his treatise, Moreau distinguishes between "the *morale* of love and sexual appetite" (original emphasis). Accordingly, he determines that two biological systems are at play in "normal passion" or "love" wherein there is a "reciprocal influence of the sexual organs and cerebral organs." Conversely, the separation of the two systems results in pathological conditions where increased sexual appetite and arousal overcome the mental faculties and cause a psychophysical imbalance. As Moreau states, if the former condition abides by nature's laws, the latter condition proves "unnatural."[37]

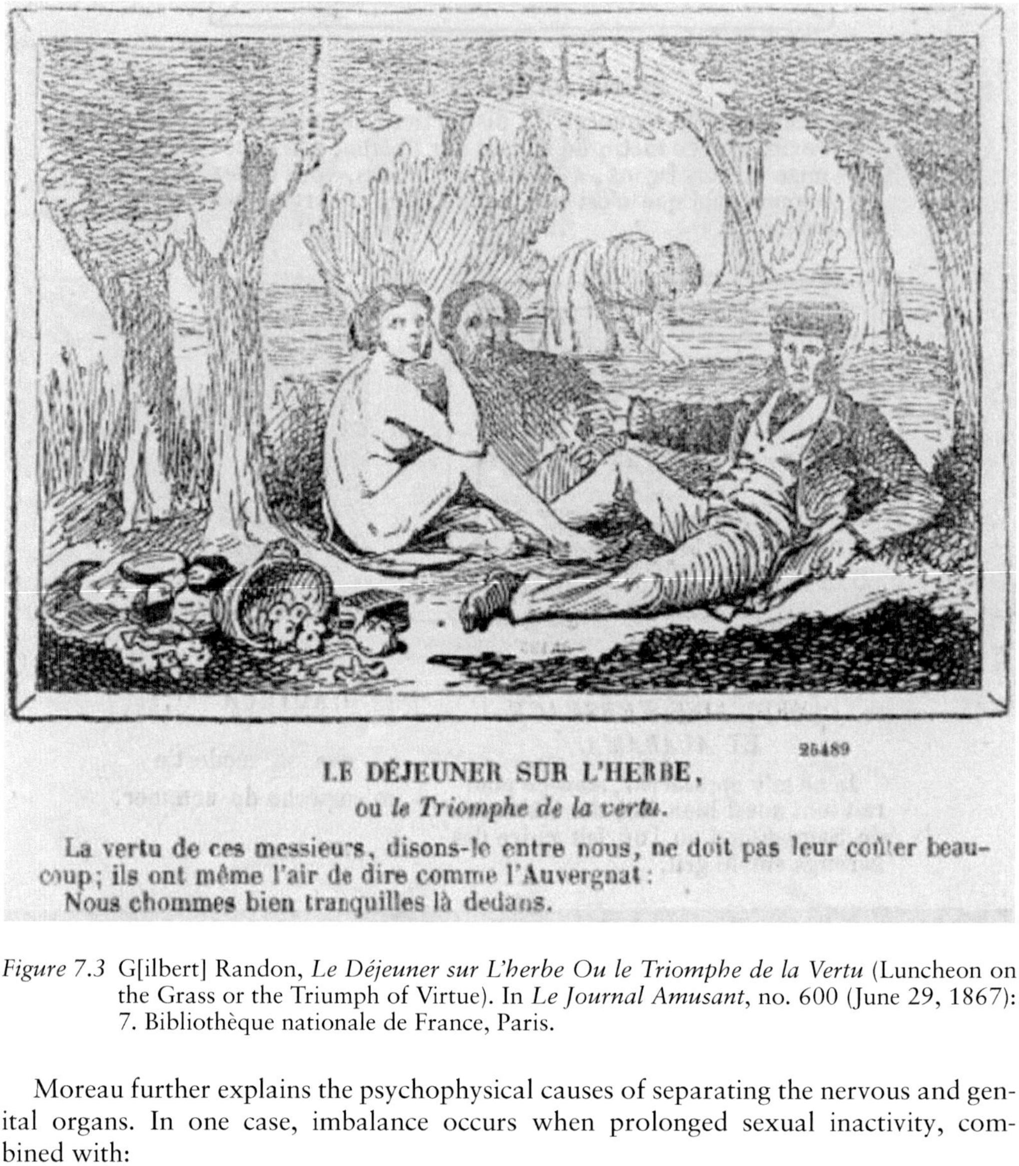

Figure 7.3 G[ilbert] Randon, *Le Déjeuner sur L'herbe Ou le Triomphe de la Vertu* (Luncheon on the Grass or the Triumph of Virtue). In *Le Journal Amusant*, no. 600 (June 29, 1867): 7. Bibliothèque nationale de France, Paris.

Moreau further explains the psychophysical causes of separating the nervous and genital organs. In one case, imbalance occurs when prolonged sexual inactivity, combined with:

> the accumulation of sensibility or nervous force, which follows, have increased its energy, and stimulates the cerebral organs to the point of bringing disorder to it exactly as a too strong dose of an excitant would do.[38]

While Moreau is most interested in the internal subjective causes of nervous excitation, we can assume that "too strong a dose of excitant" from external stimulants would include pornographic images or texts.

Moreau's description of the outward manifestations of sexual arousal in the absence of amorous sentiment asserts his belief that uncontrolled sexual activity contaminates the social sphere. The individual succumbs "to the kind of stupor, to the moral void, to the lassitude of heart which they cannot conceal." In another instance, Moreau describes a condition of excessive sexual appetite driven by "too frequent repetition of these same excitations and the exhaustion of sensibility" and its outward social manifestations. These individuals "must have recourse to extraordinary, unheard-of, impossible sensations, the very idea of which would not previously have been presented, nor would the thought of a crime, murder or theft, in the mind of an honest man." He adds, "It is no longer the moral sense of love that is at stake, it is intelligence itself[.]"[39]

Claims that nude photography—and by extension, paintings and literature deemed obscene—had a pathological and contaminating effect on society issue from an 1861 report published in the *Gazette Judiciaire*. Addressing the commerce of nude photography (and echoing Scholl's and Lecour's rhetoric), the reporter warns against its corrupting influence in the public sphere:

> Today, it is no longer a philosophical idea, not even a wrong one; it is no longer art, not even totally photography misunderstood: it is a trade, the most dangerous of them all, the trading of nudities. This commerce has a corrupting effect on everyone, from the unfortunate women who sell the image of their body so cheaply, to the young men whom these images, even in educational establishments, are going to initiate and lure into debauchery![40]

In the following section, I delve more deeply into the social sphere of the discourse on contamination and immorality and how it shaped *Olympia*'s critical reception. To do so, I map the embodied/linguistic continuum from arousal/disgust to anger as manifested in the social and censorial response to obscenity and pornography.

Mapping Anger in the Social Sphere

As noted earlier, I view the inflamed rhetoric around Manet's art as coherent with the legal response to literary and visual obscenity, as well as the psychopathology of arousal. In his mapping of linguistic expressions of emotions, Lakoff extends the metaphorical model to "a general metonymic principle," according to which "[t]he physiological effects of an emotion stand for the emotion."[41] No less embodied than metaphor, metonymy is further mediated and enriched by a conceptual understanding of the source domain. Taking "anger" as an example of metonymy's emotion-concept continuum, Lakoff decodes its "very elaborate conceptual structure," ranging "from righteous indignation, to wrath, to cold anger, and the like."[42] Lakoff's modeling of anger is particularly pertinent to connecting emotional hostility toward obscenity and the concepts that shaped the lexicon of art criticism and adjudication. However, I take liberty with Lakoff's model in my lexical analysis. Whereas Lakoff maps a direct correspondence or analogy between emotions and concepts, I propose a more nuanced mapping of metonymies wherein the emotion-concept continuum is mediated by an "understanding" of social prohibitions. For this more nuanced continuum, I draw upon cognitive and neuroanatomical research to expand Lakoff's indications of the physiological effects of anger and its metonymic descriptors.

Cognitive-Aesthetic Structure of Anger

In this regard, cognitive studies help substantiate anger's increased social dimension, especially in aesthetic matters. Cognitive psychologists Paul J. Silvia and Elizabeth M. Brown have determined that anger, disgust, and negative aesthetic emotions register with the viewer's goals and values (bottom-up processing). Using an appraisal model of emotion, they examined emotional responses when people viewed and cognitively appraised pictures, including offensive and controversial works. Such appraisals align with a person's apperception of either internal or external sources of their discomfort: that is, internal dissonance causing disgust or external offense causing anger. In this study, this internal-external dynamic shows how anger is more socially determined than disgust, but anger and disgust are mutually interactive. Accordingly, Silvia and Brown found that:

> ratings of anger increased as people appraised the pictures as more incongruent with their values,... and *as being intentionally offensive*. Not surprisingly, disgust predicted anger as well.... Ratings of disgust increased as people appraised the pictures as more incongruent with their values,... and *as more unpleasant*[.]... Not surprisingly, anger predicted disgust as well.[43]
>
> (emphasis added)

Silvia and Brown's findings not only support a socially determined emotion of anger but also the embodied metaphor of AROUSAL/DISGUST concerning pornographic images. In this regard, the male viewer appraises pornographic images, based on his sensations, as "incongruent with his values." Here, sensory disgust produces anger. We can assume that socially determined anger arises when pornographic images are perceived as "intentionally offensive," wherein such images cause sensations of disgust.

Silvia and Brown's efforts to incorporate anger, disgust, and negative emotions into the aesthetic spectrum especially pertain to an embodied linguistic modeling of pejorative critical rhetoric. As they conclude:

> [C]ontroversial art makes some people mad, and one of anger's consequences is a tendency to aggress against the perceived source of affront. Understanding censorship, funding disputes, class differences in art preferences, and the defacement of art requires understanding negative reactions to the arts.[44]

In the following explanation, I map the rhetorical metaphors of LUST/ANGER→CONTAMINATION/IMMORALITY→CENSORSHIP/MORALITY. This mapping demonstrates that hostile criticism and censorship against the obscene manifests as anger's "tendency to aggress against the perceived source of affront."

Physiology, Psychology, and Neuroanatomy of Anger

As Lakoff shows, the psychophysiological effects of anger also support embodied linguistic modeling of pejorative rhetoric. Lakoff's mapping of anger is based on "the folk theory"[45] of the physiological effects of anger that yield "a [rich] system of metonymies for anger." These effects include "increased body heat, increased internal pressure (blood pressure, muscular pressure), agitation, and interference with accurate perception. As anger increases, its physiological effects increase."[46] Empirical studies support this physiological spectrum of anger effects.[47] However, Lakoff's focus on linguistic phenomena prioritizes the folk theory to explain how "we make use of a general metonymic principle" when we describe someone as angry "on the basis of their appearance[.]"[48]

In adapting Lakoff's model to the ANGER→CONTAMINATION spectrum, I want to posit an analogy between anger and lust based on their shared physiological effects and the commonplace metaphors that equate lust with these embodied sensations.[49] That nineteenth-century viewers and readers also associated lust with such sensations can be verified by the critics' response, for example, to the first stanza of Zacharie Astruc's poem "La Fille des Îles" (Girl or Daughter of the Islands), which Manet included in the exhibition pamphlet that accompanied *Olympia*'s title.[50] The stanza's last lines, including the phrase "the august young girl in whom the fire burns," were often quoted in negative criticism.[51]

Psychologists substantiate how the physiological effects of anger further align with neuroanatomical processes.[52] For example, Brindusa Vanta elucidates the interaction between anger's physiological effects and neuroanatomical processing in mobilizing a fight response.[53] Vanta further explains how the brain is wired to control anger before it manifests outwardly as rage.[54] In a more detailed mapping of anger via the amygdala, LaVelle Hendricks and colleagues note that the amygdala processes anger as a threat event in preparation for its rational-qua-social control.[55] By examining emotive triggers and cognitive controls, we can better chart the emotion-concept continuum LUST/ANGER→CONTAMINATION related to the social sphere.

In the continuum model that I formulate here, the prefrontal cortex controls the perceived threat and ensuing anger that result from viewing *Olympia* and, analogously, pornographic photography. Art critic-painter Gonzague Privat, who admired Manet's sincerity and the model's authenticity, scorned the public's fear-turned-anger towards the painted image:

> Why are some people afraid of the young woman's [Olympia's] appearance? Does she laugh at others? Because she lives, that this life is sensible for everyone; because one feels that she could move, this woman who is found ugly and ill-made, not without some reason.[56]

Alternatively, photographer Eugène Disdéri disparaged the contradiction between the real and the ideal. He attacked *académies* and the live models' physical and moral corruption, fearing the degradation of art:

> How can the photographer obtain truly beautiful pictures with the Venuses and saints he finds on a street corner? All that is to be seen there is immodesty impersonating chastity, incredulity doing a parody of faith, a face bearing all the marks of our prosaic lifestyle vainly endeavoring to attain the ideal expression.[57]

Each of these critical responses demonstrates how anger triggers the fight response in the form of verbal aggression against the perceived source of affront. Thus, while psychological models offer cognitive control techniques for anger management, I contend that the volatile language of art criticism and civic litigation also functioned as modes of anger management in the public domain.

Conclusion: The Critical Rhetoric toward *Le Déjeuner sur L'herbe* and *Olympia* Reviewed

As I established in the previous chapter, the rhetoric of censorship litigation was infused with attacks against Realism as an artistic and literary practice. This rhetoric also informed social and legal attacks on *académies* that were deemed pornographic. The lexical codification of dirt and obscenity issued from the convergence of these trajectories and subsequently migrated into the critical attacks against Manet's *Le Déjeuner sur L'herbe* and *Olympia*.

A review of the oft-repeated criticism should suffice to recall the rhetoric around these two paintings through the linguistic spectrum of embodied arousal, disgust, and anger that obscenity triggered. As we have seen, when critics condemned the realistic portrayal of the nude model-cum-prostitute, they rehearsed the rhetoric used against pornographic photography displayed in the public domain. Similarly, just as the prosecutor Ernest Pinard aligned the technical deviations of Realist literature with the immoral subject matter, these critics aligned Manet's technical indiscretions with his models' vulgarity and immorality. Critics agreed that Manet's model in *Le Déjeuner sur L'herbe* was "the most vulgar of naked women."[58] Another critic extended the vulgarity to Manet's lack of drawing skills and his taste for the bizarre:

> Manet will show talent once he learns drawing and taste once he stops choosing his subjects for the sake of scandal. We cannot find it altogether a chaste enterprise, to set down under the trees, beside students in cap and gown, a female clad only in leafy shade. [M]ore than the composition itself, I lament the intention that inspires it. M. Manet plans to achieve celebrity by outraging the bourgeoisie. His taste is corrupted by infatuation with the bizarre.[59]

Yet another critic complained that he sought "in vain for the meaning of this uncouth riddle."[60]

Olympia's critics insisted on conflating Manet's abbreviated contouring with the model's actual filth: "[W]hy do the Realists... choose unclean women as their models, and having done so, reproduce even the filth which clings to their contours?"[61] The criticisms are repetitive. Singling out *Olympia* among the "terrible canvases" at the Salon of 1865, a critic rebukes: "What's this yellow-bellied Odalisque. this vile model picked up who knows where."[62] Théophile Gautier exclaims, "A wretched model. The flesh tones are dirty.... The shadows are indicated by more or less broad streaks of monochrome[.]"[63] Most expressive of the correlation between filth, lasciviousness, and deviant techniques was the critic from *Le Monde Illustré* (here referencing Astruc's poem): "The *auguste jeune fille* is a courtesan, with dirty hands and wrinkled feet;... her outlines are drawn in charcoal and her greenish, bloodshot eyes appear to be provoking the public protected all the while by a hideous Negress."[64] At least one critic associated *Olympia* with Baudelaire's *Fleurs du Mal*, perhaps invoking the naked lover in the censored poem "The Jewels." As Jean Ravenel wrote:

> Painting of the school of Baudelaire[;] the vicious strangeness of the little *faubourienne*,... [h]er look has the sourness of someone prematurely aged, her face the disturbing perfume of a *fleur du mal*; her body fatigued, corrupted, but painted under a single transparent light[.][65]

As this chapter has shown, the embodied experience of pornography is directly related to the time and place of its exposure. So, too, is the language of anger and disgust that pornography elicited in France during the 1850s and 1860s. In the next chapter, I have retained the experiential-affective linguistic model developed here to examine how the rupture and fragmentation of Haussmannization shaped Baudelaire's prose poems, published posthumously as *Le Spleen de Paris*. However, I do not foreground the neurobiological processes detailed here. Instead, this knowledge provides the foundation of my literary analysis of the ironic structures of the poems and the parodic devices that elicit a mirroring effect, affecting the reader's ironic self-reflection.

Notes

1 See also Charles Bernheimer, "Manet's *Olympia*: The Figuration of Scandal," *Poetics Today* 10, no. 2 (1989): 261.
2 Paul de Saint-Victor, "Salon de 1865," *La Presse* (May 28, 1865): n.p; trans. in Hamilton, *Manet and His Critics*, 514–15. "Scarron's hell" refers to Paul Scarron's *Le Virgile Travesty*, a parody of the *Aeneid*. Written in eight volumes between 1648 and 1659, it was a lengthy poem that initiated the age of burlesque literature in France; Anna Marie Roos, "*Le Virgile Travesti*, Paul Scarron, Paris, 1648," http://lister.history.ox.ac.uk/index-page_id=709.html.
3 Scholl, "Les Coulisses," 5–6; qtd. and trans., in part, in Robert, "The Photographer, the Model and Body Trafficking," in *Splendours & Miseries*, 75.
4 C. J. Lecour, *La Prostitution à Paris et à Londres 1789–1870* (Paris: P. Asselin, 1870) : 145; qtd. and trans. in Clark, *Painting of Modern Life*, 105.
5 Clark, *Painting of Modern Life*, 98.
6 Todd Love et al., "Neuroscience of Internet Pornography Addiction," 388–433. Historical parallels also justify the applicability of neural research on internet pornography addiction to the nineteenth-century practice and prevention of pornography. First, nineteenth-century "obscenities" and internet pornography are comparable as embodied modes of popular culture performed largely in the private sphere. Second, both internet and nineteenth-century pornography issued from new technologies that permeated the mass market. Now, as then, as the consumption of pornography reached epidemic proportions, official concerns for social and medical welfare likewise intensified. As Alain Vaillant asserts, "The industrialization of the production of obscenities in the middle and at the end of the nineteenth century represented a source of increasing anxiety for the public authorities and all enterprising moralists." Alain Vaillant, "Pornographie ou Obscénité: Abstract," *Romantisme* 1, no. 167 (2015): n.p.
7 Matthew Kieran, "On Obscenity: The Thrill and Repulsion of the Morally Prohibited," *Philosophy and Phenomenological Research* 64, no. 1 (January 2002): 43.
8 Kieran, "On Obscenity," 52.
9 Patrick Keilty, "Embodied Engagements with Online Pornography," *The Information Society* 32, no. 1 (December 22, 2015): 69. Part of my claim is that this process of cognitive self-reflection is not only a temporal dimension of online browsing but also registers in the processing of gazing at pornographic photography and, especially, stereocards through a stereoscope.
10 Baudelaire, "The Salon of 1859: The Modern Public and Photography," in *Selected Writings*, 295–6.
11 I have borrowed the "carnal, fleshy" phrase from Vivian Sobchack, *Carnal Thoughts: Embodiment and Moving Image Culture* (Berkeley: University of California Press, 2004), 2.
12 Keilty, "Embodied Engagements with Online Pornography," 71.
13 Keilty, "Embodied Engagements with Online Pornography," 70. See also Laurent Stern, "Pornography and Disgust," *Contemporary Aesthetics* 11 (January 2013): 7 ff.
14 R. Stark et al. "Erotic and Disgust-Inducing Pictures—Differences in the Hemodynamic Responses of the Brain," *Biological Psychology* 70, no. 1 (September 2005): 19–29.
15 "The striatum is a cluster of interconnected nuclei that form a part of the basal ganglia. It is involved in decision-making functions, such as motor control, emotion, habit formation, and reward;" Ian J. Bamford and Nigel S. Bamford, "The Striatum's Role in Executing Rational and Irrational Economic Behaviors," *The Neuroscientist* 25, no. 5 (January 24, 2019): 475, https://doi.org/10.1177/1073858418824256. According to Stark et al., "The ventral striatum was specifically activated when subjects saw highly sexually arousing pictures. This indicates the involvement of the human reward system during the processing of visual erotica" (19).

16 Vanessa Sennwald et al., "Emotional Attention for Erotic Stimuli: Cognitive and Brain Mechanisms," *Journal of Comparative Neurology* 524, no. 8 (2016): 1668.

17 Valence, as used in psychology, especially in discussing emotions, means the intrinsic attractiveness/"good"-ness (positive valence) or averseness/"bad"-ness (negative valence) of an event, object, or situation. The term also characterizes and categorizes specific emotions and describes the hedonic tone of feelings, affect, certain behaviors (for example, approach and avoidance), goal attainment or nonattainment, and conformity with or violation of the norm. "Valence (Psychology)," Wikipedia, https://en.wikipedia.org/wiki/Valence_(psychology). According to this psychological definition: "Theorists taking a valence-based approach to studying affect, judgment, and choice posit that emotions with the same valence (e.g., anger and fear or pride and surprise) produce a similar influence on judgments and choices. Stress is negative valence, and the opposite of this is pleasure or happiness. Stress can mean all unpleasant emotions." See also "valence" and "emotional valence," *APA Dictionary of Psychology* https://dictionary.apa.org/valence, https://dictionary.apa.org/emotional-valence.

18 Minming Zhang et al., "Neural Circuits of Disgust Induced by Sexual Stimuli in Homosexual and Heterosexual Men: An FMRI Study," *European Journal of Radiology* 80, no. 2 (November 2011): 418–25. Zhang's team "used fMRI to study the neural responses to disgust in homosexual and heterosexual men to investigate that issue. Thirty-two healthy male volunteers (sixteen homosexual and sixteen heterosexual) were scanned while viewing alternating blocks of three types of erotic film" (418).

19 See Sennwald et al., "Emotional Attention for Erotic Stimuli," 1668–75.

20 Zhang et al., "Neural Circuits of Disgust," 418.

21 "The thalamus translates neural impulses from various receptors to the cerebral cortex. While the thalamus is classically known for its roles as a sensory relay in visual, auditory, somatosensory, and gustatory systems, it also has significant roles in motor activity, emotion, memory, arousal, and other sensorimotor association functions," Hal Blumenfeld, "Thalamus: Definition, Anatomy, Function, & Disorders: Britannica," in *Encyclopædia Britannica*, 2020, https://www.britannica.com/science/thalamus.

22 "The frontal lobe contains most of the dopamine-sensitive neurons in the cerebral cortex. The dopaminergic system is associated with reward, attention, short-term memory tasks, planning, and motivation. Dopamine tends to limit and select sensory information arriving from the thalamus to the forebrain;" "Frontal Lobe," Wikipedia, https://en.wikipedia.org/wiki/Frontal_lobe. See also Zawn Villines, "What Does the Frontal Lobe Do?," *Medical News Today*, updated November 30, 2023, https://www.medicalnewstoday.com/articles/318139.php#Functions-of-the-frontal-lobe.

23 "Cerebellum," Wikipedia, https://en.wikipedia.org/wiki/Cerebellum.

24 "Superior Temporal Gyrus," Wikipedia, https://en.wikipedia.org/wiki/Superior_temporal_gyrus.

25 According to Zhang et al., "The insular cortex is highly interconnected with regions involved in autonomic regulation [and] might be a neural correlate of the autonomic changes associated with disgust;" "Neural Circuits of Disgust," 424.

26 Notably, the same activation was not found in the homosexual group. Regarding heterosexual responses, Zhang et al. also concluded that "Activation of the ACC suggested that emotional arousal associated with aversive stimuli engages attention resources" (423).

27 Zhang et al., "Neural Circuits of Disgust," 423.

28 Stefan Koelsch et al., "The Quartet Theory of Human Emotions: An Integrative and Neurofunctional Model," *Physics of Life Reviews* 13 (June 2015): 1. For diagrammatic modeling of the interconnectivity and bi-directionality of effector systems, affect systems, and language systems (conscious appraisal systems), see Koelsch et al., 3, fig. 2.

29 Koelsch et al., "The Quartet Theory of Human Emotions," 15.

30 Koelsch et al., "The Quartet Theory of Human Emotions," 15.

31 George Lakoff, *Women, Fire, and Dangerous Things* (Chicago: University of Chicago Press, 2008): 91.

32 In *Women, Fire, and Dangerous Things*, Lakoff variously analyzes "image schema" as a cognitive linguistic model, especially in the chapter "Case Study 2" (440–4). "An image schema is a recurring structure within our cognitive processes which establishes patterns of understanding and reasoning. Image schemas are formed from our bodily interactions, from linguistic experience, and from historical context;" "Image schema," Wikimedia Foundation,

https://en.wikipedia.org/w/index.php?title=Image_schema&oldid=1212014755. See also Beate Hampe, "Image Schemas in Cognitive Linguistics: Introduction," in *From Perception to Meaning: Image Schemas in Cognitive Linguistics*, ed. Beate Hempe (Berlin/New York: De Gruyter Mouton, 2005): 1–14. The term image schema is also foundational to cognitive art history; see Rudolf Arnheim, *Visual Thinking* (Berkeley: University of California Press, 1969); and E. H. Gombrich, *Art and Illusion: A Study in Psychology of Pictorial Representation* (New York: Pantheon Books [Bollingen series], 1960).

33 Lakoff, *Women, Fire, and Dangerous Things*, 276.

34 My use of "obscene" is consistent with the French etymology of the English word; *Oxford English Dictionary*, s.v. "obscene (adj.)," July 2023, https://doi.org/10.1093/OED/8288316530. The two English definitions of "obscene" also invoke the embodied social domain of filth and its suppression: "1. Offensively or grossly indecent, lewd; (*Law*) (of a publication) tending to deprave and corrupt those who are likely to read, see, or hear the contents. 2. Offending against moral principles, repugnant; repulsive, foul, loathsome."

35 Caption translated and annotated in Mina Curtiss, "Manet Caricatures: Olympia," *The Massachusetts Review* 7, no. 4 (1966): 733. As noted below (note 51), the first verse of the poem was published in the Salon catalogue and later by Curtiss. Five stanzas of Astruc's poem "La Fille des Îles" were inscribed on the frame of *Olympia* or attached to a sign on the frame; see Dorothee Hansen, "Manet and Astruc: Friendship and Inspiration," in *Manet and Astruc: Friendship and Inspiration*, ed. Dorothee Hansen (Madrid: Kunsthalle Bremen with Centro de Estudios Europa Hispánica, 2021): 12. Earlier, when Bertall's original caricature was published in the *Journal Amusant*, the caption launched the prostitute-dirt metaphor by labeling Olympia as a cabinetmaker's wife: "Manette, or the wife of the 'ébéniste' by Manet." [The French for cabinet-maker is derived from the word for ebony.] "How like a bouquet of flowers. [reference to popular song] This picture by Monsieur Manet is the flower of the exhibition. M. Courbet is outdistanced by the over-all length of the famous black cat. The great colorist has chosen to depict this lady just when she is about to take a highly necessary bath;" *Journal Amusant* (May 27, 1865), in Curtiss, 732. Clark also notes that Bertall's second version (*L'Illustration*) of the caricature included a chamber pot under the bed (*Painting of the Modern Life*, 144). This addition is a further reference to Olympia's insalubrious condition.

36 In the nineteenth century, Auvergne, located in south-central France, was a center of farming, cheese manufacturing, agriculture, glasswork, and ironwork. L'Auvergnat signifies the inhabitants of this region and the dialect spoken there. https://en.wikipedia.org/wiki/Auvergne and https://www.merriam-webster.com/dictionary/Auvergnat.

37 Moreau, *La Psychologie Morbide*, 264.

38 Moreau, *La Psychologie Morbide*, 265.

39 Moreau, *La Psychologie Morbide*, 266–7.

40 [Anon.], *Gazette Judiciaire*, October 2, 1861; qtd and trans. in Robert, "The Photographer, the Model and Body Trafficking," in *Splendours & Miseries*, 71.

41 Lakoff, *Women, Fire, and Dangerous Things*, 382.

42 Lakoff, *Women, Fire, and Dangerous Things*, 380.

43 Paul J. Silvia and Elizabeth M. Brown, "Anger, Disgust, and the Negative Aesthetic Emotions: Expanding an Appraisal Model of Aesthetic Experience," *Psychology of Aesthetics, Creativity, and the Arts* 1, no. 2 (May 2007): 103.

44 Silvia and Brown, "Anger, Disgust," 104.

45 While Lakoff does not define folk theory, the phrase issues from folk psychology, "a name traditionally used to denote our everyday way of understanding, or rationalizing, intentional actions in mentalistic terms" (Daniel Hutto and Ian Ravenscroft, "Folk Psychology as a Theory," in *The Stanford Encyclopedia of Philosophy*, ed. Edward N. Zalta, 2021, https://plato.stanford.edu/archives/fall2021/entries/folkpsych-theory/). Accordingly, to prove his argument regarding the emotional-conceptual structure of anger, Lakoff "presents dozens of idiomatic sayings or expressions" conventionally associated with anger (Jon Beasley-Murray, "Lakoff and the Body," *Berfrois*, published April 5, 2011, https://www.berfrois.com/2011/04/vented/).

46 Lakoff, *Women, Fire, and Dangerous Things*, 381–2.

47 See LaVelle Hendricks et al., "The Effects of Anger on the Brain and Body," *National Forum Journal of Counseling and Addiction* 2, no. 1 (2013): 1–12; and Brindusa Vanta, "Physiology of Anger," Mentalhelp.net, last updated Sept. 25, 2024. https://www.mentalhelp.net/anger/physiology/.

48 Lakoff, *Women, Fire, and Dangerous Things*, 382–3.
49 Here, arousal is replaced by lust to connote the emotive side of arousal.
50 See E. D. Lilley, "Two Notes on Manet," *The Burlington Magazine* 132, no. 1045 (1990): 266. According to Darcy G. Grigsby, "the title of Astruc's poem: 'La fille des îles' ... can be translated as 'Girl or Daughter of the Islands,' but also prostitute. For the French, 'the islands' would have been the West Indies (or Antilles);" "Still Thinking about Olympia's Maid," 436.
51 The complete stanza is translated in Clark, *Painting of Modern Life* (462): "When, weary of dreaming, Olympia wakes, Spring enters in the arms of a gentle black messenger; it is the slave, like the amorous night, who comes to make the day bloom, delicious to see: the august young girl in whom the fire burns." According to Sharon Flescher, "When Olympia was exhibited at the Salon, the contrast between the '*auguste jeune fille*' of the poem and Manet's brazenly nude model, obviously a prostitute, inflamed contemporary critics' already violent reactions to the painting;" "More on a Name: Manet's 'Olympia' and the Defiant Heroine in Mid-Nineteenth-Century France," *Art Journal*, 45, no. 1, Manet (Spring 1985): 29; see also Floyd, "The Puzzle of Olympia."
52 The simplified descriptions of the physiology of anger that follow concur with the more research-intensive "constructionist" model of emotions developed by neuropsychologist Lisa Feldman Barrett in "The Theory of Constructed Emotion: An Active Inference Account of Interoception and Categorization," *Social Cognitive and Affective Neuroscience* 12, no. 1 (January 2017): 1–23, https://doi.org/10.1093/scan/nsw154.
53 Significantly, the neuroanatomical processing of anger activates the amygdala and prefrontal cortex, the same brain regions activated by arousal and disgust. As Vanta ("Physiology of Anger") states, "Although it is possible for your emotions to rage out of control, the prefrontal cortex of your brain, which is located just behind your forehead, can keep your emotions in proportion. If the amygdala handles emotion, the prefrontal cortex handles judgment. The left prefrontal cortex can switch off your emotions. It serves in an executive role to keep things under control" (n.p.).
54 Vanta, "Physiology of Anger."
55 LaVelle Hendricks et al. provide a cogent explanation of the mediating function of the amygdala between psychical and social spheres: "The amygdala is an excellent indicator of threats. Its main purpose is emotional and social processing. We can react to the threat before the prefrontal cortex, which is responsible for the brain's thoughts and judgments, can assess the reaction's rationality. In other words, the amygdala causes the brain to react to the threat or fear before the prefrontal cortex can consider the consequences" (6).
56 Privat concluded with his endorsement of Manet: "But finally, in actual fact, this picture [*Olympia*] is a hundred times better than the yellow, coarse [academic] *Odalisque* which swells up on the opposite wall," Gonzague Privat, *Place aux Jeunes, Causeries Critiques sur Le Salon de 1865* (Paris: F. Cournol, 1865), 136–7; qtd. and trans. in Joseph C. Sloane, *French Painting Between the Past and the Present* (Princeton: Princeton University Press, 1973), 190.
57 [Eugène] Disdéri, *L'art de la Photographie* (Paris: Chez l'auteur, 1862): 304; trans. and qtd. in Robert, *Splendors & Miseries*, 70.
58 Paul, "Salon de 1863: Les Refusés," [2]; qtd. in Krell, "Manet's *Déjeuner sur L'herbe* in the *Salon des Refusés*: A Re-Appraisal," 317. Krell extensively catalogs the critical reviews of Manet's *Le Déjeuner sur L'herbe*. As demonstrated in this chapter, I disagree with Krell's argument that the criticism was not all negative and that the scandal was related to Manet's technique rather than the perception of immorality (319).
59 Ernest Chesneau, *L'art et les Artistes Modernes en France et en Angleterre* (Paris: Didier, 1864): 188–9; qtd. and trans. in Cachin et al., *Manet, 1832–1883*, 165–66.
60 Etienne, *Le Jury et les Exposants*, 30; qtd. and trans. in Cachin et al., *Manet, 1832–1883*, 166.
61 Laincel, *Promenade aux Champs-Élysées*, 13–14; qtd. and trans. in Clark, *Painting of Modern Life*, 96.
62 Jules Clarétie, "Deux Heures au Salon," *L'Artiste* 1 (January 1, 1865): 226; qtd. and trans. in Cachin et al., *Manet, 1832–1883*, 181.
63 Théophile Gautier, "Le Salon de 1865," *Le Moniteur Universel*, no. 185 (June 24, 1865); qt. and trans. in Cachin et al., *Manet, 1832–1883*, 181.
64 Ego, "Courrier de Paris," 291; qtd. and trans. in Clark, *Painting of Modern Life*, 156.
65 Jean Ravenel, [review of *Olympia*], *L'Epoque* (June 7, 1865); qtd. and trans. in Clark, *Painting of Modern Life*, 142. According to Clark, Ravenel is the pseudonym of Alfred Sensier (143).

8 Baudelaire's Prose Poems

A Paradigm for Ironic Parody

In the Introduction and Chapter 1 of this volume, I established how visual artists emulated and adapted Baudelaire's paradigm of "the painter of modern life," for whom "self-doubling" (*se dédoubler, dédoublement*) characterized the creative process in the form of ironic self-reflection. This "self-doubling" is embedded in the completed work of art and extends to the aesthetic experience itself. Parody and its ironic aesthetic effects stem from the artist's critical reflection on and inversion of artistic conventions.[1] In Chapter 7 I examined more closely how Manet parodied both the classical nude and nineteenth-century erotic photography, in *Le Déjeuner sur L'herbe* (1863) and *Olympia* (1863), for ironic effects (Figures 6.1 and 6.2). In the present chapter and those that follow I expand on the continuum between irony and parody and their fusion as ironic parody. In doing so, I circle back to and round out the overarching theme of this volume—the neuroarthistory of Baudelairean modernity. Accordingly, in this chapter I delve into Baudelaire's prose poems as the model that gave rise to not just verbal parody but, crucially for this study, visual parody. I examine here the formal attributes of literary parody that also manifest in visual parody, and analogously, I further develop the functions of the mirror as the medium and mediator of parody and its ironic effects. My emphasis on the mirror's functions and parody's effects deepens our understanding of these verbal/visual devices and their embodied effects, namely the cognitive self-reflection and transformation that ensue from verbal/visual rupture and fragmentation. This analysis also serves as a foundation for the next chapter, which investigates the dialogic relation between Zola's literary and critical texts and Manet's pictorial representations, placing Baudelaire's model as their shared contact for extending ironic parody within their respective mediums.

Baudelaire's *Lettre-dédicace* to Arsène Houssaye: Launching the Prose Poems as Ironic Parody

In their studies of Baudelaire's translation of graphic caricature into poetic form, Sonya Stephens and Maria Scott establish parody and its ironic effects as integral to Baudelaire's development of his prose poems. In his 1857 essay "Quelques Caricaturistes Français" ("Some French Caricaturists"), Baudelaire praises caricature as an ideal, modern art form because these "trivial images, sketches of the crowd and the street scene are often the most faithful mirrors of life."[2] By adapting caricature's "engagement with the everyday [as] a way of portraying and taking issue with the values, beliefs, and social practices of the bourgeoisie," Baudelaire extended caricature's "discursive duplicity, artistic paradox," and self-irony to his prose poems.[3] As Stephens, Scott, and others argue, caricature enabled Baudelaire to conceive the prose poem as a particularly modern form, employing

DOI: 10.4324/9780429442315-9

parody as a subversive device to challenge and overturn conventional art forms. My discussion builds upon Stephens' and Scott's analyses to show how parody in the prose poems is both the apparatus and product of ironic self-reflection during the writer's creative process. As I argue in Chapter 1, the mirror and mirroring are central to the act of ironic self-reflection, and I contend that ironic parody manifests as a mirroring device in both the prose poems' binary structures and critical functions.

The publishing history of the prose poems provides a time frame against which we can speculate about the creative and conceptual trajectory that extends from Baudelaire to the painter's and writer's portrayal of modern life.[4] In 1855, Baudelaire wrote his first (two) prose poems, published in an obscure anthology titled *Hommage à C. F. Denecourt. Fontainebleau. Paysages Légendes Souvenirs Fantaisies.*[5] Beginning in 1857, his prose poems would have been more widely known among his circle of writer and artist friends who coalesced in a culture of wit.[6] That year, Baudelaire also began writing a series of poems in prose under the title *Poémes Nocturnes* and published the poems piecemeal in literary journals. A group of six 'nocturnal' poems appeared in *Le Présent: Revue Hebdomadaire de la Littérature et des Beaux-arts* in August 1857,[7] and a further group of nine poems, of which six were new, was published in *La Revue Fantasiste* in 1861. Anxious to publish the whole series, Baudelaire found an opportunity when his friend Arsène Houssaye became editor of the renowned *La Presse*. In August and September 1862, Houssaye's journal published three groups of twenty prose poems, fourteen new and six reprints. Due to a misunderstanding between the editor and poet, in which Houssaye believed all the poems to be new, another series was announced but never published. Once again, Baudelaire's subsequent poems were published piecemeal in various journals during the next few years. Upon his death in 1867, Baudelaire had published 45 of his prose poems but could not secure a book publisher for the complete collection. These items and five additional poems found after his death were posthumously published in a single volume in 1869. The collection was titled *Le Spleen de Paris*, *Poémes en Prose* (Paris Spleen, Prose Poems), a title first used in the February 7 and 14, 1864 edition of *Le Figaro*.[8]

It was in the August 26, 1862 issue of *La Presse* that Baudelaire introduced the first series of poems appearing in that journal with a dedicatory letter (hereafter *lettre-dédicace*) to his then-editor, Houssaye.[9] In the letter, Baudelaire presented the prose poem as a means of expressing the experiences of modern life and overcoming the limits of verse poetry. Because many scholars consider the *lettre-dédicace* as both a manifesto on the prose poem and a parody of the same genre, I quote the text in its entirety. In doing so, I insert my own thematic headings that align with Baudelaire's practice of ironic parody and, subsequently, with the Manet-Zola dialogic exchanges.

Para. 1. In praise of fragments and fragmentation, non-linear narrative

> To Arsène Houssaye[10]
>
> My dear friend, I'm sending you a little piece of work of which it would be quite unjust to say that it has neither head nor tail, since everything, on the contrary, is both head and tail, alternatively and reciprocally. Just think what admirable advantages such a combination offers to everyone, to you [the editor], to me, and to the reader. We can interrupt wherever we wish, I, my reverie, you, the manuscript, the reader, his reading; for I am not tying a reader's recalcitrant will to the unending thread of a superfluous plot. Remove one vertebra, and the two halves of this tortuous fantasy will have no

difficulty in reuniting. Chop it into a number of fragments and you will see that each can exist on its own. In the hope that some of these stumps will be lively enough to please and amuse you, I'm taking the liberty of dedicating to you the entire snake.

Para. 2: Naming/modernizing (Romantic) sources

I have a small confession to make to you. It was in leafing through, for at least the twentieth time, Aloysius Bertrand's famous *Gaspard de la nuit* [1831] (doesn't a book known by you, by me, and by a few of our friends have every right to be called *famous?)* that I had the idea of attempting something similar, and applying to the description of modern life, or rather of a modern and more abstract life, the procedure he had applied to the depiction of ancient life, which is so strangely picturesque.[11]

Para. 3: Juxtapositions/dédoublement: old poetic conventions and new techniques

Who among us has not, in moments of ambition, dreamt of the miracle of a form of poetic prose, musical but without rhythm and rhyme, both supple and staccato enough to adapt itself to the lyrical movements of our souls, the undulating movements of our reveries, and the convulsive movements of our consciences?

Para. 4: City as inspiration; targets Romanticism of Houssaye's urban lyric poem

This obsessive ideal springs above all from frequent contact with enormous cities, from the junction of their innumerable connections. Isn't it true that you yourself, my dear friend, have attempted to translate in a *song* the strident cry of the *Glazier*, and to express in lyrical prose all the distressing suggestions that cry sends to the very attics, through the highest fogs of the street?

Para. 5: Literary invention and ironic twist; self-mocking and genre-mocking; self-surprise

But, to tell the truth, I fear my jealousy has not brought me happiness. No sooner had I begun my work than I realized not only that I remained very far from my mysterious and brilliant model, but also that I was doing something (if it can be called *something*) which was curiously different, an accident in which anyone but I would doubtless take pride, but which can bring only profound humiliation to a mind which considers the poet's greatest honour to lie in accomplishing *exactly* what he planned to do.

Your most affectionate,

C. B.

Stephens' studies of Baudelaire's prose poems sharpen our understanding of their parodic structures and ironic effects from the perspective of literary genres that the poet combined to construct a hybrid genre with subversive functions.[12] According to Stephens, parody and its dualities, "and sometimes even duplicity,"[13] are central to the prose poems' hybridity. Because this perspective substantiates my intermedia approach to Baudelaire's parodic practices, it is worth reviewing Stephens' analysis of Baudelaire's *lettre-dédicace* to Houssaye, which exemplifies both the parodic irony and dualities that characterize the entire *poémes en prose* series.

Stephens contends that the letter is a prose poem that parodies archaic literary conventions. Her analysis highlights two ways that Baudelaire subverted old forms within his new hybrid genre. First:

> the generic superscription ['*petits poémes en prose*' which headed the series published in *La Presse* in 1862] plays with a convention which, by the mid-1850s, was falling into disuse, [suggesting] an ironic appropriation of an archaic tradition in the service of an innovative and experimental enterprise.[14]

Second, Baudelaire's reference to the precedent of Aloysius Bertrand's *Gaspard de la Nuit*[15] and Houssaye's attempt at prose poem writing implies a literary debate with both authors' Romantic themes and tropes. As Baudelaire confessed in the letter's closing, he failed in his original plan to emulate Bertrand's text and, instead, he unexpectedly created something entirely new. Thus, as evidenced in the published prose poems, the poet's refusal to conform to these conventions registers Baudelaire's insincerity regarding his "profound humiliation" and forecloses any debt or influence. Given this ironic twist, Baudelaire's textual performance is doubly parodic in referencing and rejecting the literary prototypes. As Stephens concludes, the *lettre-dédicace* "is, in short, a parody of influence performed by a further parody of the prose poetry genre's referential [Romantic] practice." This parody within a parody justifies Stephens' claim that the *lettre-dédicace* is a poetic text in its own right: "It is, in other words, a prose poem on the prose poem which provides a commentary on the title and the dedicatory act and, in form and function, speaks of [its own] genre."[16]

Taking the *lettre-dédicace* as our point of departure, we can also pinpoint the textual devices and structure common to the prose poems I examine in this book. These features include surprise, genre distortion, fragmentation, and self-doubling. As Stephens argues, the generic shifting within each prose poem aligns with genre distortion and the aesthetic of surprise. Baudelaire sets up generic expectations by directly or indirectly referencing literary prototypes and conventions and then transgresses them. In doing so, he defamiliarizes the familiar,[17] whether conveyed by the transgressed genre or the often-banal subjects of the prose poems. The defamiliarization process also shapes the narrative structure of the prose poems—each "short story" ends with an unexpected, ironic turn, as seen in the last paragraph of the *lettre-dédicace*. The text is ironic in a self-doubling, mirroring way. As writer and narrator, Baudelaire reflects upon his creative process and his surprise when the final product exceeds his expectations. The aesthetic of surprise extends to the reader's self-reflexive response. Once the familiar generic patterns are disrupted, the reader must witness and register their dislocation within the psychic displacement of the plot's ironic snare.

For Baudelaire, fragmentation within the prose poems series was a means of liberating his editor, the readers of his texts, and his creative process from "the unending thread of a superfluous plot." Baudelaire introduces this feature in the first paragraph of the *lettre-dédicace* with the snake metaphor as a prescription for randomly selecting poems from head to tail or center, from one vertebra "cut" to another. The snake metaphor also refers to Baudelaire's creative process and intent since fragmentation also subverts a narrative sequence within the series, ensuring that each poem is complete in itself. Here, the cut-up snake metaphor captures the dynamic effects of navigating the modern city. This "snapshot"[18] effect of fragmentation also resonates in "The Painter of Modern Life," where Baudelaire praised Guys' renderings of the urban populace captured in non-sequential moments of city life. No less important than this literary effect, Baudelaire's aesthetic of the fragment also aligns with his advocacy of fragmented painterly techniques. Baudelaire interpreted Guys' broken contours and rapidly executed markings in his ink-and-watercolor sketches as capturing the fleetingness of modern life.

The fragmentation of the prose poem series also corresponds to Baudelaire's characterization of Guys as a "kaleidoscope with a consciousness." That this kaleidoscopic effect links Baudelaire's visual-critical thinking with his creative practice is verified by the title he proposed for the book publication of the prose poems. As he wrote in the first draft of the *lettre-dédicace*, "I have searched for titles; The 66. Although however, as this work resembles *a screw or a kaleidoscope*... it could easily be pushed as far as the cabalistic 666 or even 6666"[19] (emphasis added). Baudelaire was right to abandon mystical numerology in favor of the kaleidoscopic metaphor. Just as the colored glass fragments constantly realign in ever-new patterns, resulting from faceted mirrored reflections within the kaleidoscope cylinder, so too can the prose poem series be "chop[ped] into a number of fragments and you will see that each can exist on its own." This analogy between the kaleidoscope and the prose poem series correlates with the poet's non-linear creative process, what he referred to in the *lettre-dédicace* as his "reverie" (dream).

The kaleidoscope and screw analogies also pertain to both the ironic structure of the prose poems and the ensuing reading experience. On the one hand, the kaleidoscopic reading of the poems allows for an open work, giving the reader the freedom of interpretation and meaning-making.[20] However, as in the *lettre-dédicace*, the prose-poem propels the reader along a seemingly logical "storyline" that, upon reflection, becomes convoluted by the ironic "twist" of its ending, like the spiraling threads of a screw. Scott aptly conjoins the two analogies, marking a continuum between the poet's ironic self-reflection and the reader's fragmented experience. She attests that: "The self-reflexivity of Baudelaire's text opens up an abyss into which all certainty as to the direction of its irony must necessarily sink."[21]

Scott analyzes the prose poems from the perspective of Baudelaire's 1855 essay "Of the Essence of Laughter," which I address in Chapter 1 as foundational to Baudelaire's theory and practice of irony. According to Scott's literary analysis, Baudelaire's theory of humor and the comic and his privileging of '*le comique absolu*' (the absolute comical), as opposed to '*le comique significatif*' or '*le comique ordinaire*' (the significant or ordinary comical), correlates with his interest in caricature[22] and the grotesque and with the prose poems' self-reflexivity. Although Scott argues that *Le Spleen de Paris* is the product of Baudelaire's theory of humor and the comic, constituting a ruse or a hoax, she presents this theoretical framework with the ironic features relevant to the parodic practices formulated here.[23] In that light, it is revealing to review "Of the Essence of Laughter" from Scott's literary perspective and the recurring themes of fragmentation, self-doubling, and mirroring effects manifested in the poems' caricatural subject matter.

Scott asserts that '*le comique absolu*' results in ironic self-doubling, what Baudelaire calls "*se dédoubler rapidement*," and what she calls a "fragmentation of the ego." Scott further explains this doubling process from the perspective of reader-response theory:

> while the viewer or reader of a caricatural representation is far more likely to laugh at the expense of his or her fellow (wo)man than to recognize himself or herself as the figure of fun, [Baudelaire also] suggests that there is another kind of laughter, presumably that of the *absolute comical*, which is actually compatible with *attacks on one's self-image*.[24]
>
> (emphasis added)

Baudelaire describes this self-attack in "reference to the case of a man who laughs at his own fall, this so-called '*philosophe*' (philosopher) having '*acquis, par habitude, la force de se dédoubler rapidement et d'assister comme spectateur désintéresse aux phénomènes de son moi*' (acquired, by force of habit, the strength to split in two quickly and observe, as a disinterested spectator, the phenomena of his *self*)."[25] Scott further explains self-doubling from the writer's perspective, which aligns with my earlier analysis: Baudelaire attributes the philosopher's self-doubling to the artist who is "capable of being '*à la fois soi et un autre*' (at the same time self and other)."

Given the centrality of self-doubling in *Le Spleen de Paris*, Scott concludes that "the capacity for self-doubling [is] '[t]he necessary feature of the comic artist'"[26] and that this capacity provides the conceptual framework of the prose poem collection. Accordingly, Scott casts Baudelaire as the figure of the comic artist who presents his work as a "ruse." Just as Stephens has shown, Scott describes Baudelaire's feigned ignorance as intrinsic to his literary self-doubling and its ironic effect, a performance initiated by the *lettre-dédicace*.[27] What is significant for the present study is how Scott marks the self-transformative function of the ruse and Baudelaire's ironic tactics therein as affecting the reader of *Le Spleen de Paris*. As she explains, "to the extent that *Le Spleen de Paris* is related to the absolute comical, a particular kind of self-doubling [,]... one that presents *the reader* with a portrait of a highly self-aware author pretending to be self-ignorant" (emphasis added). She further asserts:

> One aspect of this duality is the poet's apparent mastery of ironic dissimulation. On one level, his pretenses seem to have been designed purely as a means of experiencing '*la joie de sa propre superiorité*' (the joy of his superiority); however, it would seem that the poet also sought an audience for his ruses.[28]

According to Scott, Baudelaire construed his audience as the bourgeois reader, for whom his feigned self-ignorance is an ironic trap. These readers become complicit with the prose poems' dominant speaker, who cruelly mocks their caricatured subjects. By the end of their perusal, the reader becomes the subject of Baudelaire's "narrative manipulation"[29] as he "exploit[s] the reader's credulity and narcissism"[30] and holds them up to ridicule.[31] The prose poem "Le Mauvais Vitrier" ("The Bad Glazier"), which I examine further below, exemplifies this turn-of-the-screw. Here, the narrator ridicules a sheet-glass street vendor for the untinted transparency of his wares, launching a verbal and then physical assault on his victim. For the "initiated" reader,[32] this discrepancy between cruelty and innocence conforms to the dynamic of the texts that compels the reader's self-transformation. Yet, as if suspended within a kaleidoscopic snare, "the reader of the prose poems can never be entirely sure as to the direction of Baudelaire's irony."[33] Given Scott's analysis of the reader's precarious response, we can determine the mirroring effect of Baudelaire's caricatural prose poems and their ironic turns: the reader's pleasure in enjoying their superiority at the expense of the subject of ridicule is disrupted when they reflect upon the sudden discovery of unexpected correspondences with themselves, as well as with an awareness and acceptance of one's foolishness.[34]

With Scott's and Stephens' analyses in mind, how do the prose poems mirror a "kaleidoscope with a conscience," as per Baudelaire's "The Painter of Modern Life," whether determined by the poet's creative process or the reader's transformative reading process? To explore this, I have selected texts that best represent the doubling/mirroring effects of ironic parody and the cultural/literary conventions that parody targets. Michèle Hannoosh has elaborated on parody's self-reflexivity as an act of self-criticism, attending to the writer who "parodies the act of writing."[35] She models her analysis on Richard Poirier,[36] whose definition of the genre also supports my analysis of Baudelaire's prose poems and the regenerative function that parody performs in modernist creative practices. As Poirier states, "In self-parody the 'inferred standards' behind the parody are not allowed to become authoritative, and that self-parody does not merely question the validity of a text but proposes 'the unimpeded opportunity for making new ones.'"[37] As this analysis pertains to my study, parody's targets are two-fold—both the "standards" of writing and the writer's act of writing.

Ironic Parody as a Mirroring Device in the Prose Poems

I begin with "Perte D'auréole" ("Loss of a Halo" [1869][38]) as exemplary of Baudelaire's self-reflection upon Romantic literary conventions that he targets through the lens of ironic parody. "Loss of a Halo" also best parallels "The Painter of Modern Life," since both texts manifest Baudelaire's idea of modernity comprising "the quest for presentness, a paradoxical, inconceivable, attempt to convey through art the immediacy of life outside."[39] As Edward Kaplan asserts, "'Loss of [a] Halo' fulfills Baudelaire's earliest ambition to capture 'the transitory, the ephemeral, the contingent, that half of art, whose other half is the eternal and the immutable.'"[40] The prose poem also echoes "Of the Heroism of Modern Life" from the *Salon of 1846*, where Baudelaire rejects the heroism of ancient myths and modern warfare and extols:

> [t]he pageant of fashionable life and the thousands of floating existences—criminals and kept women—which drift about in the underworld of a great city; [and] prove to us that we have only to open our eyes to recognize our heroism.[41]

What better way to achieve these goals than demystifying and dethroning the Romantic poet-prophet within the turbulence and tackiness of Haussmann's new boulevards?

"Loss of a Halo"[42] begins with a surprise—a random meeting in the middle of a new boulevard between the poet and a friend, which then ensues in the form of a dialogue and a dialogic mirroring of the ideal and the real. The friend is astonished to encounter the poet in this seedy part of the city, and he is so because of the Romantic myth of the artist as a 'seer' who writes about transcendent existence: "Hey what! You here, dear fellow! You, in a house of ill fame! You, the drinker of quintessences! You, the ambrosia eater! Really, this takes me by surprise." The poet, however, has just experienced a physical surprise—fighting the boulevard traffic, he loses his halo, which is muddied by the freshly laid tarmac:

> My dear fellow, you know my terror of horses and carriages. Just now, as I was crossing the boulevard, and hopping in the mud, in quite a hurry, through the shifting chaos where death comes galloping from all sides at once, my halo slipped off my head, in one abrupt movement, into the mire of the macadam.

Recognizing the halo's empty signification, he did not pick it up. Instead, he chose to avoid injury in the process (that is, he opts for physical relief over feigned spiritual release): "I didn't have the guts to pick it up. I considered it less disagreeable to lose my insignia than to break my bones." At the end of this brief report, the poet revels in the newfound freedom that the loss of the halo—and the friend-cum-reader and their social expectations—had imposed: "And anyway, I said to myself, misfortune is good for something. Now I can walk about incognito, commit foul acts, and indulge in debauchery like ordinary mortals. So here I am, just like you, as you can see!" This freedom notwithstanding, the friend insists that the poet report the loss to the police or post a public notice. The poet rejects these ideas: "Good God no! I'm fine here. You're the only one who recognized me. Besides, dignity irks me." In the ironic spirit of the "comic artist," the poet cheerfully casts off all social trappings of the poet-ideal and envisions his binary opposite. This "bad" poet will retrieve and don the halo, sanctifying the "other"[43] to write horrible poetry and perpetuate the ignorance of readers who consume it. The poet concludes:

> And I'm glad to think that some bad poet will pick it up and insolently stick it on his head. Make someone happy, what a delight! And especially a happy someone I can laugh at! What about X, or Z! Right! Wouldn't that be funny!

This superiority that comes with mocking an "other," here a standard-bearer of Romantic practice, compels the poet's subversive delight.

"The Bad Glazier" (1862) reiterates the self-reflective theme in which Baudelaire ridicules and parodies conventional literary tropes and the Romantic poet-ideal.[44] Although literary scholars of Baudelaire's ironic practices provide extensive analyses of this text, I approach it from the perspective of Baudelaire's art criticism and the ironic parody it cultivates. In this way, we can determine that the subjects' *dédoublement* (of self and other) is multidimensional, primarily by examining the varied narrative voice. However, the narrator also personifies both Baudelaire's "ignorant" reader (what I call the narrator-reader) and the author-poet's alter ego (who lashes out at the reader's philistine expectations). Simultaneously, the glazier personifies the same 'rejected' author-poet who refuses to conform to poetic conventions and pursues "the miracle of a poetic prose" from "frequenting great cities." As such, the glazier-cum-poet (or glazier-poet) becomes the target of the narrator-reader's attack.

The story begins with the narrator's meditation on customarily controlled and contemplative people who are then suddenly and irrationally compelled to take extreme, violent action for no other reason than capriciousness. We are also introduced to the narrator-poet. In the first paragraph, comprising a single sentence, the narrator-poet immediately discounts actions motivated by "mysterious and unknown impulses," the elixir of Romantic writers: "There exist characters, purely contemplative and completely unsuited for action, who, however, influenced by a mysterious and unknown impulse, sometimes act with a speed of which they would not have believed themselves capable."[45] Here we may recall the *lettre-dédicace* where Baudelaire ironically lamented his failure to emulate his "mysterious and brilliant model," Aloysius Bertrand. In the following five paragraphs, the narrator continues this meditation, providing examples of "friends" who have experienced "this crazy energy which hits these lazy and voluptuous souls so suddenly,... [and] find an excess courage for executing the most absurd and often even the most dangerous acts." For the author-poet, do these "lazy and voluptuous souls" personify the Romantic poet who relies upon dreams to compose his poems?[46]

Following this inventory, the narrator admits, "More than once I have been victim of such attacks and outbursts." He recounts, "One morning I had awakened sullen, sad, and worn out with idleness, and I felt impelled to do something great, a brilliant action.

And I opened the window, alas!" The first person he sees and hears becomes his target for "the most absurd and dangerous acts" and the subject of his disdain—the street peddler of windowpanes:

> The first person I noticed on the street was a glazier, whose piercing, discordant cry reached me through the heavy and dirty Parisian atmosphere. Moreover, it would be impossible for me to say why I was seized by a hatred for this pitiful man as sudden as it was despotic.

At this point in the story, the narrator-poet parodies his readers' hostility toward his presentation of the Paris underworld and its sordid streets. Upon recognizing the glazier and his fragile ware, the narrator invites him to his seventh-floor room through a narrow staircase.

Now the prose-poem reader is caught in a multidimensional snare of the author-poet's *dédoublement*; on the one hand, we recognize the author-poet's [Baudelaire's] impoverished condition; on the other hand, we realize the narrator-reader's moral depravity. The narrator admits to setting up a nearly impossible challenge for the glazier and feeling "gaiety" as he watched him "experiencing some difficulty in effecting his ascent and bumping the corners of his fragile merchandise many times." Once the glazier has arrived, the narrator verbally attacks him for the lack of colored glass among his wares: "What? You have no colored panes? no pink panes, no red, no blue, no magic panes, no panes of paradise? You are shameless! You dare walk through poor neighborhoods, and you don't even have panes which make life beautiful!" To be sure, the narrator's verbal scorn is that of the author-poet for his philistine critics who support the Romanic poets and their "mysterious" worldview. This tirade, which extends to the story's conclusion, turns into physical violence when the narrator "vigorously... pushed [the glazier] to the stairs, where he staggered grumbling." Still not satisfied, the narrator waited for the glazier to appear below at the door entrance, whereupon he went to his balcony, grabbed "a little pot of flowers," and "let it drop on the back edge of his pack." As a result of this impact, "The shock knocked him over, and he ended by breaking his entire poor itinerant fortune under his back, which produced the brilliant sound of a crystal palace smashed by lightning."

Ironically, the sound of a shattering crystal palace induces a sensorial "gaiety" (as opposed to the shrill street cry and the absence of colored glass) for the depraved narrator-reader, just as their rage against the obstinate glazier-poet induces a manic thrill: "Drunk with my madness, I shouted at him furiously, 'Make life beautiful! Make life beautiful!'" For the moment, the narrator pauses to reflect upon his hideous action, along with his reader: "Such neurotic pranks are not without peril, and one can often pay dearly for them. But what does an eternity of damnation matter to someone who has experienced for one second the infinity of delight?" Ultimately, the author-poet sustains his and the reader's *dédoublement*. This double-sided dilemma, of which the narrator speaks, challenges the reader to reflect upon their self-imposed superiority as both bourgeois and consumer of literary commodities. No less critical, the poet-alter ego undermines his reader's expectations, dismissing the poet's duty to beautify the world and directing him to portray instead "the thousands of floating existences... which drift about in the underworld of a great city."

I conclude my analysis of the prose poems with "The Mirror" to further substantiate Baudelaire's ironic self-reflection upon his literary practices, the conventions he parodies, and the ideal poet's role within the popular imaginary. Since I have already elaborated upon these strategies, a summary review of the poem will suffice to augment the foundation upon which Zola and Manet constructed their ironic parody devices, to be explored in the next chapter.

"The Mirror" (1864)[47] is the shortest of the prose poems, comprising five sentences, two of which are the lines of a dialogue between a homely man gazing at his mirrored reflection and his brutal interlocutor. I interpret this prose poem as a quintessential, self-contained episode of ironic self-reflection. As such, it presents irony and its parodic effects as the thing-in-itself—a perpetual continuum of irresolvable contradictions from which there is no escape.

> A frightful man enters and looks at himself in the mirror.
>
> "Why do you look at yourself in the mirror, since you must see yourself there only with displeasure?"
>
> The frightful man replies, "Sir, according to the immortal principles of '89, all men are by right equal. Thus I possess the right to see my reflection; with pleasure or displeasure, that only concerns my conscience."
>
> According to common sense, I was probably right; but from the legal viewpoint, he was not wrong.[48]

Literary scholars have interpreted this text as representing Baudelaire's political and ethical beliefs and his comic and caricatural strategies. These approaches notwithstanding, I regard "The Mirror" as a reflection of Baudelaire's ongoing conflict with the guardians of conventional literary aesthetics and his retort to critical attacks on his work. In this scenario, the poet's *dédoublement* is both sequential and simultaneous. Baudelaire assumes the dual personas, first that of his critics—the narrator-poet as brutal interlocutor—and then that of his poet-alter ego—the "frightful man" under assault for breaking aesthetic conventions. This brief dialogue between the interlocutor and his target sustains a constant seesawing between two rights and two wrongs (both/and). But this dialogue also extends to the reader and the poet himself, compelling ironic self-reflection and *dédoublement* for readers of Baudelaire's prose poems and mirroring Baudelaire's lived experience open to public scrutiny.

That Baudelaire figured the "frightful man" as a reflection of his public image is supported by the critical rhetoric and censoring that *Les Fleurs du Mal* unleashed, as examined in Chapter 6's review of the 1857 censorship trial of the volume and the five verse poems judged to be "obscene." As Claude Pichois asserts, the negative reviews published in *Le Figaro* on July 5 set the stage for bringing the book to trial. Here the critic (the publisher's son-in-law) drew attention to "four masterpieces of passion, art, and poetry" ("Le Reniement de' Saint Pierre," "Lesbos," and the two poems entitled "Femmes Damnées"), claiming his "duty" to "say... that whilst one might be able to understand a twenty-year-old poet allowing his imagination to run away with him and exploiting these themes, nothing can be said in defense of a man of over thirty who advertises such monstrosities."[49] A similar attack followed in *Le Figaro* on July 12.[50] During the obscenity trial, the prosecuting attorney, Ernest Pinard, embellished such monstrosities, resulting in Baudelaire's conviction for violating "public morality" and the poems' censure. Similar attacks on the 1861 edition of *Les Fleurs du Mal* suggest that this rhetoric of immorality and monstrosity endured in the public discourse. Again, *Le Figaro* launched

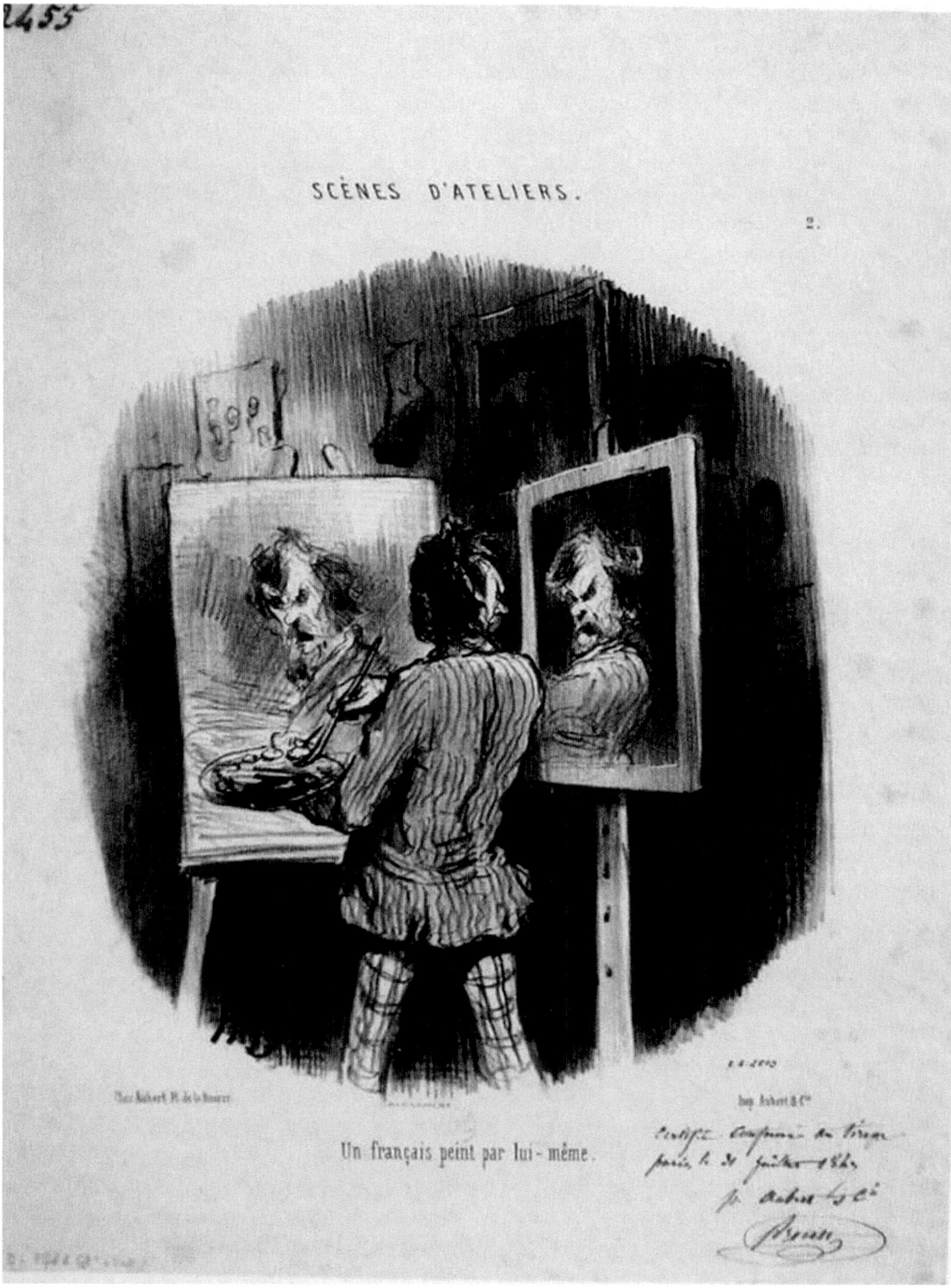

Figure 8.1 Honoré Daumier, "Un Français Peint par Lui-même" (Self-portrait of a Frenchman). In *Le Charivari*, March 29, 1849. National Gallery of Art, Washington, DC. Corcoran Collection. Gift of Dr. Armand Hammer.

the attacks. As Pichois reports, "Their critic recognized Baudelaire's 'vast, sometimes very lofty intelligence,' whilst reproaching him with his 'unbiblical,' macabre poetry and its creation of 'an unholy mixture of pagan corruption and excessive, Catholic austerity.'"[51] Another critic followed this reproach, declaring, "Never did a man more violent sing of more worthless things in a more impossible style."[52] The *Revue des Deux Mondes* was most revealing of the rift between "those who, like Baudelaire called for the autonomy of poetry and those who stood for bourgeois values in literature."[53] The critic concluded, "What sort of society or literature would adopt M. Charles Baudelaire as its poet?"[54]

These critical reviews serve to personify the interloper who attacks the frightful man for remaining focused on his own "real" self-image, just as Baudelaire portrayed the erotic and transgressive subjects of his poems, drawn from both his imagination and lived experiences. In both cases, the imagery and circumstances are disturbing and unpleasant, if not repugnant. Yet, as author-poet, Baudelaire parodies his critical adversaries and his critical adversity. The interloper represents the bourgeois culture that refuses to accommodate the poet's boundary-breaking lyrical and prose poetry. Also, by way of ironic self-reflection, Baudelaire parodies the suffering poet-artist who exposed his creative freedom in a moronic bourgeois world. As noted earlier in this chapter, Baudelaire translates the graphic art of caricature into literary form, here, specifically, Honoré Daumier's portrayal of an artist rendering his self-portrait (Figure 8.1).[55] The depicted artist is not deceived by his own "frightful" reflection that mirrors his actual physiognomy. Baudelaire's composition shares with Daumier's the incessant snare of "frightful" reflections from which the artist/poet and viewer/reader cannot escape; both personae are forced to reconcile with their own 'ideal' expectations.

Conclusion

This chapter examines Baudelaire's prose poems to underscore the performative and formal attributes of literary parody that manifest in visual parody. These features include fragments and fragmentation, non-linear narrative, juxtapositions/*dédoublement* of old poetic conventions and new techniques, and ironic twists/genre-mocking. Each of these incurs mirroring effects of ironic self-reflection, a paradigm Baudelaire sustained from German Romantic irony.

By concluding this chapter with "The Mirror," I wish to emphasize the prominence of the mirror in the intermedia trajectory from Baudelaire to Manet and Zola, the next chapter's subject. Therein, I conduct a comparative analysis between Zola's novels *L'Assommoir* (1877) and *Nana* (1880) and Manet's mirror-paintings, notably *A Bar at the Folies-Bergère* (1882), to further establish the four mirror functions that link Manet and Zola's ironic parody with Baudelaire's model: metaphoric, structural/compositional, mimetic (the presence of mirrors and mirror reflections), and discursive (critical). Each of these mirror functions engages the reader/viewer in self-reflective snares, playing upon displacements of the ideal with the real, and the disparities between expectations for closure and new encounters with ambiguous, irresolvable outcomes. As witnessed in Baudelaire's prose poems, these subversive mirrored disruptions compel readers' experiences of psychical immersion and cognitive and temporal delay. Such experiential aesthetics are intrinsic to Manet's modernity, denoted by mirrors and mirrored spaces, and to Zola's Naturalism, characterized by his novels' antithetical plot designs.

Notes

1 As discussed in the Introduction, I am guided by Linda Hutcheon's *A Theory of Parody*.
2 Baudelaire, "Some French Caricaturists," in *Selected Writings*, 209; originally published as "Quelques Caricaturistes Français" in *Le Présent* II, 12 (October 1, 1857): 77–95; rpt. in *Oeuvres Complètes*, II, 389–419. Baudelaire's accompanying essay, "Quelques Caricaturistes Étrangers" ("Some Foreign Caricaturists"), was published in the same issue of *Le Présent* (185–93).
3 Sonya Stephens, *Baudelaire's Prose Poems: The Practice and Politics of Irony* (Oxford/New York: Oxford University Press, 1999): 119, 159.
4 This chronology is adapted from Francis Scarfe, "Introduction," in Baudelaire, *Baudelaire: The Poems in Prose with "La Fanfarlo,"* vol. II, 11–12.
5 The first two prose poems, "Le Crépuscule du Soir" and "La Solitude," were revised when published in 1857. This publishing chronology is from Scarfe, "Introduction," 11; and "Chronologie," in Baudelaire, *Oeuvres Complètes*, 935. For the anthology publication, see Auguste Luchet and Claude-François Denecourt, *Hommage à C.-F. Denecourt. "Fontainebleau", Paysages, Légendes, Souvenirs, Fantaisies…* (Paris: Hachette, 1855).
6 According to Ainslie Amstrong McLees, Baudelaire was at the center of "the cult of wit [that] permeated every level of society, from the intellectual to the uneducated" (90); *Baudelaire's "Argot Plastique": Poetic Caricature and Modernism* (Athens: University of Georgia Press, 1989): 88–90.
7 See "Chronologie," in Baudelaire, *Oeuvres Complètes*, 936; and Baudelaire, "Poemes Nocturne," *Le Présent: Revue Hebdomadaire de la Littérature et des Beaux-Arts* 1, no. 8 (August 24, 1857): 284–90. Publication of *Le Présent* was short-lived, appearing from July through October 1857 (https://catalogue.bnf.fr/ark:/12148/cb328424055).
8 The CNRTL (portal) provides the literary definition of *Spleen*: "État affectif, plus ou moins durable, de mélancolie sans cause apparente et pouvant aller de l'ennui, la tristesse vague au dégoût de l'existence" (*Liter*. Affective state, more or less lasting, of melancholy without apparent cause and which can range from boredom, vague sadness to disgust with existence.) "SPLEEN: Définition de SPLEEN," CNRTL, accessed September 15, 2022, https://www.cnrtl.fr/definition/spleen. According to Scarfe, Baudelaire could not settle on a suitable title, having proposed and published a range of options. Scarfe opines that *Le Spleen de Paris, Poémes en Prose* was used only once and should not be regarded as "the main or definitive title," but rather a "subtitle;" "Introduction," *Baudelaire: The Poems In Prose*, 12. In the introduction to the poems published in *Le Figaro* (February 7, 1864), G. Bourdin states "'*Le Spleen de Paris'* est le titre adopté par M. C. Baudelaire pour un livre qu'il prépare, et dont il faire un digne pendant aux *Fleurs du Mal*," 3; ("'Le Spleen de Paris' is the title adopted by M. C. Baudelaire for a book that he is preparing, and of which he is making a worthy counterpart to *Les Fleurs du Mal*"), 3.
9 G. Bourdin echoed these ideas in *Le Figaro* (February 7, 1864): 3.
10 "To Arsène Houssaye," *The Prose Poems*, 30–31 (original emphases).
11 According to Rosemary Lloyd, *Gaspard de la Nuit* is a collection of prose poems published posthumously in 1842. "Bertrand (1807–41) drew inspiration for his prose poems from medieval Dijon and created a prose form that is reminiscent of the ballad in its structure," *The Prose Poems*, 121 n. 30.
12 According to Sonya Stephens, Baudelaire's literary genres include the short story, lyric, allegory, and popular argot, blague, and boutade forms. See Stephens, "Argot Littéraire, Argot Plastique: Caricature in Baudelaire's Prose Poetry," *Australian Journal of French Studies* 30, no. 2 (1993): 197–206; Stephens, "Boundaries, Limits and Limitations: Baudelaire's Poèmes-Boutades," *French Studies* 52, no. 1 (January 1, 1998): 28–41; and Stephens, *Baudelaire's Prose Poems*, 21, 70, 78. 105–6, 152.
13 Stephens, *Baudelaire's Prose Poems*, 2.
14 Stephens, *Baudelaire's Prose Poems*, 8.
15 Baudelaire's letter to Houssaye refers to Aloysius Bertrand's *Gaspard de la Nuit*, a work subtitled *Fantaisies a la Maniere de Rembrandt et de Callot*. According to Stephens, "Though known only to a few cognoscenti (only forty copies were sold), the text was becoming increasingly renowned [within Baudelaire's inner circle]" (*Baudelaire's Prose Poems*, 8).

16 Stephens, *Baudelaire's Prose Poems*, 16.

17 A phrase borrowed from Russian formalists, coined by Viktor Shklovsky in his essay "Art as Device" (alternate translation: "Art as Technique"); see Lawrence Crawford, "Viktor Shklovskij: Différance in Defamiliarization," *Comparative Literature* 36, no. 3 (1984): 209–19.

18 For an analysis of the snapshot effect in Baudelaire's lyric poetry, see Susan Blood, "The Sonnet as Snapshot: Seizing the Instant in Baudelaire's 'A une Passante,'" *Nineteenth-Century French Studies* 36, no. 3–4 (2008): 255–69.

19 Baudelaire, "Plans and Projects," *The Prose Poems*, 109. The Kabalistic numbers 666 (and 6666) have an esoteric, messianic meaning as opposed to the New Testament association with the devil. As summarized in "Ask the Rabbi," a database for Ohr Somayach (a yeshiva based in Jerusalem), "The number 666 could thus represent the strength and perfection of the physical world, which Judaism teaches will occur in the messianic era, when the physical world will reach its ultimate purpose, to be a vehicle through which the *created* experience the Creator;" "Six Six Six," Ohr Somayach, accessed July 10, 2020, https://ohr.edu/ask_db/ask_main.php/277/Q1/. This article further explains that 666 is recorded in the Old Testament, in Jewish commentaries on the Bible and the Torah.

20 I adapt the concept and phrase "the open work" from Umberto Eco, *The Open Work* (Cambridge: Harvard University Press, 1989).

21 Maria C. Scott, *Baudelaire's Le Spleen de Paris: Shifting Perspectives* (Aldershot, Hampshire: Ashgate, 2005): 41.

22 Baudelaire originally wrote "Of the Essence of Laughter" to introduce his study devoted to caricature. In the end, he only wrote two chapters: "Some French Caricaturists" and "Some Foreign Caricaturists," both published in 1857 (see note 2 above). Stephens uses "Of the Essence of Laughter" to frame her comparative analysis of Baudelaire's prose poems and graphic caricature in *Baudelaire's Prose Poems*. On dating the essays on laughter and caricature, see Baudelaire, *Oeuvres Complètes*, II, 1342.

23 In labeling *Le Spleen de Paris* "a ruse," Scott argues that parody functions in the collection of prose poems as an attack on Baudelaire's audience, given the poet's "resentment at the public's confusion of his person with the subject matter of the verse [*Les Fleurs du Mal*] collection," *Baudelaire's Le Spleen de Paris*, 207–8. Stephens examines Baudelaire's criticism of bourgeois-artist and bourgeois taste in art in the chapter "The Politics of Form" in *Baudelaire's Prose Poems*, 72–107.

24 Scott, *Baudelaire's Le Spleen de Paris*, 19.

25 Scott, *Baudelaire's Le Spleen de Paris*, 19; original emphases.

26 Scott, *Baudelaire's Le Spleen de Paris*, 20; Scott here quotes Michèle Hannoosh, *Baudelaire and Caricature: From the Comic to an Art of Modernity* (University Park, PA: Pennsylvania State University Press, 1992): 252.

27 Scott agrees with Stephens, among others, that the letter is a parody of the prose poetry genre and, specifically, that "the obsequious acknowledgment of a debt to Houssaye is 'clearly ironic' [Stephens]," *Baudelaire's Le Spleen de Paris*, 24.

28 Scott, *Baudelaire's Le Spleen de Paris*, 20.

29 Scott, *Baudelaire's Le Spleen de Paris*, 43.

30 Scott, *Baudelaire's Le Spleen de Paris*, 48.

31 Scott, *Baudelaire's Le Spleen de Paris*, 43.

32 Scott argues that Houssaye was not among Baudelaire's "initiated readers" but instead duped by Baudelaire's feigned sincerity in the *letter-dédicace*. Scott's phrase "initiated reader" refers to the "informed reader" used in standard reader-response literary theory; see *Baudelaire's Le Spleen de Paris*, 24.

33 Scott, *Baudelaire's Le Spleen de Paris*, 44.

34 Scott, *Baudelaire's Le Spleen de Paris*, 48.

35 Michèle Hannoosh, "The Reflexive Function of Parody," *Comparative Literature* 41, no. 2 (1989): 115. While Hannoosh focuses on the "comic" effects of parody as a critical device (113–15, 117), I suspend the comic to focus directly on parody's critical functions.

36 See Richard Poirier, "The Politics of Self-Parody," *Partisan Review* 35 (1968): 339–53.

37 Poirier, "The Politics of Self-Parody," 352; qtd. in Hannoosh, "The Reflexive Function of Parody," 115.

38 According to Edward K. Kaplan, "Loss of a Halo" was published for the first time in the posthumous edition, vol. 4 of Baudelaire's *Oeuvres Complètes* (Paris: Michel Levy Frères, 1868–1870); see "Appendix 2," in Charles Baudelaire, *The Parisian Prowler: Le Spleen de Paris. Petits Poèmes en Prose*, 2nd ed., trans. Edward K. Kaplan, (Athens: University of Georgia Press, 1997): 181.
39 Edward K. Kaplan, *Baudelaire's Prose Poems: The Esthetic, the Ethical, and the Religious in "The Parisian Prowler"* (Athens/London: University of Georgia Press, 2009): 142.
40 Kaplan, *Baudelaire's Prose Poems*, 142.
41 Baudelaire, "Of the Heroism of Modern Life," in *Selected Writings*, 106.
42 See "Loss of a Halo," trans. in Baudelaire, *The Parisian Prowler*, 113.
43 Here, I use the "other" as the objectified, inferior subject of the "sovereign Subject," a binary structure codified in critical literary theory; see Dino Franco Felluga, *Critical Theory: The Key Concepts* (London/New York: Routledge, Taylor & Francis Group, 2015): 202.
44 Stephens interprets Baudelaire's "The Bad Glazier" as a parody and critique of Arsène Houssaye's Romantic version of the glazier, "La Chanson du Vitrier" (The Glazier's Song; 1850), and Scott, among others, concurs. See Stephens, *Baudelaire's Prose Poems*, 68–9; and Scott, *Baudelaire's Le Spleen de Paris*, 25–6.
45 Baudelaire, "The Bad Glazier," trans. in Baudelaire, *The Parisian Prowler*, 13–5.
46 See, for example, Safoora Torkladani and Pyeaam Abbasi, "Dream: A Brief Comparative Study of Nerval and Keats," *International Letters of Social and Humanistic Sciences* 55 (July 2015): 140–46, https://doi.org/10.18052/www.scipress.com/ilshs.55.140.
47 "The Mirror" was published in the December 25, 1864 issue of the *Revue de Paris* (Kaplan, *Baudelaire's Prose Poems*, 132).
48 Baudelaire, "The Mirror," trans. in *The Parisian Prowler*, 101.
49 Gustave Bourdin, qtd. and trans. in Pichois with Ziegler, *Baudelaire*, 224.
50 Baudelaire solicited positive reviews from within his literary circle (Pichois with Ziegler, *Baudelaire*, 225).
51 Pichois with Ziegler, *Baudelaire*, 275.
52 Eugene Moret, *La Causerie*, qtd. and trans. in Pichois with Ziegler, *Baudelaire*, 275.
53 Pichois with Ziegler, *Baudelaire*, 276.
54 Armand de Pontmartin, qtd. and trans. in Pichois with Ziegler, *Baudelaire*, 276.
55 In "Some French Caricaturists," Baudelaire extols Honoré Daumier as the best example of the art of caricature and a modern artist par excellence (215). He praises the artist for his ironic treatment of contemporary Parisians, exaggerating their gestures, postures, and situational contrasts, to depict their moral incongruity and physical and social vulgarity. For a comparative analysis of Baudelaire's prose poems and graphic caricature, particularly with Daumier, see Stephens, *Baudelaire's Prose Poems*, 114–25.

9 Verbal and Visual Parody

Zola's and Manet's Ironic Encounters

This chapter meticulously traces the transformation of ironic parody from a literary device to a pictorial device, focusing on the significant contributions of Manet. I delve into his parodic practices, which are strongly influenced by Baudelaire's art criticism and aligned with Zola's Naturalist novels, particularly *L'Assommoir* (1877) and *Nana* (1880). Baudelaire's "The Painter of Modern Life" (1863) is a starting point, illustrating how the mirror functions as a metaphor and metonymy of modernity and its dualities—simultaneously fleeting and fragmentary yet eternal and immutable. "The Painter of Modern Life" also exemplifies how artistic self-doubling becomes a mandate for the artist's response to "multifarious" modern urban conditions. Manet's *A Bar at the Folies-Bergère* (1882) vividly demonstrates Baudelairean parodying devices and their ironic effects (Figure 1.9). Interpreting this painting within the dynamic context of a renovated Parisian metropolis, I argue that the painter, like the novelist, used the mirror as a parodic device to stimulate viewers' critical self-reflection upon modern conditions and artistic norms. I use this painting, among others, to reveal how a word-and-image approach to visual and verbal parody restores the signifying functions of artistic and literary techniques to the realm of sensory and cognitive urban experiences, thereby illuminating the transformative power of ironic parody.

It is essential to reiterate the methodology that guides my word-and-image analysis. A comparative study of Manet's visual parody with verbal parody provides a case study to demonstrate a discursive method in which word and image function as interactive yet independent signifying agents. Manet's paintings do not illustrate literary texts, which is the method of iconographic analysis, but they participate in a shared discourse on modernity and its experiential effects. Thus, while I argue that *A Bar at the Folies-Bergère* is a manifesto of Baudelairean modernity, I also maintain that Manet's painting is not simply a graphic demonstration of Baudelaire's essay "The Painter of Modern Life." Rather, in this painting Manet transcribed his own formulation of the visual, technical means by which to represent modernity. Just as the mirror-motif subverts literary conventions and narrative structures in Zola's novels, in Manet's paintings, the painted mirror and mirrored imagery subvert pictorial conventions, inviting the viewer to engage actively in an ironic encounter with aesthetic expectations.

While the mirror is not explicitly present in "The Painter of Modern Life" text, the artist and his artworks are depicted as agents of mirror reflections. To substantiate this inferred presence, it is instructive to review the trajectory from earlier German Romantic irony to the formation of ironic parody in Baudelaire's prose poems and his aesthetic criticism, as discussed in Chapter 1. As formulated by Friedrich Schlegel, among others, Romantic irony consists of an active state of critical self-reflection upon the constant

DOI: 10.4324/9780429442315-10

contradictions of the finite and infinite, the self and the non-self, and the past and present. This ironic state likewise entails a self-induced mirroring effect: a subject's recognition of a double self. As Schlegel attests, irony arises from "the insoluble conflict of the absolute and the relative." It provides "the freest of all liberties, for it enables us to rise above our self [and engage in] constant self-parody."[1]

The Romantic ironists held that these oppositions are reconcilable but irresolvable, making dynamic tension and paradox imperative in the finished work of art. Yet irony provides a unifying perspective in the work of art and its reception. Accordingly, Schlegel further defined a poetics of irony as "the artistic reflection and beautiful self-mirroring"[2] in a work of art. The ironic perspective is thus "a glance back over the artifact which hovers over the entire work,... namely the object and the response to the object."[3]

During the nineteenth century, the object of ironic self-criticism and self-reflection shifted from the subjective self to the artistic medium, and the context of self-criticism shifted from the metaphysical to the material realm. As demonstrated in Baudelaire's prose poems, the subjective ironic attitude converges with the formal techniques of parody. As an aesthetically motivated strategy aimed at inverting or subverting literary or artistic conventions, parody triggers ironic insight in the reader/viewer.

Linda Hutcheon's *A Theory of Parody: The Teachings of Twentieth-Century Art* is foundational for my intermedial approach to modern parody and its subversive results. According to Hutcheon, modern parody is inherently double-coded since it inscribes both "continuity" with and "critical distance" from the old (targeted) text in the new text. Hutcheon further shows how irony is both the cause and effect of this parodic technique of double coding. When parody produces a doubling effect between the new text and the (old) targeted text, this effect induces an ironic attitude or situation for the viewer/reader. It is just this irresolvable opposition between tradition and its inversion that makes irony, as critical self-reflection (or mirroring), contiguous with parody. This reciprocity of technique and effect constitutes ironic parody, which became consonant with modernity during the nineteenth century.

At this defining moment of modernism, Manet's paintings deserve to be re-viewed. The four sections in this chapter explore this question from various angles. First, I review the doubling devices and their mirror functions in Baudelaire's prose poems, the subject of the preceding chapter, but now through the lens of "The Painter of Modern Life." The second and third sections examine Manet's paintings in which mirrors are either implied or represented, followed by a discussion of the mirror-motif in Zola's novels. The last section demonstrates how Zola used the mirror motif as a critical standard to champion Manet's art and how Manet, in turn, devised a critical, pictorial response to that interpretation. In each case, the mirror engages the reader/viewer in ironic parody as a device for aesthetic disruption and critical self-reflection.

The Prose Poems through the Lens of "The Painter of Modern Life"

To support the word-and-image mediation of ironic parody and Manet's importance within this discursive space, I re-view Baudelaire's prose poems through the lens of "The Painter of Modern Life." I do so to demonstrate Manet's reception of and artistic response to these texts within the experiential context of modernity.

As explored in Chapter 8, from a chronological perspective it is noteworthy that Manet created his most explicitly parodic works, *Le Dejeuner sur L'herbe* (1863) and *Olympia* (1863), coincident with Baudelaire's seminal literary publications: the *lettre-dédicace* that introduced the series of 20 prose poems, and with his drafting (1859–60) and publication of "The Painter of Modern Life" (1863) (Figures 6.1 and 6.2). From a biographical perspective, we can assume that Manet was aware of these works before their publications. At least since 1858, the poet-critic and the painter were close, and they engaged with mutual friends in a "cult of wit."[4] Until Baudelaire departed for Brussels in 1864, and resuming upon his return to Paris in 1866, the poet-critic and the painter met daily.[5] During this time, Manet frequently painted and sketched portraits of Baudelaire, an example of which appears in his 1862 *Music in the Tuileries*,[6] the painter's first manifesto of modern life (Figure 1.5).[7] Here, Manet also portrayed himself among their like-minded friends—artists and illuminati—gathered in the Tuileries Gardens to listen to the twice-weekly concerts performed there. Ostensibly, these gatherings provided occasions to discuss the paradoxes of modern life.

To align "The Painter of Modern Life" with the prose poems, I extend the premise of Chapter 8, presenting the *lettre-dédicace* as a manifesto on the prose poem and a model of ironic parody that targets the same genre. In this comparative analysis, I follow the intertwined thematic axes outlined in the previous chapter, namely: 1) the city as inspiration; 2) the fragment and fragmentation as formal and technical devices; 3) ironic self-doubling as a mirroring device; 4) parodying old artistic forms in the new texts; and 5) the kaleidoscopic effect. Here, however, I have reordered and condensed these motifs to match the tenets that Baudelaire mandates for the painter's creative process and their artistic product as mirrors of the modern city.

From this double word and image perspective, my analysis focuses on the dualities and doublings that the city generates. This approach is based on the form and content of "The Painter of Modern Life." A close reading of this text reveals a lexicon of contrasts with which Baudelaire distinguishes between conventional artistic practices and standards and newly formed responses to the modern city. Furthermore, whereas dualities are implicit in the *lettre-dédicace* and, by extension, in the prose poems, artistic self-doubling and aesthetic dualities/juxtapositions are at the core of "The Painter of Modern Life" and, consequently, of Baudelaire's aesthetic principles and critical standards. This more explicit array of dualities facilitates the transposition of ironic parody from the literary text to the painted surface. Specifically, the essay marks Baudelaire's shift of focus from the subjective self as the site for psychical doubling, the subject of the prose poems, to the technical strategies and material medium as the sites for artistic and aesthetic doubling and the staging of ironic parody.

Aligning Manifestos: Baudelaire, the City, and Doubling

As argued throughout this study, beginning with Baudelaire's essay "Of the Heroism of Modern Life," written for his 1846 Salon review, and subsequently in his essay "The Painter of Modern Life," written from November 1859 to February 1860[8] and published in 1863, Baudelaire prescribed an aesthetic of dualities and acts of self-doubling as the visceral response to the urban experience and its manifestation in artistic form. The latter essay especially mirrored the actual social and physical disruptions that Manet and Zola experienced during the Second Empire's modernization of Paris under Haussmannization. Indeed, the experience of the modern city is what foremost binds "The Painter of Modern Life" with the prose poems. This correlation unifies Baudelaire's two authorial voices as art critic and poet. (Hereafter, I refer to the poetic voice as "Baudelaire-poet" to distinguish it from his critical voice, referred to simply as "Baudelaire.") From these correspondences, Baudelaire's articulation of the affective experience of modernity is central to his aesthetic mandate for artistic renewal and his motivation for reconceiving the prose poem as a modern art form. As Baudelaire-poet "confessed" in the *lettre-dédicace*, the same embodied urban experience spawned his prose poems and their efficacy for generic/literary change:

> Who among us has not, in moments of ambition, dreamt of the miracle of a form of poetic prose, musical but without rhythm and rhyme, both supple and staccato enough to adapt itself to the lyrical movements of our souls, the undulating movements of our reveries, and *the convulsive movements of our consciences*?
>
> This obsessive ideal springs above all from frequent contact with enormous cities, from the junction of their innumerable connections.
>
> (emphasis added)

Likewise, in "The Painter of Modern Life," Baudelaire asserts that the artist must submerge himself in the urban crowd and embody its vital essence. Here, Baudelaire attributes these visceral sensations to the watercolor painter and illustrator Constantin Guys, for whom "the crowd is his domain, [he] establish[es] his dwelling in the throng, in the ebb and flow, the bustle, the fleeting and the infinite."[9]

In "The Painter of Modern Life," Baudelaire ascribes to Guys the dual attributes of the prose poet—he is both the subject of self-doubling and the creator of artworks inscribed with doubling (ironic) effects. For Baudelaire, Guys' self-doubling issues from the dynamic encounter between the self and the non-self—the artist is both a man of the urban crowd and separate from it. In the first instance, the artist is a "passionate observer," "[he] is an ego athirst for the non-ego, and reflecting it at every moment in energies, more vivid than life itself, always inconstant and fleeting."[10] Secondly, this dual existence propels artistic intervention upon aesthetic norms and their doubling effects. Just as Baudelaire-poet proposed the modern prose poem as a conflation of the old Romantic lyricism and the new upheavals of the city, Baudelaire defined the "double composition" of beauty as a constant tension between "an eternal and invariable element" and "a relative circumstantial element [of] contemporaneity."[11] This aesthetic doubling echoes Friedrich Schlegel's ironic tenets and correlates with Baudelaire's earlier review of the 1855 Universal Exposition, where he suggests how the painter of modern life can achieve aesthetic dualism and, ultimately, cultural renewal through parodic means. The artist does so by presenting "the ever-new which eternally elud[e]s," but still alludes to "the rules and analyses of the school," a dualism that triggers in the viewer the "shock of surprise."[12] This axiom is tantamount to the parodic techniques and ironic effects that Baudelaire implemented in his prose poems.

Baudelaire's self-referencing between his texts also creates a mirroring between his lyrical and critical writing. These cross-references also undergird the efficacy of his ironic-parody practices for writers and painters alike. Simultaneously, the cross-references reveal Baudelaire's shifting perspective from ironic self-reflection to ironic reflection upon the medium and techniques of representation. The kaleidoscopic trope best represents this intermedial exchange.

In "The Painter of Modern Life," Baudelaire restores the kaleidoscopic effect from the first draft of the *lettre-dédicace* but with a different host body. In the *lettre-dédicace*, Baudelaire projects the kaleidoscopic effect onto the body of texts that comprises the prose poem series. Here, the non-linear series of poems produce a kaleidoscopic effect wherein the individual reader can order the reading sequence of poems, just as turning the kaleidoscope's mirrored cylinder of colored glass fragments creates ever-new patterns. In the final version of the *lettre-dédicace*, Baudelaire replaced the kaleidoscope metaphor with the snake metaphor but still inferred the kaleidoscopic effect. Regarding the snake's regenerative skeletal system, Baudelaire asserts that the prose poem series can be "chop[ped] into a number of fragments and you will see that each can exist on its own." Each poem, in turn, induces a kaleidoscopic effect by fracturing the linear sequence of events with irresolvable, unforeseen outcomes, affecting the shock of surprise.

Baudelaire shifts the kaleidoscopic effect from the literary body to the artist's body and the body of the artwork in "The Painter of Modern Life." In this context, Baudelaire brings the kaleidoscope's mirroring function to the fore. As the embodiment of the crowd, Baudelaire explains, the artist "may also be compared to [...] a kaleidoscope endowed with consciousness, which with every one of its movements presents a pattern of life, in all its multiplicity." As "a kaleidoscope with a consciousness," Guys not only immerses himself in the "ebb and flow" of the boulevards but absorbs its tumults and shocks, just as Baudelaire-poet mediates "the convulsive movements of our consciences" into the "staccato" patterns of his non-rhyming prose poems. At this point, the kaleidoscopic effect conforms to the mirroring functions of the artist, his technique, and the finished artwork.

In "The Painter of Modern Life," Baudelaire suggests how the artist mirrors his lived experience through an artistic technique that is itself fleeting and fragmented: such a technique must "reflect [the artist's consciousness] at every moment in energies, more vivid than life itself, always inconstant and fleeting." As detailed in Chapter 1, in Baudelaire's assessment of Guys' modernity, the sketch-like pen-and-ink drawing and transparency of watercolor washes convey fleeting cognitive experiences of the dynamic cityscape and physical encounters with the myriad urban populace. Here, Baudelaire extends the fragment and fragmentation, which are intrinsic to the kaleidoscopic structure of prose poems, the artist's technique, and the signification of that technique as a break with tradition. Just as the prose poems' irresolvable endings subvert conventional plot development, Guys' sketch-like techniques counter expectations for the 'finished' work of art, comprising closed contours, precise details, and smooth surfaces. Guys' snapshot-like capture of assembled or moving figures also forecloses narrative readings. These convention-breaking techniques became Manet's signifiers for mirroring his Baudelairean modernity as a "kaleidoscope with a consciousness."

Manet's Mirrors: Agents of Ironic Parody

This section explores the three ways in which Manet used the mirror and its reflections: first, as a corollary of the viewer's gaze (whereby vision is an optical mirror) in *Music in the Tuileries* and *Olympia*; and second, as a painted mirror that reflects physical surroundings, material/painterly facture, and social spaces in *Nana* (1877), *The Café-Concert* (1879), and *A Bar at the Folies-Bergère* (Figures 6.9, 9.2, and 1.9). The third kind of mirroring, as a means of critical-ironic self-reflection upon the pictorial medium, is embedded in the three paintings.

Mirror as a Corollary of the Viewer's Gaze: Music in the Tuileries

As earlier noted, and acknowledged by Manet scholars, *Music in the Tuileries* counts as Manet's first manifesto of modern life, and more specifically, a tribute to Baudelaire and his soon-to-be-published "The Painter of Modern Life." I begin with this medium-sized group portrait painting to add to the discussion of the viewer's location in relation to the crowd observed. My contribution focuses on the absent-but-implied mirror and its historical reference, a reference that demonstrates Manet's "self-doubling"—of his own person and the viewer—and his parodic doubling of "eternal" museum masterpieces and contemporary cultural practices and art forms.[13]

Scholars have focused on Diego Velázquez as Manet's model for *Music in the Tuileries*, especially given the latter's intense engagement with the Spanish masters between 1850–62.[14] Accordingly, since Manet copied *Les Petits Cavaliers* in 1859 (Chrysler Museum of Art, Norfolk, Virginia), then attributed to Velázquez and exhibited in the Musée du Louvre,[15] the original and the copy have been established as his model for and precursor to *Music in the Tuileries*.[16] To be sure, Manet's painting modernized the outdoor gathering of seventeenth-century Spanish artists portrayed in *Les Petits Cavaliers*.[17] This comparison includes Manet's self-portrait among his artist and writer friends that mimics (what was assumed to be) Velázquez's self-portrait among his own artist friends. Even more telling is that both artists are positioned at the left edge of the canvases and gaze toward the viewer. Yet, Manet is not the only person gazing outward. While others' gazes may be turned toward the viewer, their eyes are obscured by lowered eyelids and patchy brushstrokes. Mme. Lejosne, seated in the foreground, is the only figure looking directly at the viewer with her eyes wide open (more about these two gazers shortly).

The importance of Velázquez notwithstanding, Manet's outwardly gazing figures engage the viewer/observer in a mirroring process, whereby vision is an optical mirror of what is viewed. This analogy is especially relevant to the doubling effect of reciprocal gazes between the observer and the observed. Such an exchange issues from the viewer's spatial relation to both the assembled group of listeners (located before the viewer) and the (unseen) musicians performing atop a gazebo pavilion (located in front of the listeners but behind the viewer).[18] In this encounter, the viewer's position is doubly marked: the viewer is sandwiched between the listeners and the performers so that the listeners are also observers of the person viewing both the painting and the assemblage of listeners.

In analyzing this mirroring effect as a function of ironic parody, I want to bring attention to Manet's dialogue with French masterworks, namely Jacques-Louis David's *The Intervention of the Sabine Women* (1796–9) (Figure 9.1).[19] Not only was David's painting prominently displayed in the Louvre,[20] but Manet's atelier master, Thomas Couture, lauded the painting in his 1867 publication, *Méthode et Entretiens D'atelier*.[21] At first glance, we notice a compositional similarity between David's painting and Manet's *Music in the Tuileries*: each work comprises a multitude of figures horizontally ordered in a frieze-like arrangement. In this basic comparison, we can already ascertain that Manet had followed Baudelaire's mandate to replace the ancient [Roman] heroes with the heroes of modern life. From this Baudelairean perspective, we recognize multiple ways Manet targeted and thereby inverted or parodied David's Neo-classical subject, figures, and accessories.[22]

First, on the right, Manet replaced the upraised lances and the trophies they hold with tree trunks and their verdant boughs. In this comparison, Manet seems to have exaggerated the slant of otherwise upright trunks (as recorded in his ink-wash sketch, *Corner of the Tuileries* (1862[?]; Bibliothèque nationale, Paris)[23] and thereby inverted ancient urban warfare into modern urbane leisure within a tamed natural setting. These exact replacements also indicate Manet's carry-over of David's one-point perspectival orthogonal lines, delineated by the middle-ground lances and building contours. Manet obscured but

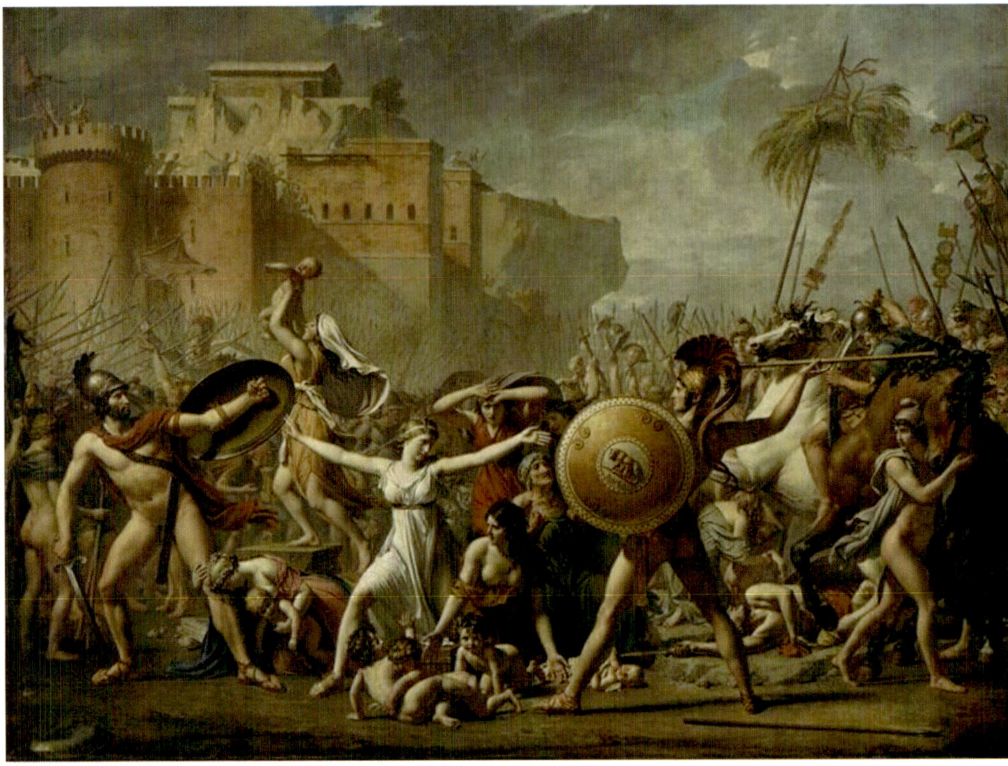

Figure 9.1 Jacques-Louis David, *Les Sabines* (*The Intervention of the Sabine Women*), 1796–9, oil on canvas, 152 x 206 in. Louvre Museum, Paris. © RMN-Grand Palais/Art Resource, NY. Photo: Michel Urtado.

retained these spatial markers with the painterly rendering of trees and figures that diminish in scale as they recede at mid-point along the central vertical axis of the canvas.

Manet not only subverted the pictorial precision of David's painting. He also subverted the multitude of male warriors and female pawns and their children with convivial persons in modern dress, and he did so with direct echoes of battle gear and individual figures. For example, he demoted the warrior's shield, right of Hersilia and frontal to the picture plane, to a swash of grey brushstrokes, to the right of center, that both denotes and obscures the veiled visage of Manet's mother.[24] Similarly, in the center foreground of Manet's painting, two girls, adorned in white dresses and colorful sashes, shovel dirt into their pails, engaging in a playful activity that undercuts the nude, frightened children in the center foreground who spill out of David's canvas. Two of these children gaze out of the canvas, appealing to the viewer's compassion, which, in turn, echoes the gaze of the forward-charging woman to the right of Hersilia. Located just above and behind the children, this woman is the only other figure who beholds the viewer, with her gaze and slightly open mouth (as if about to cry out) and upraised bent arms shielding her forehead. She, too, is a target of Manet's subversion since the grey shawl that unfolds over her head dissolves into Mme. Manet's blurred veil (at the canvas's center), disrupting the pictorial surface and spatial depth.

Gazes and the Mirror: Presence and Absence

The woman's gaze and those of the children not only behold the viewer of David's painting. As the *Sabine Women* was initially exhibited in the Louvre,[25] the figures' gazes were reflected back upon the viewer in a mirror placed on the gallery wall opposite the almost life-size canvas. In this setting, the viewer's gaze was doubly activated by the outwardly gazing figures and their bodily pivot away from the painting to face the mirror. From this position, the viewer's reflection thrusts them into the multitude. Yet, the beholder's stillness would side with Hersilia and the Sabine women to become an agent of national reconciliation, which the Sabine women symbolized in post-revolutionary France.[26] Aesthetically considered, viewers of David's painting "found themselves hemmed in between the two planes and referred from one to the other."[27] Although the mirror is not physically present in the viewing of *Music in the Tuileries*, I argue that the mirror is inferred by the viewer's placement in relation to the painted space and the dialogic gazes between observer and observed. As implied above, in this mirroring scheme, the viewer, like the self-portrayed Manet-artist, is both of the crowd and separate from it. Likewise, just as the viewer of David's painting felt hemmed in between painting and mirror, the viewer of Manet's painting is sandwiched between the listeners and the band players. We can further speculate that if Manet's viewers imagined pivoting to view the band behind them, that illusion would enhance the mirroring effect of being both part of the crowd and distinct from it.

Affinity or Precedent?

Did Manet know about David's mirrored installation of *The Sabine Women*? Do the analogies between David's and Manet's mirroring effects pertain to affinity or known precedent? In the preceding discussion, these analogies seem to be matters of affinity. I now want to address issues of precedent, suggesting how David's example endured. For several reasons, I suggest that there may have been an oral legacy of David's innovative exhibition that circulated among artists, within their studios, and in their discussions about exhibition strategies. First, this legacy may have been established at the time of David's exhibition and survived in Manet's time;[28] at least two contemporary critics described, in printed texts, the mirror's presence and the viewer's actively embodied engagement with the painting's reflection.[29] Second, *The Sabine Women* exhibition was not the only occasion that David integrated a mirror into his exhibition space; he subsequently used a mirror in 1807 to exhibit the *Coronation of Napoleon* (Louvre Museum, Paris) and again in 1824 for *Mars Being Disarmed by Venus* (Musées royaux des Beaux-Arts de Belgique, Brussels).[30] As Frédérique Desbuissons states, David's "systematic use [of the mirror] indicates a role that was more than fortuitous: it was essential."[31] If this essential integrity of mirror and painting mattered, it would have set a precedent, if not in practice, then in word. David's students, who also became atelier masters, may have circulated memories and mentions of David's radical exhibition practice to the next generation of artists. This discourse may have endured in Manet's time. Third, the perennial presence of actual mirrors in artists' studios[32] provides a material link between David, his students-cum-masters, and Manet's integration of mirroring effects in *Music in the Tuileries* and subsequent paintings.

Even if we cannot establish that Manet knew the history of David's exhibitions with mirrors, Manet invokes the presence of his studio mirror in *Music in the Tuileries*.

His self-portrait signals his mirror reflection and asserts his gaze directly at the viewer standing before the center of his painting. As previously noted, only Mme. Lejosne shares this outward gaze and the viewer's returned gaze. In this social setting, their gazes are mutually tied to Baudelaire's presence: Mme. Lejosne, and her husband Commandant Lejosne, introduced Manet to Baudelaire at their home in 1858.[33] This alliance is also mirrored in Baudelaire's shrouded presence, as seen in the profile directly above Mme. Lejosne's blue bonnet. Ironically, although Baudelaire's visage is obscured, Manet compels our engagement in this triadic union and, in turn, his testament to Baudelairean modernity that *Music in the Tuileries* embodies.

That Manet invoked the mirror in his first masterpiece was a pivotal strategy for his subsequent forays into mastering the painting of modern life and the mirroring of ironic (self-) reflections.

Viewer as Voyeur: Mirroring Ironic Self-Reflection

Although the mirror is absent in *Olympia* and is non-functional in *Nana*, a mirroring effect is activated by the directness of the female gaze. Each figure engages the viewer in the optical mirroring of her self-image: we return the perceived image to the figure that poses before us. By engaging the viewer with the female gaze, Manet inverted or parodied the male gaze that traditionally beheld the passive female nude as the embodiment of ideal beauty.

As is well known, Manet's painting of Olympia is a modernization of Titian's *Venus of Urbino* (1538; The Uffizi, Florence), which had become a standard for subsequent classical nudes. As seen in Titian's example, the figure's tilted head and sideways glance render her passive in her encounter with the male viewer. As demonstrated in Chapter 6 on pornography and censorship, when Manet turned the model's gaze directly to the viewer, he further modernized the classical nude by evoking the photographic nude. Indeed, during the 1850s, prior to Manet's painting, photographers had already inverted the male gaze by having the model look head-on into the camera aperture. Olympia's modernist affinity with the photographic model goes beyond her pose: both are inscribed with the specific social identity of a prostitute.

While Manet mirrored his bourgeois friends and family in *Music of the Tuileries* as heroes of modern life, he inverted the bourgeois male's status in his painting of Olympia. Here, the heroine-as-prostitute resides in the Paris underworld, as touted in Baudelaire's 1846 "Of the Heroism of Modern Life." By posing Olympia to mirror the bourgeois viewer's gaze, Manet effectively launched a critique of bourgeois values. This same cultivated viewer would have immediately recognized the dialogic relationship between Titian's Venus, the eroticized Venus of contemporary photography, and the sex worker in her brothel setting.

The photographic intervention and the portrayal of a modern urban type transform Manet's interpretation of Titian's Venus into a visual parody. Manet not only removed the artifice of an idealized, mythological figure to depict a recognizable prostitute,[34] he also left intact visual references to Titian's painting—the model's gestures and pose, the cuddly dog replaced by a seductive black cat, and the two-part division of the background screen. By doing so, he forced the viewer to reflect backward upon the sixteenth-century museum masterpiece and experience the transformation of the traditional ideal into the contemporary real. In this cognitive response, memory is also engaged as a mirroring device. This mnemonic event, in turn, compels the mirroring effect of ironic self-reflection.

In his painting of *Nana*, Manet further endowed the model[35] with photographic references and multiplied mirrored reflections. As discussed in Chapter 6, Nana posing before the mirror recalls studio photography, both in her stance and the presence of the mirror. For example, Louis-Camille d'Olivier often placed the models before a psyche mirror (Figure 6.7). This swiveling apparatus was a commonplace studio prop for painters and especially useful for photographers. The latter creators exploited the multiple views of the figure that the mirror reflected, producing images that served as both the painter's model and the pornographer's erotic spectacle. Olivier's more ironic stereograph ([Self-portrait With Two Nudes], 1856) signals how Manet borrowed photographic techniques to activate the viewer's gaze as a mirror reflection (Figure 6.8). In this image, the photographer's presence is made real by his self-portrait framed by the psyche mirror,[36] showing him in the mirror while gazing at his model. Yet his reflected presence makes his physical absence behind the camera equally real. Instead, the viewer is now in front of the model, so their gaze replaces the photographer's gaze through the camera apparatus. In this stance, the viewer reflects upon their complicity in producing the model's erotic appeal.

The viewer of Manet's painting also occupies the photographer's position directly in front of his model.[37] Nana stands before a psyche mirror,[38] apparently applying makeup, but turns her head to pose before the implied camera. Nana's gaze is doubly reflected. First, the camera's mirroring is synonymous with the viewer's optical mirroring. Second, the viewer's and her top-hatted companion's gazes mirror Nana's brazen self-image. In each beholding, our/his admiring gaze reproduces Nana's self-image as it reflects back to her, creating a narcissistic circuit of self-absorption and self-satisfaction. It is for this reason that Manet rendered the actual mirror useless. Instead of reflecting Nana's image, the oval surface reflects the painterly medium, color, and brushstrokes through which the artist has made Nana visually present. The mirror thus becomes an ironic device to reflect upon the production and reception of the painted image: is painting a mimetic/mirror reflection (like a photograph) or a creative re-presentation of the real? It is precisely this question that Manet addresses in his mirrored reflections of urban social spaces.

Mirroring the Painted Surface and/in the Social Sphere

In *The Café-Concert*, Manet creates a pictorial space that induces experiential associations with the modern condition (Figure 9.2). In doing so, he uses the mirror to parody pictorial perspective. To begin with, our initial encounter with the painting is ironic. At first glance, the singer performing on stage appears to be situated in the receding space of the concert hall. However, upon closer viewing, we realize that the image of the singer is a mirror reflection. The mirror's gilt frame partially defines this reflected image, while the staged performance occurs before the foreground figures. Thus, where we expect to see into pictorial depth beyond the painted surface, as in Jean Béraud's *La Pâtisserie Cloppe* (1889; Musée Carnavalet, Paris), Manet's mirror prohibits our entry into the pictorial space. The mirror's painted surface reinforces this negation—broadly laid brushwork renders the mirror as flat as the grey wall beside it. Even more so than Nana's mirror, the *Café-Concert* mirror is a critical reflection upon the painter's materials with which experiences of modernity are made real. Accordingly, Manet parodies perspective by using the mirror to place the viewer within rather than outside the picture space. If the singer's

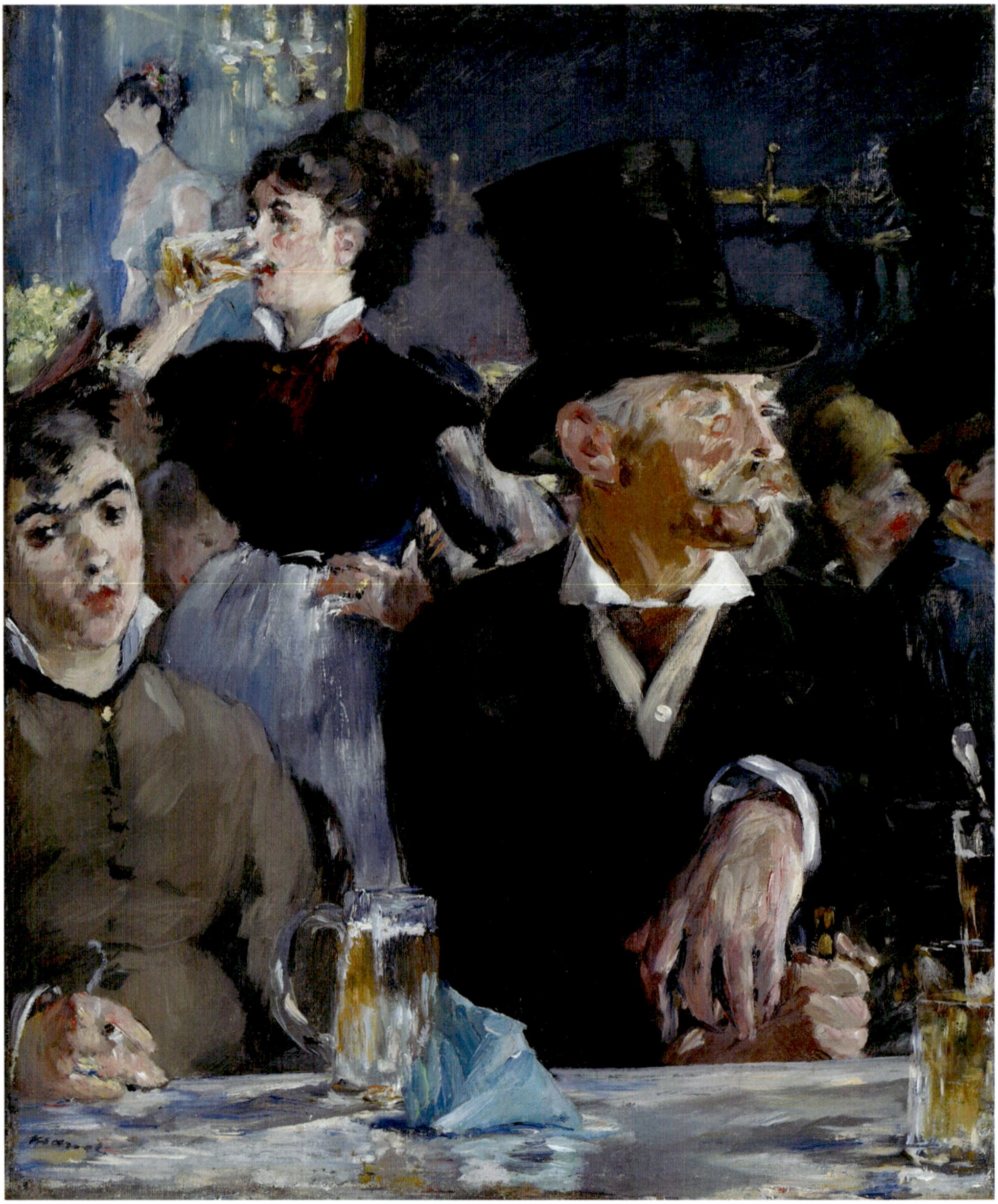

Figure 9.2 Édouard Manet, *The Café-Concert*, c. 1879, oil on canvas, 18 5/8 x 15 3/8 in. The Walters Art Museum, Baltimore. Acquired by Henry Walters, 1909–10.

performance takes place in front of the foreground figures, then the viewer of Manet's painting is situated between those figures and the stage. As encountered in *Music in the Tuileries*, even though the spectacle occurs behind us, we directly experience the multifarious ambiance of the café-concert. By placing the viewer in the middle of the picture space, Manet also compels the viewer to reflect upon the fleeting and fragmented spaces of modernity that subvert expectations with ironic encounters.

Ironic encounters are also triggered by how Manet uses the mirror to reflect on modern social conditions. For one thing, the man and woman in the foreground are not engaged with each other, and only the man observes the performance before them. Furthermore, while the café-concert was a modern site where upper and lower classes mingled, Manet has ruptured this alliance by inserting a still-life between the foreground figures who gaze in opposite directions. The rupture signifies the working-class woman, recognized by her brownish-grey garb, and the top-hatted man as her potential lover. While contemporaries would label her a *grisette* or *lorette*, they would assign him to the upper-class bourgeoisie or lower-level aristocracy.[39] Manet intensified the chasm between social types by aligning the *grisette/lorette* with the other working women on the left side of the composition.

In *The Café-Concert*, Manet has depicted three urban types of working woman: the prostitute, the waitress, and the singer.[40] Aligned from bottom to top, each woman's head leads the viewer's eye along an arrow-shaped path and into the "supposed" pictorial space. However, although their bodies connect in this formal arrangement, their gazes and gestures are disconnected. Here, too, Manet has registered a social scale among the working-class women: the *grisette/lorette* remains at the lowest rung, while the waitress can aspire to the middle class (note her wedding band). The singer, who may or may not rise above prostitution, gains social privilege among the upper classes.[41] Thus, *The Café-Concert* mirrors the social order of modern Paris, at once unified and divided.

In *A Bar at the Folies-Bergère*, Manet extended the mirror as a parodic device to invoke ironic social reflection. In doing so, he combined and perfected those mirroring effects he achieved in his earlier paintings. However, the mirror almost covers the entire picture surface, suggesting that Manet intended the painting to celebrate Baudelairean modernity. Even more so than the *Café-Concert*, Manet's parodic formal devices in *A Bar at the Folies-Bergère* conform to Baudelaire's celebration of modernism's multifariousness, a state of constant paradox, surprise, and change. In *A Bar at the Folies-Bergère* Manet also achieves the "double composition" of modern beauty and thereby invokes critical self-reflection by rendering the familiar unfamiliar and by presenting "the ever-new which eternally elud[e]s," but still alludes to "the rules and analyses of the school."[42]

As in *Olympia*, Manet's ironic approach can first be identified in how he parodies "the rules of the school," that is, museum masterpieces and academic techniques. In *A Bar at the Folies-Bergère*, the painted mirror provokes contrastive associations with the tradition of mirrored feminine beauty and the smooth academic finish in which this cultural ideal was portrayed. These include mythological types, such as Titian's *Venus with a Mirror* (c. 1555; National Gallery of Art, London) and Ingres' *Portrait of Comtesse d'Haussonville* (1845; The Frick Collection, New York). Manet subverts these conventions by rupturing expectations for mimetic representation of mirrored reflections. Unlike the contiguity between the figure and her reflection in the traditional compositions, in Manet's painting, the barmaid's reflection is too far removed from her centered, frontal pose. This disjunction was intentionally ironic, as confirmed by Manet's study for *A Bar at the Folies-Bergère* (Figure 9.3). The barmaid's reflection directly aligns with her frontal stance in this rendering. In the final painting, the sketch-like painterly brushwork also subverts our expectations for smooth mirror surfaces and reflections in academic paintings.

Like the *Café-Concert*, in *A Bar at the Folies-Bergère* Manet depicts modernity through the fragmented form. First, we encounter fragments of colors and pigments that do not physically cohere. Likewise, the mirror image is a fragment of an overall ambiance and

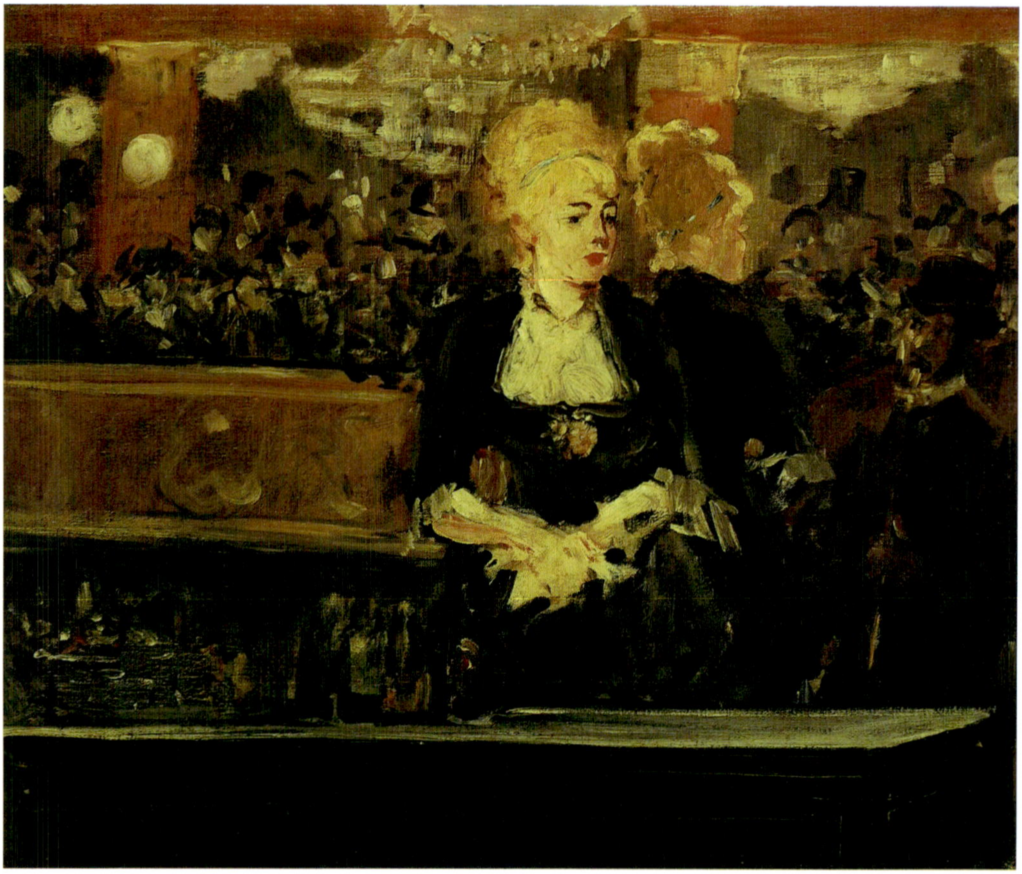

Figure 9.3 Édouard Manet. Study for *A bar at the Folies-Bergère*, 1881, oil on canvas, 18 1/2 x 22 in. Private Collection. Image: Wikimedia Commons.

fragments of the objects and figures reflected therein (note, for example, the cropped legs of the trapeze artist). The disconnected gazes between the barmaid and her bourgeois customer, who is reflected in the mirror, add to a psychological fragmentation of the whole, distinct from the preparatory sketch in which she does look at the man. At the same time, these painterly and figurative fragments cohere as the mirror image of the contradictory, ironic condition apropos of modern life and the corollary experience of endless transformations.

As a place of popular entertainment and social class mingling, the Folies-Bergère theater signified a microcosm of the urban whole.[43] Yet, in the painting of this Parisian site, as in *The Café-Concert*, Manet inserted social rupture within the disparate crowd, especially in the figure of the barmaid. Given her status as a working-class woman, she is denied the pleasures that surround her. An isolated, static figure, the barmaid stands in contrast to the dynamic spectacle of revelers, shimmering light works, and lively paint marks. Her psychological isolation is compounded by her physical position, for the mirrored spectacle is only a reflection of the action that unfolds before her. The following section will consider where the viewer stands in relation to that mirrored space.

Mirrors in Zola's Novels

Given the centrality of the mirror in *A Bar at the Folies-Bergère*, I now want to restore Manet to an enduring discourse on Baudelairean modernity and his dialogic encounter with Zola within that discursive space. I do so to demonstrate the parodic devices and ironic effects that *A Bar at the Folies-Bergère* shares with the Naturalist novel.

Why Zola? Zola provides a vibrant resource for this discursive analysis.[44] First, Manet and Zola were personal friends.[45] Second, Zola championed Manet in his art criticism, and, in acknowledgment, Manet painted Zola's 1868 portrait. In the same year, Zola dedicated his fourth novel, *Madeleine Férat*, to Manet.[46] This latter exchange reveals their camaraderie as convention-breakers in line with Baudelairean modernity. Zola's dedicatory statement signals their shared strategies for disrupting the status quo:

> To Edouard Manet
>
> The day when, with an indignant voice, I undertook the defence [sic] of your talent, I did not know you. There were fools who then dared to say that we were two friends in search of scandal. Since these fools placed our hands one in the other, may our hands remain for ever [sic] united. The crowd willed that you should have my friendship; this friendship is now complete and durable, and as a public proof of it, I dedicate to you this book.[47]

Midway through the novel, Manet hailed Zola for breaking the codes against obscenity (as the painter had done in *Olympia* and *Le Déjeuner sur L'herbe*):

> My dear friend, I'm in the middle of *Madeleine Férat* and don't want to wait till I've finished to send my congratulations. You paint the redheaded woman [Madeleine] so well it makes one jealous, and *the expressions you find to render the love scenes would deflower a virgin just to read them.*[48]
>
> (emphasis added)

Considering their alliance, I ground the verbal-visual comparisons between the painter's images and the writer's texts in the Baudelairean themes of modernity: 1) the creative and experiential acts of self-doubling; 2) the familiar—or conventional—made unfamiliar; and 3) the portrayal of "heroes" of high life and low life in modern Paris.

The sustained doubleness that ironic parody engenders "through the playing with multiple conventions"[49] initially connects Manet's paintings with Zola's Naturalist novels. The mirror is central to achieving these effects for the writer and painter. Specifically, the following discussion centers on the mirror as a visual form and formal structure in two Zola novels (which Manet knew well[50]), *L'Assommoir* (1877) and *Nana* (1880). Like Manet, Zola uses actual mirrors as commonplace interior furnishings and as vehicles for multiple reflections within an interior space.[51] More importantly for this study, Zola also uses the mirror indirectly as an ordering device to contrast reciprocal yet opposing events within the narrative structure. This mirroring device aligns Zola's Naturalist novels with Baudelaire's prose poems discussed in the previous chapter. Here, I wish to extend the formal analysis that exposed the mirroring structure of Baudelaire's poems to a structural analysis of Zola's novels. Building on critical literary theory, this structural analysis will show how the writer and painter used formal patterns and technical means to enforce a mirroring dialogue between the reader/viewer and the text or work of art. Each medium invites the reader's/viewer's anticipation of a conventional series of events—such as illusionism in painting or idealist plot development in the novel—and then denies these expectations. This section explores how the mirror-motif, embedded in the novels' texts, functions as a parodic device for critical self-reflection.

Literary historian David Baguley provides a model for the structural analysis of formal patterns and their ironic effects within the Naturalist novel. As he observes, the heroic character types and idealized plot development of earlier nineteenth-century Romantic novels are targeted for parodic inversions in the Naturalist novel. In the earlier narratives, conflicts and misunderstandings are resolved in fortuitous outcomes for the characters and their personal developments. However, in the Naturalist novel, an ironic narrative sets up "contrast[s] between a character's hopes, ideals, aspirations and ensuing plight."[52] While the reader's expectations for edifying character and plot development are met at the novel's inception, its "banal" or "deprivation" ending subverts those expectations by denying resolution or redemptive closure.[53] Heroes are replaced by anti-heroes in the Naturalist novel, and edifying evolution is replaced by degeneration. This degenerative course of events, what Baguley calls the "entropic" metatheme of the Naturalist novel, narrates the moral and social decay incurred by modernization. The narrative structure of entropy thus deploys the banal ending as a parodic strategy of inversion and subversion.[54]

In the narrative space between the novel's inception, where the reader's expectations are met, and the ensuing events, where expectations are denied, the reader is caught in a textual "snare."[55] Just such a liminal reading space is comparable to the ironic snare induced in Baudelaire's prose poems. *L'Assommoir* provides an example of the entropic metatheme in Zola's novels and the snare it engenders. At the novel's beginning, the main character, Gervaise Macquet, rises above her rural origins and achieves bourgeois prosperity as a laundress-proprietor in Paris. But her success is short-lived. Midway through the novel, the author incisively details the anti-heroine's slow degeneration; consumed by material indulgence and poverty, Gervaise dies from alcoholism and starvation at the end. Baguley's structural analysis of the entropic metatheme in *L'Assommoir* further demonstrates how the mirror-motif is embedded in Zola's novels. In a geometric diagram, Baguley charts Gervaise's rise and decline on either side of a triangle. Here, the narrative is structured by mirroring reciprocity since each chapter that records these events is paired symmetrically with its opposite.

Analogous to the Naturalist novel's subversive narrative structure, in *A Bar at the Folies-Bergère* Manet produces an ironic situation through a parodic inversion of pictorial structural form. Here, as in the novel's textual snare, the viewer is engaged in a visual one. By presenting a painting of a mirrored reflection, Manet initiates and then disrupts a conventional way of "entering" the picture. The impasto brushwork on the painted surface contravenes pictorial depth. The painted mirror frustrates the viewer by inviting and denying (or delaying) an illusionistic extension of our space or the painted objects' space. Likewise, the male figure's presence and absence are central to the viewer's ironic aesthetic experience. Although we see the man's frontal view reflected in the right side of the mirror, the direction of his gaze and the absent view of his back in front of the bar deny our conventional expectations for entering an illusionary space. We would expect a mediating *repoussoir* figure to direct the viewer into the picture space, as seen in Renoir's *Luncheon of the Boating Party* (1881; The Phillips Collection, Washington, D.C.), where the figure in the lower right turns inward. In the absence of this figure in *A Bar at the Folies-Bergère*, we are placed directly in the middle of the spectacle, facing the barmaid and the inescapable reflectivity and materiality of the mirror.

The novel *Nana* offers another example of the textual mirroring that Zola embeds in his texts. The novel also provides historical links with his earlier novel *L'Assommoir* and with Manet's painting of Nana. In *L'Assommoir*, Nana appears at the novel's end as Gervaise's daughter. Zola tracks her growth from a playful infant and then an abused[56] and delinquent young girl, to a respectable artisan as a hat maker, and finally to a seductive young woman. Blond and buxom at the age of fifteen, she procures a wealthy fifty-five-year-old bourgeois lover, whom she quickly abandons to pursue the life of an upscale prostitute.[57] In the novel *Nana*, we are introduced to Nana when she is eighteen.[58] In the first chapter, she takes Paris by storm as the thinly veiled star of "The Blond Venus," the title of an operetta performed at the Théâtre des Variétés. Nana's career as a prostitute is also the story of how she destroys every man who pursues her. Count Muffat is the most injured—he loses his fortune to Nana's voracious appetite for material riches, commits adultery with Nana, and is humiliated by Nana's betrayals in extending her affection to other men. The entropic metatheme also prevails over the narrative. Nana's rise to stardom on the stage is brief, replaced by her achieving entry into the aristocracy. However, Nana's insatiable appetite for expensive finery and food ruins men's fortunes, leading to her downfall. In the last chapter, Nana leaves Paris to "foreign exotic parts," having become bored with her abundance of riches—of objects and men. When she returns months later, she gets smallpox and dies a horrible, disfiguring death.

As in *L'Assommoir*, the entropic metatheme in *Nana* is structured by mirrored textual symmetries. As Pierre Petit has shown, mirroring occurs at the level of localization: the sites where the action takes place. Petit divides the fourteen chapters into three locations, as his diagram illustrates. Chapters labeled A take place in "Au Spectacle," Chapters labeled B take place in "Chez Nana," and Chapters C take place in "Chez Muffat." In the first five chapters, the scenes alternate between B and C, Nana's and Muffat's residences, and are framed by A, the Théâtre des Variétés.[59] These scenes form the pattern A-BCB-A. At the novel's center, the following three chapters take place elsewhere. Chapter VI takes place in the peaceful countryside at Nana's villa, while the next two, Chapters VII and VIII, return to Paris. These two chapters form a textual snare and are pivotal to Nana's degeneration. Immersed in a frenzy of extravagance, Nana traverses the city to various locations. Symmetry returns to the last six chapters, again forming the pattern A-BCB-A.[60] Except for Chapter XI, another outdoor scene at the racetracks (where Nana herself becomes the spectacle), the same alternating sequence of locations between Nana's residence (B) and Muffat's residence (C) is framed by A, the world of the spectacle; Chapter IX takes place at the Théâtre des Variétés, and Chapter XIV at the Grand Hotel. In this last scene, the spectacle resumes with Nana's death: she dies surrounded by female spectators in her hotel room while male spectators wait for her demise in the hotel lobby.[61]

The *mise en abyme*[62] is another mirroring device that Zola uses in *Nana*. According to Lucien Dällenbach, the literary *mise en abyme* is "a means by which the work turns back on itself, appears to be a kind of reflection"[63] or, more simply, it is "the mirror in the text."[64] In the case of *Nana*, an actual mirror is at the center of the mirror in the text. Also, the *mise en abyme* appears as a story within the story, which occurs in Chapter VII, strategically placed at the exact center of the novel. Nana and Muffat are in her dressing room at the theater, and she is gazing at her naked self in the full-length mirror, having "grown absorbed in her ecstatic self-contemplation."[65] Muffat, however, is troubled by the story published in a journal titled "The Golden Fly," which he reads to Nana. The story describes "the life of a harlot descended from generations of drunkards," who now, "[w]ith her... rottenness that is allowed to ferment among the populace[,] is carried upward and rots the aristocracy... [and] corrupts and disorganizes all Paris." The harlot is compared to "a fly of sunny hue which has flown up out of the dung a golden fly,... [and] enters the windows of palaces and poisons the men within by merely settling on them in her flight." This story, of course, mirrors both Nana's history and Muffat's corruption. Realizing this, Muffat, at first, is repulsed. He then gazes at Nana's reflection in the mirror. He becomes seduced again by "her satin skin and the supple contours of her shape," even more pronounced by her erotic movements impelled by her self-induced ecstasy.[66]

If this textual scene reminds you of Manet's *Nana*, it is not by accident. Manet's 1877 painting was named after the young woman Nana, who became a prostitute at the end of *L'Assommoir*.[67] This novel was published in the same year Manet created the painting. Similarly, Zola used Manet's painting as a *mise-en-scène* in Chapter VII of *Nana*, providing a stage-like setting and describing the characters' positions in Nana's dressing room. As in the painting, the reader of Zola's text is caught in the narcissistic circuit of Nana's gaze, which is returned to her by the reader's and Muffat's gazes. This creative exchange between painter and writer is an important historical event and invites an investigation of their critical exchanges. While Zola championed Manet in his critical reviews of the artist's works, he did so in the name of Naturalism and the mimetic reflections Naturalism promoted. Manet, in turn, responded with pictorial retorts to the limitations that Naturalism imposed. I will treat this critical exchange and the mirror's pivotal place therein in the remainder of this chapter.

Manet's Retort: Mirror as a Critical Device

When Zola began writing about Manet in 1866, he invoked the mirror-motif as a critical tool for promoting Manet and his work as a modern way of seeing.[68] Zola regarded Manet as a mirroring agent and his art as a mirror reflection of his subjects' physical appearance and social being. In doing so, the critic transposed his Naturalist literary theory onto the artist and his works.[69] The following section examines how the mirror-motif joins Zola's literary theory and his art criticism and, in turn, provides Zola with the means for presenting Manet as a kindred spirit in his battle for the new art of Naturalism.[70]

In 1866 and 1867, Zola wrote two critical reviews of Manet's art. During these years, Zola also formulated his theoretical tenets of the Naturalist novel, which he published in the 1868 preface to the second edition of *Thérèse Raquin* (first published in 1867).[71] Here, he explained his "analytical method" as a defense against "accusation[s] of immorality" in his novel. In this defense, Zola argued that he used the analytic method, inspired by "scientific curiosity," to "simply [be] an analyst who may have become engrossed in human corruption, but who has done so as a surgeon might in an operating theater." In this way, the Naturalist writer acts as a scientist who objectively observes nature's specimens and depicts these observations with a "clear and natural language."[72] As such, Naturalist language functions as a mirror reflection of reality. In this respect, Zola's 1868 description of his literary method echoes the analytic method he ascribed to Manet two years earlier. In his 1866 Salon review, Zola praised Manet's talent for its "simplicity and accuracy," "that is to say, the exact observation of objects."[73]

This earlier assessment of Manet suggests that Zola's Naturalist theories were shaped both by science and the pictorial arts. In fact, in his 1868 preface, Zola compares his process of transcribing human nature to the painter's process of mirroring physical nature:

While I was busy writing *Thérèse Raquin*, I forgot the world and devoted myself to copying life exactly and meticulously, giving myself up entirely to precise analysis of the human being.... The human side of the models ceased to exist, just as it ceases to exist for the eye of the artist who has a naked woman sprawled in front of him but who is solely concerned with getting on to his canvas a true representation of her shape and coloration.[74]

Zola elaborated on Manet's direct observation and simple translation of nature in his more extended 1867 Salon review, where he called Manet "an analytical painter." As such, Manet is just like the modern scientist who has, in Zola's words, "returned to a precise observation of facts." The artist, he writes, is "an interpreter of *that which is*, and his works [consist] of a precise description made in a human and original language"[75] (emphasis added). For Zola, the individual artist's originality or uniqueness is only significant for the way their vision is heightened by one aspect of nature over another. Zola called this unique sensibility "temperament," about which he elsewhere explained: "A work of art is a corner of nature seen through a temperament."[76] When Zola applied this axiom to Manet, he further aligned the artist with his analytical method and the new art it engendered. For Zola, Manet's temperament renders natural light with heightened brightness, which, in turn, enhances his formal simplicity. Zola thus surmised:

[I]f I... were asked what new language Édouard Manet spoke, I would reply: he speaks a language of simplicity and exactitude. The new note that [his temperament] contributes is one of blondness [or bright tones], filling the canvas with light. The translation that he gives us is a precise and simplified one, achieving its effect through large units and broad masses.[77]

Zola here posited the painter's language as analogous with the Naturalist writer's language, both mediums functioning as an objective mirror of nature.

Zola's description of *Olympia*, written for his Salon of 1866, demonstrates how he applied his art criticism and Naturalist theory to a specific painting. In the first case, Zola explains how the formal elements signify the artist's direct observation and "nature seen through a temperament." Viewing *Olympia* directly, Zola describes how simplified forms and painterly brushwork unexpectedly cohere as a visual translation of reality. He writes, "[A]t the first glance you distinguish only... two tones played off against each other"—"Olympia, reclining on the white sheets, is a large pale spot on the black background." Similarly, he observes that when you look closely, all you see are flat brush strokes and patches of color: "details have disappeared" and "everything is simplified." These broad masses of tonal contrasts result from the artist's vision as affected by his temperament. However, Zola recommends that the viewer "step back a bit" to "reconstruct reality." Then, he discovers, "a curious thing happens. Each object falls into its proper plane. Olympia's head projects from the background in astonishing relief; the bouquet becomes marvelously fresh and brilliant." Zola attributes this transformation of paint into reality to the artist's "accurate eye and... direct hand.... The painter worked as nature works, in simple masses and large areas of light, and his work has the somewhat rude and austere appearance of nature itself."[78] Zola then explains how the artist's direct translation of physical nature extends to the artist's translation of human nature.

To extol Manet's painting as an accurate representation of social realities, Zola identified Olympia as "a girl of our times, whom we have met in the streets." As if observing a specimen of the urban environment for his novels, Zola further explained:

> She is a girl of sixteen, doubtless some model whom Edouard Manet has quietly copied just as she was.... *When other artists correct nature* by painting Venus, *they lie.* Manet asked himself,... Why not tell the truth?... As usual, the public took good care not to understand what the painter wanted. *There are even people who have looked for a philosophical meaning in the painting.*[79]
>
> (emphasis added)

This last statement deserves our attention since it denies meaningful content in Manet's paintings, a denial repeated by Zola's later claim that Manet never wanted "to put ideas in his paintings."[80] Both statements reveal Zola's attempt to enlist Manet into the ranks of Naturalism and promote his own artistic agenda.[81] However, in doing so, he rids the painter of any extra-pictorial or ironic intentions and his paintings of the parodic play with pictorial conventions.

Manet, an avid reader of Zola's novels, had read his friend's 1868 Preface to *Thérèse Raquin* shortly after its publication. In a letter dated late April or May 1868, Manet praised the Preface for advancing progressive writers and artists: "Bravo, my dear Zola, that was a formidable preface, and you're making the case not only for a group of writers but for a whole group of artists as well."[82] Manet's praise will resonate in Zola's September 1868 dedication of *Madeleine Férat* to the artist. Given their alliance, what would Manet-the-artist have thought about Zola's subjugation of painting to writing and his refusal of meaningful content in the painter's art? We cannot rely on textual evidence to answer this question since Manet wrote little about his art. Instead, we can demonstrate that when Manet painted Zola's portrait in 1868,[83] he did so less as a tribute to the critic than as a retort to the latter's misinterpretation of the painter's artistic method and pictorial intentions (Figure 9.4).[84] Manet's portrait of Zola also reveals how the artist, in turn, appropriated the writer and his writing to his own creative agenda.

As Robert Lethbridge and others have argued, Zola is "awkwardly misplaced" among objects and images alien to him.[85] Instead of placing Zola amidst his writings and literary activities (the photographic convention for portraits of literati, as seen in Nadar's photographic portrait of Zola[86]), Manet surrounded the writer with artworks and texts that shaped his artistic practices. Zola holds an open book, Charles Blanc's popular *Histoire des Peintres de Toutes les Ecoles* (1865–77),[87] which Manet would have owned and relied on for its illustrations.[88] The Japanese screen behind the sitter on the left and the Japanese print on the sitter's right refer to Manet's engagement with Japanese art. This print is mounted with a reproduction of Velázquez's *The Drinkers*, marking Manet's admiration of Spanish art, and an etching of Manet's *Olympia*.[89] Ironically, in this image, Olympia turns her gaze toward Zola to appreciate his admiration for the painting.[90] Olympia's gaze also connects the actual etching with the writer-critic (Figure 6.13). Directly below *Olympia* is a blue brochure, inscribed with Manet's name on its cover, in which the engraved *Olympia* had been published nine months earlier.[91] This brochure marks another tribute to Manet; it is a reprint of Zola's 1867 review of Manet's work, published on the occasion of the artist's one-person exhibition in May 1867.[92] However, Manet reversed the writer's appropriation of his paintings by appropriating Zola's writing: Manet replaced the pamphlet's printed title with his signature, which was used to sign the painting.[93]

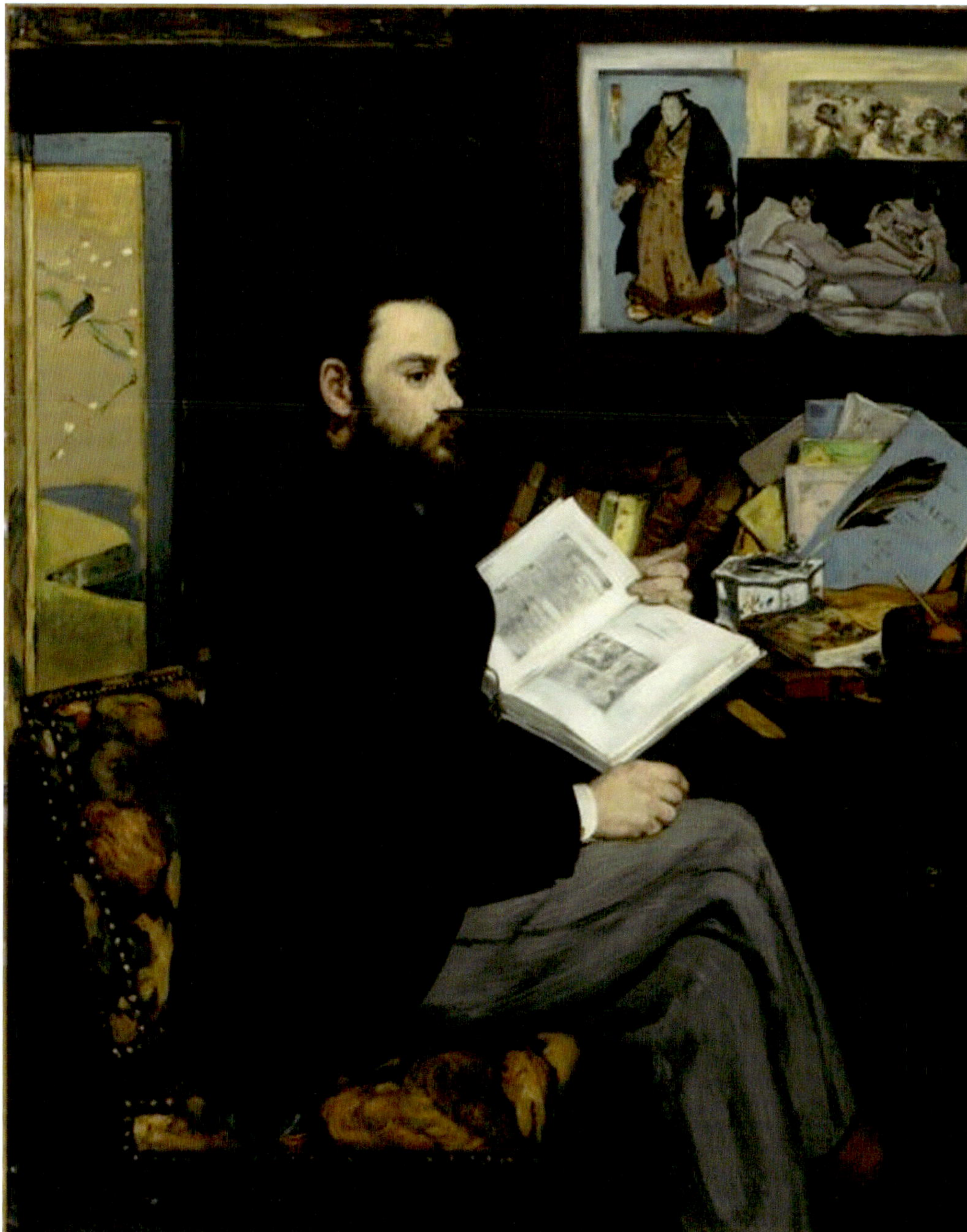

Figure 9.4 Édouard Manet, *Emile Zola*, 1868, oil on canvas, 57 3/4 x 45 in. Musée d'Orsay, Paris. © RMN-Grand Palais/Art Resource, NY. Photo: Hervé Lewandowski.

Although Zola appears oblivious to all the visual references to Manet's artistic "sources,"[94] we are not. For one thing, we see and recognize the pictorial cues from which Zola looks away. For another, compositional delineations compel the viewer to focus on pictorial images and the artist's inscription. Manet directs our gaze diagonally, scanning the illustrated pages of the open book, following Zola's thumb and fingers, finally settling on the artist's signature. Similarly, Manet uses the L-shaped contours of Zola's arm and torso to repeat the frame's outline that holds the print collection.[95] In doing so, he reinforces the pictorial, rather than analytic, methods that underlay his art. The viewer's gaze now mirrors the artist's self-reflection on his medium and the conventions specific to pictorial representation. In short, I agree with Lethbridge, who maintains that the portrait's internal pictorial references "reassert [Manet's] creative contacts with tradition that Zola had missed."[96]

Notwithstanding Manet's claim for his painting's autonomy and traditions, Zola responded to the portrait in a literary way. Indeed, he used his 1868 Salon review to reaffirm Manet's position as an analytical observer in service of the writer's Naturalist theories. Here, Zola echoed his 1868 preface to *Thérèse Raquin*, where he compared the writer's process to the painter's process, wherein each forgets his subject's physical presence during sessions of intense observation. Thus, in describing Manet's performance while painting his portrait, Zola recalls "those long hours of sitting" during which he watched the artist, standing before the canvas, absorbed in his task. He recounts: "Manet had forgotten me, he didn't know I was there, he was copying me as he would have copied any human animal, with an attention, an artistic awareness, the like of which I have never seen." Like the disengaged writer who only records what he observes, Manet painted the surrounding inanimate objects and the living model with equal intensity and color harmony. Zola thus concludes that "Manet is before all a naturalist. His eye sees and renders objects with an elegant simplicity."[97]

In the following years, Zola further developed the Naturalist novel and continued championing Manet's achievements, which were similar to his literary theory. He variously called Manet "a modern artist, a realist, a positivist"[98] (1875), "a naturalist, an analyst"[99] (1876), and "one of the tireless workers of naturalism" (1880).[100] With all these labels, Zola matched Manet's art and method with the later positivist tenets presented in "The Experimental Novel," Zola's treatise on his scientific methods published in 1880. Here, Zola explained his empirical and determinist techniques of studying humans through their relationship with the social environment. Like the scientist, the writer is free of personal feelings and moralizing and controls his "*a priori* ideas" by "observation and experiment."[101] While Zola aligned his experimental method with the scientists, he also aligned the artist with the experimental novelist. As he put it, "The personal feeling of the artist must be subject to the control of truth.... The artist has the same starting point as the scientist; he stands before nature, has an *a priori* idea, and works in line with that idea." Yet, he cautioned, the artist only "diverges[s] from the scientist if he carries his idea out to the end without verifying its exactness by observation and experiment."[102]

In conclusion, I return to *A Bar at the Folies-Bergère* to argue that this painting also represents Manet's retort against Zola's latest appropriations of his art. Manet began this painting in 1881, just one year after Zola called the painter "one of the tireless workers of naturalism" and published "The Experimental Novel." Manet's retort is not just against the assimilation of the visual by the textual[103] but also against the take-over of modernity by science. With *A Bar at the Folies-Bergère* Manet reclaimed his Baudelairean modernity and, at the same time, disassociated himself from Zola's determinist positivism.

As we have seen, because Manet's painting shares mirroring techniques and urban themes with the Naturalist novel, associations with entropic metathemes are inevitably encoded in the network of visual forms, whether figurative or abstract. For example, for the nineteenth-century viewer, the barmaid might foretell the working-class woman's inevitable "fall" into prostitution.[104] Likewise, the café-concert scene's dematerialized ambiance might evoke an incessantly decaying atmosphere. Conversely, we can argue that these surroundings evoke the constant flux and contradictions that Baudelairean modernity extolled. In short, Manet shared Baudelaire's affirmation of such modern phenomena as a source of cultural renewal. Thus, Manet used the mirror to demonstrate that his art was not just a static record of modern life but a dynamic translation of the lived experience. Baudelaire identified this cognitive experience as fundamental to the artist's creative faculties. As earlier quoted, Baudelaire praised the painter of modern life as "the lover of life, [who] may also be compared to a *mirror* as vast as this crowd; to a *kaleidoscope* endowed with consciousness, which with every one of its movements presents a pattern of life, in all its multiplicity" (emphasis added).[105]

Given his Baudelairean allegiance, Manet's choice of the Folies-Bergère as a site of modernity can be explained by the spectacle of mirrors encountered in the building's interiors. The semi-circular theatre and the adjoining rectangular hall (winter garden) were lined with mirrors, and the two rooms opened onto each other (Figures 9.5 and 9.6). As such, these interior spaces reflected the visual/cognitive experience of the newly built boulevards and the scintillating windows of buildings that lined them.

While it has been suggested that Manet depicted the two architectural spaces by painting the theater from a vantage point inside the rectangular hall,[106] the contemporary graphic illustrations of the Folies-Bergère indicate that he did much more. These documents record the presence of several bars at both ends and at the center of the semi-circular balcony of the theater and along the ground-story of the two-story winter garden. A more realistic rendering of *The Bar at the Folies-Bergère* (1878; Brooklyn Museum, New York), by Jean-Louis Forain, illustrates the actual curvature of a mirrored bar at one end of the theater balcony. The curvature also explains why Manet rendered the barmaid's reflection off-center.[107] Poster advertisements also celebrate the café-concert's festive atmosphere with a panoply of vistas, including the numerous bars. The poster depicting the staircase arcade that bridged the theater and winter garden is significant. This image suggests that the figures in Manet's mirrored balcony scene sit and stand in the two-storied rectangular winter garden. That space is further indicated by Manet's rendering of perspectival depth, here marked by the receding orthogonal lines behind and to the left of the barmaid. These trajectories are carried by piers and columns, global light fixtures, and the loosely brushed commingling men and women patrons. Yet, the cropped legs of the trapeze artist in the upper left corner indicate that we are standing within the theater where this act was performed. Given these simultaneous views, we can assume that Manet combined the two spaces into a single space to portray modernity's experiential, cognitive effects. Most important in comparing photographs[108] and graphic illustrations with Manet's painting is that the artist distorted reality and mastered his technique in representing the spectacle of the Folies-Bergère as an unending hall of reflections, mirroring, in effect, the kaleidoscopic experience of modern Paris.

Suppose we accept that Manet's bar at the Folies-Bergère is not one bar but a composite of all the bars enclosed within its walls. In that case, we will also agree that his portrayal is a product of the lived experience transformed by his imagination rather than the translation of nature by his "temperament." As such, another Baudelairean narrative

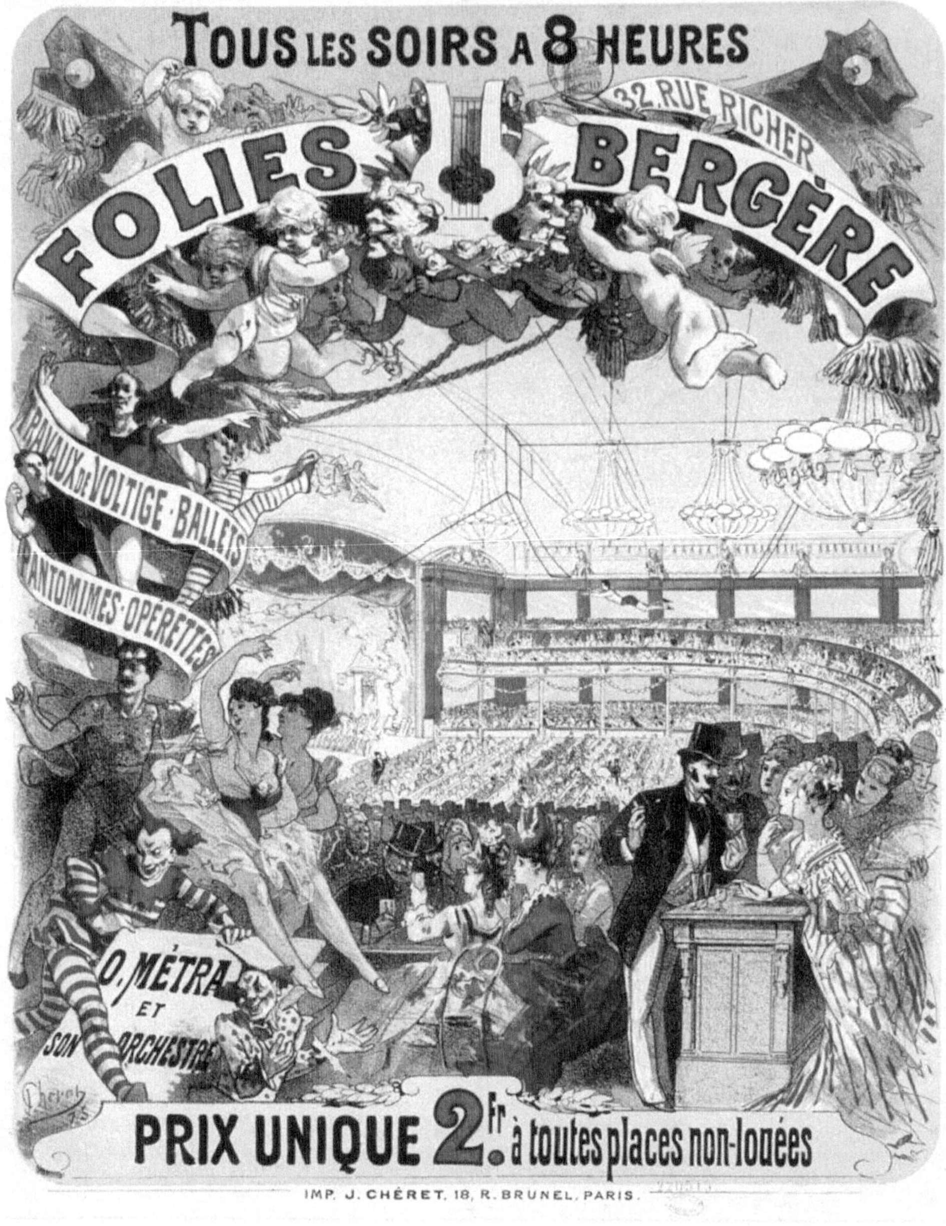

Figure 9.5 Jules Chéret, "Folies-Bergère, Tous les Soirs à 8 Heures" (Folies-Bergère, Every Night at 8), 1875, color lithograph (poster), 21 5/8 x 17 3/8 in. Bibliothèque Forney, Paris.

unfolds, one that follows the creative process prescribed in "The Painter of Modern Life." As frequently cited in this book, Baudelaire described the modern artist (Constantin Guys) as a man of the crowd from dawn to dusk who retreats from the crowd at night. Alone in his studio, the artist reflects upon his sensory experiences and renders their essence:

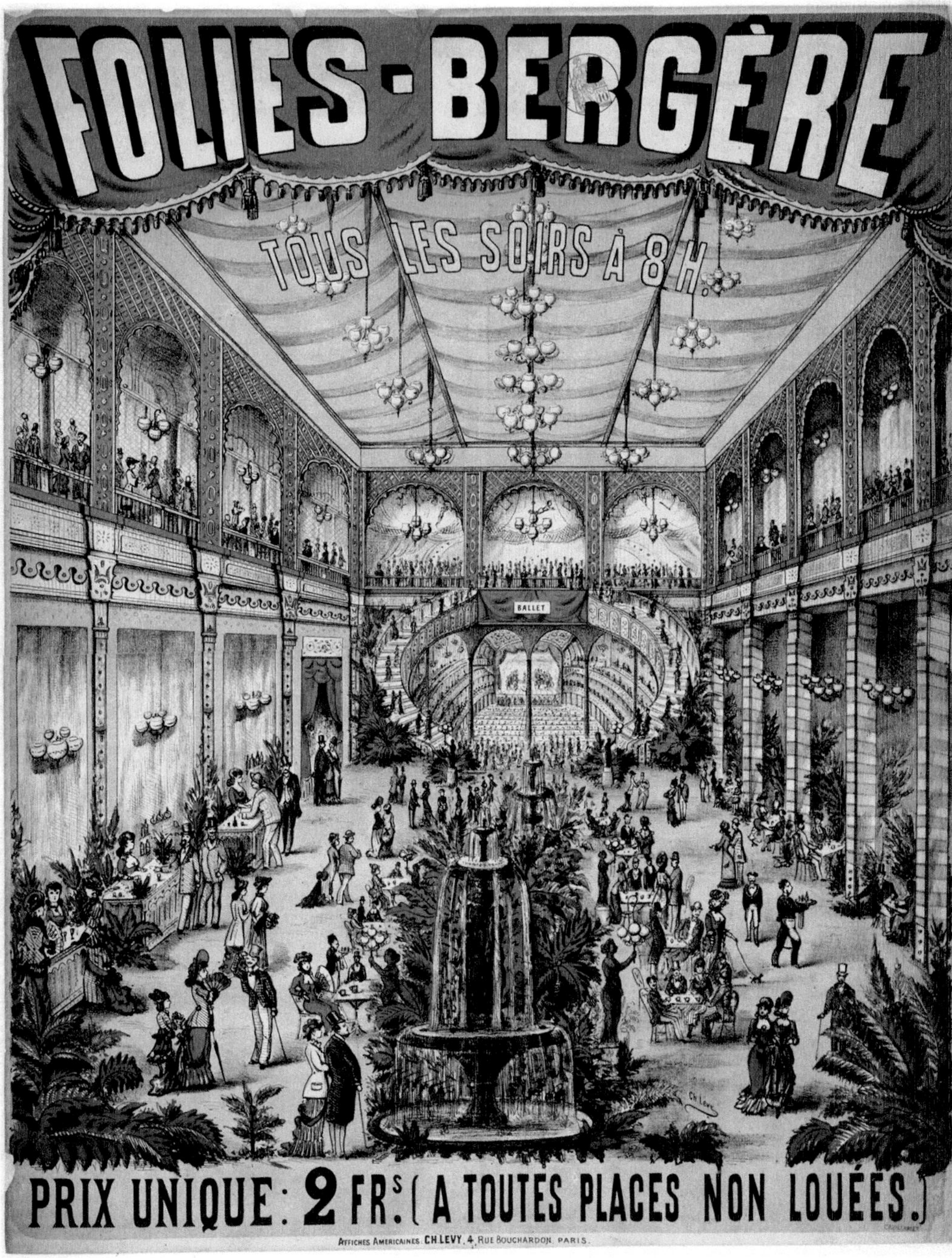

Figure 9.6 View of the winter garden. Charles Lévy, "Folies-Bergère, Tous les Soirs à 8 H." (Folies-Bergère, Every Night at 8), 1880–1, color lithograph (poster), 22 3/8 x 17 in. Paris Musées/Musée Carnavalet, Histoire de Paris.

"All the materials, stored higgledy-piggledy by memory, are classified, ordered, harmonized, and undergo that deliberate idealization, which is the product of a childlike perceptiveness that is acute and magical by its very ingeniousness."[109] Manet followed the process and means of Baudelaire's prescription for the creative translation of nature. As an ink drawing of the Folies-Bergère shows,[110] Manet rendered his first impressions on the spot. Away from the crowds in his studio, he retrieved the essence of the café-concert and re-presented its scintillating ambiance in the fragmented but unified surface of the canvas. Suppose his viewers recognized *A Bar at the Folies-Bergère* as a dynamic synthesis of low and high art, actual and remembered space, sketch-like brushwork, and visual spectacle. In Baudelaire's words, would they not have viewed Manet's painting with "the shock of surprise"? If so, Manet would have affected an ironic aesthetic response that Baudelaire celebrated as the ultimate purpose of artistic creation.

In summary, Manet responded to Zola's Naturalist appropriation of his art by demonstrating how the painted mirror (on the canvas) sustains the transitoriness of modernity and its ironic effects. In doing so, he created an experiential immediacy that the novel could not have achieved. Although Zola's novels have character and plot developments structured by mirrored symmetries, these developments depend on sequential events that unfold in real time for the reader and abstract time within the novel's structure. Ironically, the *mise en abyme* in *Nana* comes closest to a point where visual and textual mirrorings converge. The story within the story, illustrated by Muffat reading "The Golden Fly" to Nana, is also reflected in the mirroring of Nana and Muffat's gaze. To be sure, Zola adapted this scene from Manet's *Nana*. Manet, however, ironized this mimetic interpretation of his paintings. Constantly reflecting on his medium, Manet used the mirror to subvert mimetic representations and to present a painting within a painting about painting instead. Finally, the viewer is caught in a visual snare that defies narrative appropriation and prolongs ironic contemplation.

In the following chapter, I reflect upon the liminal space, captured in Manet's painting of modernity, through the lenses of cognitive poetics and neuroaesthetics. In doing so, I restore the viewer's gaze to the embodied effects of fragmentation and rupture and, hypothetically, to the embodied experience of nineteenth-century Parisians during the ongoing Haussmannization of their city.

Notes

1 Schlegel, "Critical Fragments" ("Kritische Fragmente" [1797]), in *German Aesthetic and Literary Criticism*, 48.
2 Schlegel, "Critical Fragments," 49.
3 Wheeler, "Introduction," *German Aesthetic and Literary Criticism*, 21.
4 McLees, *Baudelaire's "Argot Plastique,"* 88, 91. For Manet's characteristic "witticism," see Beth Archer Brombert, *Edouard Manet: Rebel in a Frock Coat* (Chicago: University of Chicago Press, 1997): 93, 116, 120, 190.
5 Brombert, *Edouard Manet*, 58–9.
6 On Manet and Baudelaire's friendship at the time of painting *Music in the Tuileries*, see Cachin et al., *Manet, 1832–1883*, 124.
7 Manet also painted Baudelaire's mistress, Jeanne Duval, as *Lady with a Fan* (1862). The poet and painter communicated via letter as well. In the most notorious of these (May, 1865), Baudelaire responded to Manet's complaints about people laughing at his works exhibited in the Salon of 1865 (*Le Déjeuner sur L'herbe* and *Olympia*); Baudelaire chastises Manet for his dismay, despite the critic's "fondness" for the artist; see Pichois with Ziegler, *Baudelaire*, 284.
8 See Baudelaire, *Selected Writings*, 458, note 1.

9 Baudelaire, "The Painter of Modern Life," in *Selected Writings*, 399.
10 Baudelaire, "The Painter of Modern Life," 399–400.
11 Baudelaire, "The Painter of Modern Life," 392.
12 Baudelaire, "The Exposition Universelle," in *Art in Paris 1854–1862*, 124.
13 Manet was born and raised in Paris. From the time of his youth, as a student and under his uncle Edmond's guidance, Manet was a frequent visitor to the Louvre and Luxembourg museums, where he studied and copied old masters and living artists. Manet continued these visits throughout his lifetime. At the Louvre he met his lifetime artist friends, Edgar Degas and Henri Fantin-Latour, and Berthe Morisot, his future sister-in-law. See Brombert, *Edouard Manet: Rebel in a Frock Coat*, 9, 221, 162; and Cachin et al., *Manet, 1832–1883*, 302.
14 See Tinterow and Lacambre, *Manet/Velázquez*, 205–28. Manet visited Spain in 1865, especially to see Velázquez paintings at the Prado Museum, and while there, he became enamored by the work of Goya and El Greco (228–34).
15 Manet regularly visited the Louvre (Tinterow and Lacambre, *Manet/Velázquez*, 204). *Les Petits Cavaliers* has recently been retitled *Réunion de Quatorze Personnages* (1645/50) and attributed to Juan Bautista Martínez del Mazo, who was trained in Velázquez's workshop and married the master's daughter. The painting was purchased by the Musée du Louvre in 1851. See Musée du Louvre, Juan Bautista Martinez del Mazo, "Réunion de Quatorze Personnages," Musée du Louvre, 1645, https://collections.louvre.fr/en/ark:/53355/cl010064949.
16 Cachin et al., *Manet, 1832–1883*, 126; and Tinterow and Lacambre, *Manet/Velázquez*, 207, 220.
17 Cachin, 126; and Tinterow and Lacambre, *Manet/Velázquez*, 207, 220.
18 Manet's inclusion of his outward-looking self-portrait among others in this gathering has also elicited comparisons with Velázquez's self-portrait in *Las Meninas*. Manet only later mentioned this painting in a letter to Fantin-Latour during his 1865 visit to the Prado (Tinterow and Lacambre, *Manet/Velázquez*, 231). See Ben Pollit, who argues that Manet invokes *Las Meninas* in establishing a relationship between the observer-viewer and the observed (the crowd scene); "Manet, Music in the Tuileries Gardens," Khan Academy, https://www.khanacademy.org/humanities/becoming-modern/avant-garde-france/realism/a/manet-music-in-the-tuileries-gardens.
19 Recent scholarship's focus on Manet's Spanish connections overlooks his dialogue with the French academic tradition of artistic conventions.
20 According to the Louvre Museum website, *The Intervention of the Sabine Women* was exhibited at the Louvre from 1799 to 1804 and at the Salon of 1808. The painting was acquired by the state on December 11, 1819, from M. de la Harpe (?) with *Leonidas at Thermopylae* for 100,000 francs, and then displayed in the Musée du Luxembourg (established in 1818 by Louis XVIII as a museum dedicated to living French artists). Following David's death (1825) and the museum's protocol, the *Sabine Women* was transferred to the Louvre. See Jacques-Louis David and France, "Les Sabines," Musée du Louvre, 1799, https://collections.louvre.fr/en/ark:/53355/cl010065426 and "History of the Musée Du Luxembourg," museeduluxembourg.fr, https://museeduluxembourg.fr/en/history-musee-du-luxembourg.
21 Thomas Couture, *Méthode et Entretiens D'atelier* (Studio Method and Interviews) ([Paris]: [n.p.], 1867): 313–4.
22 Notably, Couture's *The Romans of the Decadence* (1847; Musée d'Orsay, Paris) also formalistically aligns with David's *Sabine Women*, but with opposite didactic intent, perhaps inspiring Manet's subversion of David's masterpiece.
23 For an illustration of *Corner of the Tuileries*, see Cachin et al., *Manet, 1832–1883*, 127.
24 Marni Kessler, "Dusting the Surface, or the Bourgeoisie, the Veil, and Haussmann's Paris," in *The Invisible Flâneuse?: Gender, Public Space, and Visual Culture in Nineteenth-Century Paris*, eds. Aruna D'Souza and Tom McDonough (New York: Manchester University Press, 2006): 61.
25 According to Frédérique Desbuissons, *The Sabine Women* painting was exhibited in the old meeting hall of "the Académie d'Architecture which overlooked the Cour Carré and the room formerly reserved for the models was used as an antechamber;" in "A Ruin: Jacques-Louis David's *Sabine Women*," *Art History* 20, no. 3 (September 1997): 432.
26 Ewa Lajer-Burcharth was the first scholar to reconstruct David's use of the mirror and its post-revolutionary, ideological function in "David's *Sabine Women*: Body, Gender and Republican Culture under the Directory," *Art History* 14, no. 3 (September 1991): 411, 417. See also

Desbuissons ("A Ruin," 432–48), who reconstructs the discursive and critical space of David's exhibition of *The Sabine Women* and the inseparable relation between the painting and the mirror, comparable to contemporary installation art. Lajer-Burcharth identifies "a large pivoting mirror, a type called a *psyché*, a furnishing novelty under the Directory" (406). Desbuissons counters this claim, arguing that rather than "a swing-mirror (*psyché*)," David likely installed "simply a large wall mirror" and did so "to take into account the whole of the environment" (438–40, 447n45).

27 Desbuissons, "A Ruin," 440.

28 Thomas Couture may have been Manet's conduit to David's work since Couture studied at the École des Beaux-Arts with Antoine-Jean Gros, who had studied with David.

29 Lajer-Burcharth identifies several critics who commented on the effect of the mirror, which, for this study, establishes a textual archive. She relies on Pierre Chaussard, art critic of the Directory, for the ideological and social function of *The Sabine Women*, of which the mirror installation was a part ("David's *Sabine Women*," 398–406; 426n25). Chaussard's critique was published as a pamphlet titled *Sur le Tableau des Sabines par David* (Paris, an VIII – 1800) (Lajer-Burcharth, "David's *Sabine Women*," 424n1). In addition to Chaussard, Desbuissons advances Charles-Paul Landon, artist-art critic [as per Chaussard, *Sur le Tableau*, 19], for his detailed testimony regarding the mirror's reflection and its effect ("A Ruin," 437–8; C.-P. Landon, "Peinture," *Journal des Arts, de Littérature et de Commerce*, no. 32, 10 nivôse, an 8 [December 31, 1799]).

30 Desbuissons, "A Ruin," 434. For Charles-Paul Landon's description of the exhibition of *Coronation* and mirror effect, see Desbuissons, "A Ruin," 446n28; Landon praises the mirror's reflection for effecting an "animated" scene.

31 Desbuissons, "A Ruin," 439.

32 Desbuissons, "A Ruin," 442.

33 Alan Krell, *Manet and the Painters of Contemporary Life* (New York: Thames and Hudson, 1996): 15; and Cachin et al., *Manet, 1832–1883*, 123. See also Beth Brombert who cites Antonin Proust, Manet's biographer, to suggest that Manet and Baudelaire "may have met earlier, [when seated together] over lunch, often in the company of Proust and other writers, at the Rôtisserie Pavard and at Dinocheau's, where one spent little and talked a lot;" *Edouard Manet*, 59.

34 As noted in Chapter 6, note 86, Manet's model, Victorine Meurent was a recognizable *parisienne* within Manet's artists' circle as a model and artist.

35 Henriette Hauser, a young actress, posed for *Nana*. Hauser was then the mistress of the Prince of Orange, who nicknamed her "Citron." See Cachin et al., *Manet, 1832–1883*, 393.

36 The first photographer's self-portrait dates from 1840: Hippolyte Bayard, *Self-Portrait as a Drowned Man*, (direct paper positive; Société Française de Photographie, Paris); see "Bayard, Hippolyte 1801–1887 – Self-portrait as a Drowned Man 1840," n.d., https://commons.wikimedia.org/wiki/File:Bayard,_Hippolyte_1801-1887_-_Selfportrait_as_a_Drowned_man_1840.jpg.

37 Hippolyte Bayard, *Autoportrait Avec Appareil sur Pied*, 1850–1855, albumen print, 13.6 x 10.8 in. http://www.artnet.com/artists/hippolyte-bayard/autoportrait-avec-appareil-sur-pied-2Aj3N9Cg8TFbFH6UNbgdEQ2.

38 Manet owned the psyche mirror depicted in Nana, part of his studio furnishings; see Brombert, *Edouard Manet*, 385.

39 According to the tradition of nineteenth-century *physiologies*, social types were primarily marked by their clothing. From the 1840s to the 1890s, small book publications titled "*Physiologie de…*" (*L'étudiant, le Bourgeois, le Dandy*, etc.) flourished in popular culture, providing a means to lock Paris's burgeoning society into typologies or stereotypes. The *grisette* and *lorette* were among the most popular social types, figuring in visual arts and literature; see, for example, Constantin Guys, "Two Grisettes" (1840s–70s; https://www.metmuseum.org/art/collection/search/337264) and Baudelaire, "To a Passerby"; orig. 1857, in *Les Fleurs du Mal/The Flowers of Evil* [2nd ed. 1861], https://fleursdumal.org/poem/224. Both working-class social types sought financial security in prostitution. Libby Murphy and Greggor Mattson note, "The main differences between the *grisette* and *lorette* stem from their relationships with men. The name *grisette* is believed to be derived from the French word for grey or *gris*. This dull

color was worn by seamstresses when they were at work. The grisette is the companion of a young, male college student. She is one part caretaker--doing his laundry and cooking--and one part lover." The *grisette* is often working as a seamstress, embroiderer, or tailor. "The *lorette*, however, is frequently depicted as having older bourgeois *souteneurs* or 'caretakers.' This relationship is much closer to what we [today] think of a socially ambitious 'kept woman'" ("Types of Prostitutes in 19th Century France," *A la Recherche des Femmes Perdues* [online exhibition], Oberlin, OH: Oberlin College Library, 2016, https://onprostitution.oberlincollegelibrary.org/exhibits/show/prostitutes-in-19th-century-fr.). For the culture of and about the *grisette* during Honoré Balzac's, Guys', and Baudelaire's time, see Claire Scamaroni and Maison de Balzac, *Elle Coud, elle Court, la Grisette: 14 octobre 2011–15 janvier 2012*. Paris: Paris-Musées, 2011.

40 Hollis Clayson examines the popular and actual associations of café-concert and brasserie barmaids and waitresses with "clandestine" prostitution in Clayson, *Painted Love*, 133–53.

41 Such was the trajectory of Nana, the titular main character of Zola's 1880 novel, discussed further below.

42 Baudelaire, "The Painter of Modern Life," 11; and "The Exposition Universelle 1855," in Baudelaire, *Art in Paris 1845–1862*, 124–5.

43 The Folies-Bergère was referred to as a "*café-spectacle*" in 1868; see Raoul Muriand, *Les Folies Bergère* (Sèvres: Editions La Sirène, 1994): 10.

44 As Robert Lethbridge states, "The creative relationship between [Zola and Manet] has many mirroring dimensions," which he lists in "Zola and Contemporary Painting," in *The Cambridge Companion to Emile Zola*, ed. Brian Nelson (Cambridge: Cambridge University Press, 2007): 73.

45 Manet and Zola's friendship endured for seventeen years. In 1866, Manet wrote fifty letters and notes to Zola, published in "Appendix I: Letters from Manet to Zola," in Cachin et al., *Manet, 1832–1883*, 518–30. For Zola and Manet's friendship and Zola's critical endorsement of Manet, see Frederick Brown, *Zola: A Life* (London: Papermac, 1997): 135–9, 144–7. Despite Zola's 1879 rebuke of Manet's loose painterly technique, their friendship continued until Manet's death (Zola was a pallbearer at Manet's funeral); see Brown, *Zola*, 504–7.

46 The dedication to Manet is dated "1 septembre 1868" in the French edition of *Madeleine Férat*, six months after Manet painted Zola's portrait (February 1868); see my discussion below.

47 Émile Zola, *Madeleine Férat: A Realistic Novel* (1868; Eng. trans. London: Vizetelly & Co., 1888; repr., Portland: Mint Editions, 2021), n.p.

48 Manet, letter to Zola, December 1868, in Cachin et al., *Manet, 1832–1883*, 521. As noted here, Manet extolled how Zola "painted the redheaded woman [the heroine Madeleine Férat]," who resembled Victorine Meurent, Manet's favorite model. See also Brombert, *Edouard Manet*, 196.

49 Hutcheon, *A Theory of Parody*, 7.

50 See letters from Manet to Zola in which the painter refers to attending the theatrical productions of *L'Assommoir* [after] January 18, 1879, and of *Nana* [after] January 29, 1881, in Cachin et al., *Manet, 1832–1883*, 527, 528. On Manet's liking of *L'Assommoir*, see Brombert, *Edouard Manet*, 196, 385.

51 William J. Berg discusses the physical presence of mirrors in Zola's novels in *The Visual Novel: Emile Zola and the Art of His Times* (University Park, PA: Pennsylvania State University Press, 1992): 63–9.

52 David Baguley, *Naturalist Fiction: The Entropic Vision* (Cambridge: Cambridge University Press, 1990); 160. See also Baguley, "An Essay on Naturalist Poetics," *The Journal of the British Neo-Formalist Circle* 12, no. 1 (April 1987): 41–56; and "Parody and the Realist Novel," *University of Toronto Quarterly* 55 (Fall 1985): 94–108.

53 Baguley, "An Essay on Naturalist Poetics," 50.

54 For a synopsis of the entropic metatheme, see Baguley, *Naturalist Fiction*, 216–9.

55 Baguley, "An Essay on Naturalist Poetics," 50.

56 See Brombert, *Edouard Manet*, 385.

57 Zola devotes a chapter (Chapter XI, "Little Nana") to a detailed account of Nana's upbringing and delinquency.

58 Nana is described in Chapter I as "[e]xceedingly tall, exceedingly strong, for her eighteen years." https://www.online-literature.com/view.php/nana/1?term=nana. However, there is an

age discrepancy between the novel *Nana*, which begins in 1867, and the Rougon-Macquart family tree, denoting that Nana (birthname Anna) was born in 1852; meaning she would have been 15 at the start of the novel *Nana*. For the Rougon-Macquart family tree, see "Émile Zola. Oeuvres. Manuscrits et Dossiers Préparatoires. Les Rougon-Macquart. Le Docteur Pascal. Dossier Préparatoire," Bibliothèque nationale de France, Département des Manuscrits, Paris NAF 10290. Published online June 3, 2013. gallica.bnf.fr, https://gallica.bnf.fr/ark:/12148/btv1b90797703/f173.image.

59 Pierre Petit, "Localisation Structurale d'un Roman de Zola: *Nana*," *New Zealand Journal of French Studies*, 5, no. 1 (May 1984): 19–21.

60 Petit, "Localisation Structurale," 20.

61 Another set of symmetries is constructed as places of "harmony" where Nana is the spectacle and antithetical, a matter of different points of view: Nana seen in the "*salle*" (Chapter I) of the theater or backstage (Chapter V) and Nana victorious (Chapter IX) or defeated (Chapter XIV). Petit, "Localisation Structurale," 23.

62 *Mise en abyme* (also *mise en abîme*) is a "term from heraldry, meaning the reduced reproduction of an image within itself. It was popularized by Gide to refer to a similar phenomenon in literature (play within play, novel within novel, etc.);" "Mise en Abyme," in *The New Oxford Companion to Literature in French*, ed. Peter France, (Clarendon: Oxford University Press, 1995), https://www.oxfordreference.com/view/10.1093/acref/9780198661252.001.0001/acref-9780198661252-e-3152.

63 Lucien Dällenbach, *The Mirror in the Text*, trans. Jeremy Whitely with Emma Hughes (Cambridge, UK: Polity Press, 1989): 8.

64 Dällenbach, *The Mirror*, 35.

65 Zola, *Nana: A Realistic Novel* (1880), trans. Ernest Alfred Vizetelly (London: Vizetelly, 1884), http://www.online-literature.com/view.php/nana/7?term=goldenfly.

66 Zola, *Nana*, http://www.online-literature.com/view.php/nana/7?term=golden fly.

67 Nana's adolescence to womanhood is described in Chapter XI of thirteen. Much has been written about the chronologies of Zola's novels *L'Assommoir* and *Nana* and Manet's painting *Nana*; for an overview, see Brombert, *Edouard Manet*, 385–9. Scholars have also discussed the similarities between Zola's Nana and Manet's Nana, both her physical features and status as a prostitute. For a discussion of Nana's narcissism and the relation between the novel and Manet's painting, see Berg, *The Visual Novel*, 67–8. Brombert, however, distinguishes between Zola's Nana, whom the author lades with deleterious sociological meanings, and Manet's Nana, whom the painter frees from these associations and portrays with "humor and delight" (*Edouard Manet*, 389).

68 Literary and art historians today regard Zola's art criticism of Manet as a transference of his literary theory onto the artist and his works. However, these writers have not considered how Zola used the mirror-topos as a critical device, the method on which I focus. For discussions of the intersection of Zola's literary theory and art criticism, including Manet, see Berg, *The Visual Novel*, 33–60; and Lethbridge, "Zola and Contemporary Painting," 67–85.

69 For an ekphrastic, critical analysis of Zola and Manet's verbal-visual dialogue, see Lauren S. Weingarden, "Mirroring Naturalism in Word and Image: A Critical Exchange between Émile Zola and Edouard Manet," in *Ekphrastic Encounters: New Interdisciplinary Essays on Literature and the Visual Arts*, eds. David Kennedy and Richard Meek (Manchester: Manchester University Press, 2019): 144–64. In the essay I address how the writer and artist's critical and theoretical exchanges exceed the limits of text and image and activate a third level of cross-referencing, which permeates a complex, multivalent engagement between the 'sister arts.'

70 See Lethbridge, "Zola and Contemporary Painting," and Theodore Reff, "Manet's *Portrait of Zola*," *The Burlington Magazine*, 117, no. 862 (January 1975): 35–44.

71 Zola dated the Preface to the 1868 edition of *Thérèse Raquin* as "15 avril 1868;" Émile Zola, *Thérèse Raquin*, 2nd ed. (Paris: C. Marpon et E. Flammarion, 1883): vii, https://gallica.bnf.fr/ark:/12148/bpt6k5679412r.

72 Zola, "Preface to the Second Edition (1868)," in *Thérèse Raquin*, trans. Leonard Tancock (London: Penguin Classic, 1962): 27.

73 Émile Zola, "Mon Salon: M. Manet," *L'Evénement* (May 7, 1866); repr. "Mon Salon (1866): M. Manet," in *Oeuvres Complètes Illustrées de Émile Zola. 32, Oeuvres Critiques*, t. 1 (Paris: E. Fasquelle, 1906) : 84–6, https://gallica.bnf.fr/ark:/12148/bpt6k5495479r. This text is the

fourth article of Zola's *Salon de 1866*, which is entirely dedicated to Manet; the text reappeared, amplified, and published as a brochure in 1867; see Émile Zola, *Pour Manet*, ed. Jean-Pierre Leduc-Adine, Le Regard Littéraire, 32 (Bruxelles: Editions Complexe, 1989): 65, note. Translation by author; original text: "simplicité et de justesse," "c'est-à-dire à l'observation exacte des objets."

74 Zola, "Preface (1868)," *Thérèse Raquin* (1962), 23.

75 Émile Zola, "Edouard Manet. Étude biographique et Critique, 1867," originally published in *La Revue du XIXe Siècle* (January 1, 1867) with the title "Une Nouvelle Manière en Peinture: Edouard Manet (1867)"; see Émile Zola, *Pour Manet*, 81, note; trans. "A New Style of Painting," in Nochlin, *Realism and Tradition*, 74–5. Zola's January 1867 essay on Manet was published in June 1867, as a brochure, on the occasion of Manet's private exhibition on the outskirts of the 1867 Universal Exposition; titled *Éd. Manet: Étude Biographique et Critique; Accompagnée d'un Portrait d'Éd. Manet par Bracquemond; et d'une Eau-forte d'Éd. Manet D'après "Olympia"* (Paris: E. Dentu, 1867), https://gallica.bnf.fr/ark:/12148/bpt6k6350838h. This later text was republished in 1879, with some variations, in *Mes Haines* and in *Mon Salon*; see BNF Catalogue Notice http://ark.bnf.fr/ark:/12148/cb31690884g.

76 Zola, "Mon Salon (1866): Les Réalistes du Salon," originally published in *L'Evénement* (May 11, 1866); rpt: *Les Cahiers Naturalistes. Société Littéraire des Amis d'Émile Zola*, https://www.cahiers-naturalistes.com/documents/zola-et-arts/ecrits-sur-lart-de-zola/mon-salon-1866-les-realistes-du-salon/. The full sentence reads: "La définition d'une œuvre d'art ne saurait être autre chose que celle-ci: Une œuvre d'art est un coin de la création vu à travers un tempérament."

77 Zola, "A New Style of Painting," 74.

78 Zola, "A New Style of Painting," 74–5.

79 Zola, "A New Style of Painting," 74–5.

80 Zola, "A New Style of Painting," 73.

81 When Zola aligned the new art with scientific observation, he sought to distinguish it from Romantic tendencies of emotional expression and spiritual aspiration. For this reason, he denied any relationship between "the paintings of Edouard Manet and the poems of Charles Baudelaire." While he admits "that a lively sympathy has brought the poet and the painter together, but I think that I can affirm the latter has never made the blunder, committed by so many others, of wanting to put ideas in his painting." Manet is only motivated by his talent for translating what and how he sees nature: "if he brings together several objects or several figures, he is guided in his choice only by the desire to obtain beautiful color areas, beautiful oppositions. It is silly to try to make a mystical dreamer out of an artist obedient to such a temperament" ("A New Style of Painting," 73).

82 Cachin et al., "Appendix 1," *Manet, 1832–1883*, 521.

83 The chronological proximity of Manet's execution of Zola's portrait in February 1868 (Cachin, *Manet*, 282) and Zola's 1868 Preface to *Thérèse Raquin*, penned in April 1868, suggests an unrecorded verbal debate between the artist and writer that occurred in Manet's studio. Perhaps following this extended encounter, both creators dug deeper into their respective positions, creating polemical painting and writing.

84 See Robert Lethbridge, "Manet and Textual Frames," in *Artistic Relations: Literature and the Visual Arts in Nineteenth-Century France*, eds. Peter Collier and Robert Lethbridge (New Haven: Yale University Press, 1994): 155–8; and Lethbridge, "Zola and Contemporary Painting," 74–5.

85 Lethbridge, "Manet and Textual Frames," 157.

86 See "Émile Zola," photograph by Nadar, 1895, https://commons.wikimedia.org/wiki/File:%C3%89mile_Zola_Nadar.jpg.

87 See Lethbridge, "Manet and Textual Frames," 157, and Reff, "Manet's *Portrait of Zola*," 36. According to Reff, by 1868, seven of the final 14 volumes of *Histoire des peintres* had appeared.

88 Reff, "Manet's *Portrait of Zola*," 36.

89 Reff, "Manet's *Portrait of Zola*," 39.

90 Reff, "Manet's *Portrait of Zola*," 39; see also Lethbridge, "Manet and Textual Frames," 155.

91 See Robert Lethbridge, "Zola, Manet and *Thérèse Raquin*," *French Studies* 34, no. 3 (July 1, 1980): 282. Lethbridge states that Zola already possessed the engraving *Olympia*, which Manet had provided to the writer for the 1867 brochure. Reff also acknowledges Manet's

inclusion of the engraving and that her seductive gaze was created "with a visual wit altogether characteristic of him," and with his appreciation of Olympia's acknowledgment of Zola's efforts to defend her reputation ("Manet's *Portrait of Zola*," 41).

92 Manet's 1867 one-person exhibition was organized in protest of the official Salon. For a discussion of Zola's and Manet's collaboration in staging this event, see Alan Krell, "Manet, Zola and the 'Motifs d'une Exposition Particulière' 1867," *Gazette des Beaux-Arts* 99 (1982): 109–15.

93 Lethbridge also asserts that the painting is evidence of Manet's ironic, "visual wit" and articulates "the contemporary debate about the relative autonomy of literature and painting." This interpretation suggests that Manet's portrait of Zola was a kind of joke or satire. As referenced in Chapter 8, this interpretation also aligns with McLees' analyses of the "cult of '*la blague*' (wit)" to which Baudelaire belonged, and by extension, to which Manet and likely Zola belonged as well; see *Baudelaire's "Argot Plastique,"* 88.

94 As Lethbridge observes, "dangling eyeglasses suggest the perceptual blank of Zola's gaze, oblivious not only of pictures in hand and around him but also of the portrait painter" ("Zola and Contemporary Painting," 75).

95 Lethbridge argues that Manet's framing of Zola, within interlocking frames, echoes Manet's compositions with mirror reflections; see "Manet and Textual Frames," 157.

96 Lethbridge, "Zola and Contemporary Painting," 74.

97 Zola, "Mon Salon (1868): Édouard Manet," originally published in *L'Événement Illustré*, May 10, 1868; repr: *Les Cahiers Naturalistes. Société Littéraire des Amis d'Émile Zola*, https://www.cahiers-naturalistes.com/documents/zola-et-arts/ecrits-sur-lart-de-zola/mon-salon-1868-edouard-manet/. Zola attributed Manet's voice to the dialogue between himself and the artist: "C'est en remuant ces idées que j'ai vu la toile se remplir. Ce qui m'a étonné moi-même a été la conscience extrême de l'artiste. Souvent, quand il traitait un détail secondaire, je voulais quitter la pose, je lui donnais le mauvais conseil d'inventer. *Non*, me répondait-il, *je ne puis rien faire sans la nature. Je ne sais pas inventer. Tant que j'ai voulu peindre d'après les leçons apprises, je n'ai produit rien qui vaille. Si je vaux quelque chose aujourd'hui, c'est à l'interprétation exacte, à l'analyse fidèle que je le dois.*" (original emphasis).

98 Zola, Salon de 1875 (extract), in *Pour Manet*, 158.

99 Zola, Salon de 1876 (extract), in *Pour Manet*, 167.

100 Zola, "Le Salon de 1880: Le Naturalisme au Salon," originally published in *Le Voltaire* (June 18–22, 1880); repr. *Les Cahiers Naturalistes*, https://www.cahiers-naturalistes.com/documents/zola-et-arts/ecrits-sur-lart-de-zola/index-des-ecrits-sur-lart-de-zola/le-salon-de-1880/. "M. Édouard Manet a été un des infatigables ouvriers du naturalisme."

101 In "The Experimental Novel," Zola defined the experimental novelist as "he who accepts proved facts, who shows in man and society the mechanism of the phenomena which science has mastered, and who lets his sentiments enter in only concerning those phenomena whose determinism is not yet fixed, while he tries to control this personal sentiment, this *a priori* idea as well as he can by observation and experiment" in *Documents of Modern Literary Realism*, ed. George Joseph Becker (Princeton: Princeton University Press, 1963): 195; original edition: "Le Roman Experimental," in Emile Zola, *Le Roman Experimental* (Paris: G. Charpentier, 1880): 18.

102 Zola, "The Experimental Novel," 193–4.

103 See Lethbridge, "Manet and Textual Fames," 157.

104 The "fallen" woman is one of the most prevalent themes of the Naturalist novel, which follows the anti-heroine within an entropic metatheme pattern (see Baguley, *Naturalist Fiction*, 105–6). The barmaid's lowly labor in the public realm of entertainment marks her sexual availability and, in turn, her inevitable degeneration into prostitution. Often, the anti-heroine succumbs to prostitution for material and social gain, as witnessed in *Nana*.

105 Baudelaire, "The Painter of Modern Life," 400.

106 Novelene Ross argues that Manet painted the mirror reflection of the winter garden, from which the auditorium could be viewed. While her evidence is taken from a contemporary description of the Folies Bergère, such an analysis attributes to Manet's painting a photographic or documentary "realism," against which I am arguing; see *Manet's 'A Bar at the Folies Bergère' and the Myths of Popular Illustration* (Ann Arbor: UMI Research Press, 1982).

107 Robert L. Herbert notes the curved balcony walls, covered with mirrors. However, he attributes the barmaid's displaced reflection in the mirror as a matter of "distortion," not a result

of the actual mirror's curvature. Herbert also uses Chéret's poster of the Folies-Bergère to document the balcony's mirrored walls. See *Impressionism: Art, Leisure, and Parisian Society* (New Haven: Yale University Press, 1991): 79–80. I independently located Chéret's poster and related documents to establish the mirrored curved balcony wall during my Paris-based research project in 2001.

108 I refer to photographs taken by Frères Séeberger in 1925 and 1928; the earlier date shows the interior in original decor before being refitted in the Art Deco style (documented on the later date). These photographs show the pervasive use of mirror paneling throughout the interior: in the entrance, a street-level bar, the recessed alcoves of the hall, and the rear walls of the semi-circular balcony auditorium. A photograph of the auditorium, taken from the stage, shows the balcony to one side and the camera's flash reflecting from the curved mirrored wall ("Théâtre des Folies Bergères/Scène: Vue prise des coulisses;" Référence APFSK01-07). I viewed these photographs in 2001 at the (former) Caisse Nationale des Monuments Historiques et des Sites (Paris; now accessible via the Ministère de la Culture database, https://www.pop.culture.gouv.fr/).

109 Baudelaire, "The Painter of Modern Life," 402.

110 See Manet, *At the Folies-Bergère*, ca. 1880, pen and brown ink over graphite, 51 1/2 x 21 1/8 in, private collection; illustrated in Ann Dumas et al., *The Private Collection of Edgar Degas* (New York: Metropolitan Museum of Art, 1997): 188, fig. 265.

10 Reflections on Baudelairean Modernity

Introduction

In this final chapter, I formulate a model for reifying the embodied aesthetic of nineteenth-century Paris during Haussmannization. In doing so, I draw from empirical psychology models that track the sensory, affective, and cognitive events that comprise aesthetic processing.[1] However, my focus extends beyond these discrete events. I explore the liminal spaces that connect physiological and neural processes as sites that register rupture, ambiguity, and dissonance, attributes consonant with Baudelairean modernity. Manet's *A Bar at the Folies-Bergère* is at the heart of this exploration, a painting that has yielded compelling findings in my collaborative eye-tracking study (Figure 1.9).

By placing *A Bar at the Folies-Bergère* at the center of this chapter, I also establish the painting as a framing device for this book. Introduced in Chapter 1 and examined at length in Chapter 9, *A Bar at the Folies-Bergère* combines the themes and methods presented in previous chapters. A word-and-image approach has provided the overarching methodology to integrate Baudelaire and Baudelairean artists, notably Manet, into a network of shared responses to Haussmannization. At the same time, I have aligned this methodology with the modeling of experiential and embodied aesthetics and neuroaesthetics to reimagine the lived experience of rupture and fragmentation during the demolition and rebuilding of Paris in the second half of the nineteenth century. As I have argued, this lived experience manifested in writers' and painters' use of irony and ironic parody as literary and pictorial devices. Within this network, the mirror emerges as a literal and metaphoric conveyor of rupture, fragmentation, and ironic self-reflection, playing a pivotal role in both the formal structures and thematic representations of Baudelairean modernity. I have also identified contemporaneous scientific and critical texts to substantiate this experiential and embodied aesthetics. Taken together, the reimagined lived experience of Haussmannization, its literary and pictorial representations, and its psychical and formal attributes of irony and ironic parody provide a platform for systematizing a neuroarthistory of nineteenth-century Parisian culture.

Mirror as Liminal Space

Based on the standard English and psychological definitions of liminality, I aim to connect Baudelairean modernity and the liminal spaces of aesthetic processing. In the first case, liminality is "[c]haracterized by being on a boundary or threshold, especially by being transitional or intermediate between two states, situations, etc."[2] In psychological terms, liminality is defined as "relating to the threshold of a sensation," with threshold

DOI: 10.4324/9780429442315-11

defined as "the minimum intensity of a stimulus that is necessary to evoke a [neural] response[.]"[3] Given these definitions, liminality and the liminal spaces that connect the physiological and neural processes of aesthetic experience are consistent with Baudelaire's definition of modernity as transitory and the painterly translation of modernity as transformative.

The temporal dimensions of liminality also evoke the mirror as a liminal space, particularly regarding the mirror and its reflections in *A Bar at the Folies-Bergère*. Postmodernist Michel Foucault's phenomenological description of mirroring effects, as articulated in his definition of heterotopia, (inadvertently) succeeds in rendering our physical and psychical immersion in the liminal space of Manet's painted mirror. For Foucault, heterotopias (real sites) are the binary opposite of utopia (non-sites) that pertain to (imperceptible) socio- and geo-cultural spheres.[4] Yet the mirror manifests the alternation between these two sites in the viewer's existential self-reflection. As Foucault states, "I believe that between utopias and... heterotopias, there might be a sort of mixed, joint experience, which would be the mirror." On the one hand, "[t]he mirror is... a utopia since it is a placeless place.... I see myself there where I am not, in an unreal, virtual space that opens behind the surface." At this point, an existential doubling of presence and absence occurs: "I am over there, there where I am not, a sort of shadow that gives my own visibility to myself, that enables me to see myself there where I am absent: such is the utopia of the mirror." On the other hand, the mirror "is also a heterotopia in so far as the mirror does exist in reality" and reflects—in reverse—an existential doubling of absence and presence. Foucault continues:

> From the standpoint of the mirror, I discover my absence from the place where I am since I see myself over there. Starting from this gaze that is, as it were, directed toward me, [and] that is on the other side of the glass, I come back toward myself; I begin again to direct my eyes toward myself and to reconstitute myself there where I am. The mirror functions as a heterotopia in this respect.

But this (heterotopian) real-site simultaneously fluctuates with the (utopian) non-site:

> [the mirror] makes this place that I occupy at the moment when I look at myself in the glass at once absolutely real, connected with all the space that surrounds it, and absolutely unreal since to be perceived it has to pass through this virtual point which is over there.[5]

Foucault's description of self-mirroring as a dynamic event not only translates into the aesthetic experience of Manet's painted mirror. The existential doubleness, the shifting spatio-temporal dynamics between the self and the non-self and between the familiar and unfamiliar, also mimics the self-reflective mirror-functions in Baudelairean irony and Manet's ironic parody. As explored in Chapters 1 and 9, the painted mirror in *A Bar at the Folies-Bergère* represents and affects the processing of irony and the dialogic relation of the parodied "text" within the new "text," thereby making visible that which is invisible. In this context, the (temporal) liminal spaces of literary and pictorial mirroring are analogous with the sensory-affective-conceptual trajectories triggered during the viewer's embodied aesthetic experience. In the following discussion, I adapt models and methods from cognitive poetics, psychology, and neuroaesthetics to track the temporal processing of visual irony and parody.

The Problem of Beauty, Reward, and Rupture

This section maps the temporal-dynamic space of liminality as integral to experiential aesthetics and, accordingly, reviews scientific research that examines neural and psychical responses to artworks deemed ambiguous, indeterminate, and "not beautiful." From the inception of neuroaesthetics (dating from the 1990s), researchers have not attended to such artwork. Instead, they have limited neuroaesthetics to the neural bases of beauty perception in art, focusing on brain areas that activate pleasure and reward systems. As noted in Chapter 7, scientists have recently addressed neural responses to dissonance and the difference between assessing beauty and ugliness.[6] According to Camilo J. Cela-Conde and colleagues:

> the aesthetic experience that encompasses the concept of 'beauty' constitutes a narrow realm in modern aesthetics.... [I]n many modern art forms... the prime role of beauty has diminished and is even actively challenged. Thus, an ugly object may still be a great work of art.[7]

While today we may not regard Manet's *A Bar at the Folies-Bergère* as "ugly," it does constitute an aesthetic of rupture and ambiguity that transgresses beauty and problematizes "reward."

To establish the liminal spaces wherein rupture and ambiguity occur during aesthetic processing, I first adapt models from cognitive poetics. Cognitive poetics is a branch of literary theory that analyzes the mental processes at work during literary reading.[8] Reuven Tsur, David Miall, and Philip Davis represent varying degrees to which literary theorists of cognitive poetics have collaborated with neuroscientists. I have extracted cognitive and neurological data from their studies that apply to experiential neuroaesthetics as manifested by literary irony and visual ironic parody. The studies discussed here focus on the mental processing of rupture and defamiliarization as modalities unique to literary rather than non-literary reading. These modalities also distinguish many modern artworks, such as *A Bar at the Folies-Bergère*, from quotidian appearances and phenomena.

Tsur's work on the grotesque exemplifies how he uses cognitive psychology to "systematically account for the relationship between the [literary devices] and their perceived effects" in "information-processing."[9] The grotesque is a valuable model for the cognitive experience of defamiliarization, which correlates with distortion and disorientation in *A Bar at the Folies-Bergère*. As Tsur explains, the reader "perceives the image as grotesque, [as] something that is simultaneously laughable and horrifying or disgusting." This initial perception issues from "the grotesque poetic devices that [subsequently] produce an emotional disorientation which is experienced as a shock, perplexity [or] surprise."[10] To analyze the "emotional disorientation" that the grotesque induces, Tsur developed the "rapid and delayed categorization" principle derived from neurological information processing models.[11] Tsur aligns the cognitive processing of the grotesque (and the unfamiliar in general) with the right cerebral hemisphere, where delayed categorization occurs in a diffuse state, as with affects and emotions. Conversely, rapid categorization occurs in the left hemisphere, where sequential processing forms logical information (for example, everyday non-literary language). Delayed categorization is prolonged because it engages the memory system, sensory loading, and extensive mental processing space.[12] As Tsur states, such a delay "involve[s] a period of uncertainty that may be quite unpleasant or even intolerable[.]"[13] Delayed categorization facilitates a mapping of the brain's increased activity in the right hemisphere that registers diffuse input when applied to our emotionally or cognitively disorienting experience of the mirror distortion/dislocation in *A Bar at the Folies-Bergère*.

Tsur's neurological model of delayed categorization also substantiates the locus and the temporal dynamics of liminal space produced by Manet's mirroring devices and Baudelaire's exemplary ironic literary devices. As we have seen in Chapter 8, the defamiliarization process is embedded in Baudelaire's prose poems themselves. In this context, defamiliarization arises from the ironic turns with which the narrative structures end, causing the reader to be shocked and surprised. Similarly, in Manet's painting, the mirror is the agent of defamiliarization, affected by the distortion of the mirror's actual (in situ) curve, as described in Chapter 9. Here, Manet exaggerates the distortion by misaligning and eliminating objects and figures from the mirror's reflection. Viewers of Manet's painting are first caught in a visual snare, transitioning to a dynamic liminal space marked by prolonged delayed categorization as they struggle to find spatial orientation. We can assume that this cognitive processing is followed by shock, perplexity, or surprise.

The fact that contemporaneous viewers were perplexed by this defamiliarization process is evidenced by a caricature of *A Bar at the Folies-Bergère*. In the graphic caricature, the illustrator inserted the male patron in front of the bar and barmaid, viewed from behind, thereby conforming to expected spatial (and social) relations (Figure 10.1). As the caption indicates, the man reflected in the mirror does not exist in the "*tableau*" (scene or painting); because of the painter's distraction, the illustrator had to complete—albeit correct—the painting.[14]

Literary historian David Miall has collaborated with psychologist Don Kuiken to measure the duration of "delayed categorization" in real-time, using "foregrounding" as the literary device for tracking the "defamiliarization" process.[15] Foregrounding refers to making something stand out from the surrounding words or images.[16] Its defamiliarizing function occurs in two related ways: "as unexpected regularity (...), and as unexpected irregularity."[17] Based on their review of neuroscience research on literary reading (using event-related potentials [ERPs] and functional magnetic resonance imaging [fMRI])[18] and their empirical studies (using surveys and timed reading tasks), Miall and Kuiken have found that foregrounding inhabits a complex cognitive process, beginning with pre-cognitive responses to its unfamiliarity.[19] As a temporally unfolding experience, foregrounding evokes the mental processing of liminal spaces affected by mirrored reflections. As Miall explains:

> Foregrounding thus appears to initiate a rich response prior to awareness: as early as 155 msec, the detection of unfamiliar stylistic features may initiate a response process that includes prosodic, affective, and semantic aspects. This [process] is followed by an inhibitory phase during which feeling is contextualizing the unfamiliar, leading to the emergence of new meanings or shifts in perspective a number of seconds downstream from the foregrounding onset.[20]

Empirical psychologists Matthew Pelowski and Fuminori Akiba's analysis of processing aesthetic disruption provides a visual art model for assessing "foregrounding" and comparable pre-cognitive phases in viewing *A Bar at the Folies-Bergère*. As reviewed in empirical Chapter 3, Pelowski and Akiba's five-stage model of art perception is "organized around initial disruption and subsequent meta-cognitive reflection and

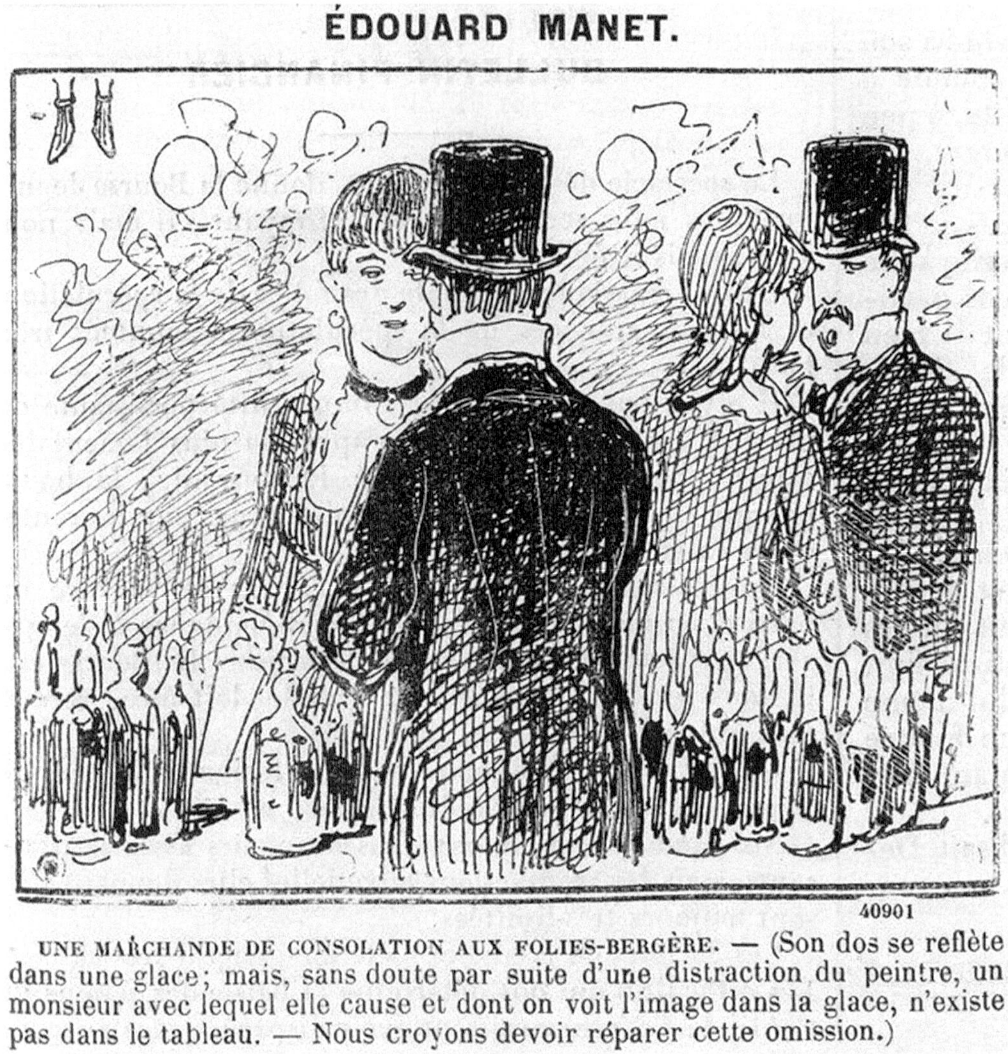

Figure 10.1 Stop (Louis Pierre Gabriel Bernard Morel-Retz), "Edouard Manet. Une Marchande de Consolation aux Folies-Bergère" (Edouard Manet. A Consolation Seller at the Folies-Bergère). In *Le Journal Amusant*, no. 1343 (May 27, 1882): 5. Bibliothèque nationale de France, Paris. Image: gallica.bnf.fr/Bnf.

self-transformation."[21] Based on physiological assessments and viewer surveys, their findings pertain to two kinds of viewer responses to discrepancy. First, there is the viewer who either succumbs to "the human instinct of escape or to assimilation through surface evaluation or self-withdrawal."[22] This viewer, for example, would be repulsed by Manet's pictorial distortions and withdraw from further engagement or enter a satirical (caricatural) dialogue with the artwork. The second kind of viewer is open to schema change. This viewer completes the five stages of the art-perception model. In the first stage, when confronted by unfamiliar (or defamiliarized) artworks, this viewer overcomes the escapist/withdrawal instinct and instead "enters an intractable position whereby one might use their own disruptive encounter as a means of self-reflection and transformation and therefore come to behave or see something new[.]" In this case, the viewer may also "feel pleasure, epiphany, and harmony."[23]

At this later stage, Pelowski and Akiba's aesthetic model of self-reflection aligns with Miall and Kuiken's assessment of the reader's transformative experience that leads to new meanings or shifts in perspective. Like Pelowski and Akiba, Miall and Kuiken also used psychological and quantitative methods (surveys) to identify the "self-modifying" effects of foregrounding as an agent of defamiliarization.[24] Based on these experiments, the latter team has redefined "delayed categorization" as the "defamiliarization-reconceptualization process," wherein the two poles are mediated by feeling.[25] Because Pelowski and Akiba and Miall and Kuiken have broadened the affective spectrum of the defamiliarization process, their research offers a model for making delayed categorization a concrete and measurable temporal exponent of experiential neuroaesthetics. These empirical analyses substantiate the temporal dynamics of liminal space activated by mirror reflections.

Philip Davis's research focuses directly on brain anatomy. Davis collaborated with linguists, psychologists, and neuroscientists to capture the various brain regions activated by reading Shakespeare's texts that deploy the "functional shift," another form of disruption and defamiliarization. This rhetorical device changes "the grammatical status of words, e.g., using nouns as verbs" or verbs as nouns.[26] Davis's team finds that the functional shift "triggers a surprise effect inviting mental re-evaluation, seemingly independent of semantic processing."[27] This research is most helpful in translating the liminal space of viewing mirror reflections into experiential neuroaesthetics. First, it demonstrates how literary and, coincidently, pictorial techniques of rupture can become the subject of neuroscientific research. Second, the team's neurological findings might also apply to the transformative effects triggered by the defamiliarization processing of the mirror in *A Bar at the Folies-Bergère*.[28]

Beginning in 2008, Davis's team used three computer-based technologies to measure brain activity and the effects of the functional shift: electroencephalographic (EEG), magnetoencephalography (MEG), and magnetic resonance imaging (MRI). Participants were instructed to read four sentence stimuli based upon Shakespeare's texts ranging from normative syntax and grammar to texts in which the functional shift (in syntax or grammar) distorted meaning or made it nonsensical. People undergoing the experiment had to press a button if the sentence roughly made sense to them. During the reading process, the team used the EEG to locate that functional shift's "unique effect on the brain." They used MEG to record "rapid, 'peak' [brain] activity during the comprehension of functional shift[.]" They used MRI to pinpoint "brain-activation by location."[29] Their results showed that functional shift processing increased activity in both the left and right hemispheres. Most importantly, they found that this activity was much more widely dispersed throughout the brain's anatomy "in regions not typically activated in everyday language processing."[30] Their neuroimaging findings "show how Shakespeare's grammatical exploration forces the listener to take a more active role in integrating the meaning of what is said."[31] Taken together, the Davis team's data support and make concrete the psychological and theoretical models of delayed categorization, foregrounding, and the five-stage modeling that puts discrepancy and self-transformation at the heart of an aesthetic experience.

A Bar at the Folies-Bergère and a Neuroaesthetic of Rupture

We can now adapt the preceding studies to speculate about the temporal, affective, and cognitive processes of viewing *A Bar at the Folies-Bergère* from the perspective of a nineteenth-century Parisian.[32] Initially, the viewer might recognize the popular entertainment venue as conferred by the painting's title. However, as per Pelowski and Akiba's five-stage model of art perception, during the first stage of direct evaluation, the viewer's memory-based recognition would be immediately disrupted by a perceived "discrepancy" among the formal elements—the painterly, fragmented surface, the frontal figure's disengaged demeanor, and the mirror distortions and omissions. Now, the viewer enters the second stage of Pelowski and Akiba's model, "a period of meaninglessness and loss of understanding of the artist's motive," reverting to pre-cognitive brain functions. We can amplify this disruptive pre-cognitive stage with literary models of neurocognitive rupture. For example, foregrounding is experienced as "unexpected irregularity," wherein the barmaid, her mirror reflection, and the client are misaligned. This assessment might result in a visual functional shift, in which pictorial "grammar" of perspective and spatial recession is replaced by paint marks that simultaneously cohere as objects and figures and dissolve into assorted pigments, impastoed surface texture, and sketch-like brushwork. Concomitantly, the viewer experiences delayed categorization, affecting surprise or confusion.

As noted above, stage three of the Pelowski/Akiba five-stage model is a matter of self-evaluation, determining whether the viewer escapes the artwork, cannot tolerate the discrepancies, or remains engaged. This stage "also entails a switch from 'lower order and often unconscious' perception to a 'higher order process involving conscious assessment' and greater cognitive involvement."[33] Viewers who sustain engagement enter a fourth stage wherein the direct evaluation of the artwork shifts from self-evaluation (when the viewer assesses their success or failure to understand) to:

> a meta-cognitive approach to their interpretation: acknowledging discrepancy, revisiting their previous expectations or perceptual/conceptual schema, and eventually discarding or changing these schema [...]—often, essentially 'giving up' their attempt at direct control[.] Thereby, viewers alter their own hierarchical self-image, achieving a relative 'self-transformation.'[34]

However, this schema change entails discrepancy processing and "appears to temporarily intensify the preceding tension and emotion," affecting the autonomic nervous system (ANS).[35]

Pelowski and Akiba's model ends with the fifth and final stage, constituting the "aesthetic phase" that incorporates the "schema change or self-transformation."[36] The viewer now experiences "the arousal of 'aesthetic' harmony, pleasure and emotion itself" and an adjustment of the viewer's "self-image or relationship with the world."[37] This deeply embodied aesthetic experience notwithstanding, I have circumvented stage five to reify the stage four discrepancy processing as more enduring and meaningful for the nineteenth-century viewer. According to Pelowski and Akiba, discrepancy processing engages both subsystems of the ANS, the "coordinated activation of parasympathetic [accelerated nervous] response and sympathetic [slowed down nervous] response [which makes the processing] such a potent psychological state."[38] I propose that this psycho-neural processing parallels and mirrors the viewer's affective embodiment of the modern urban experience inscribed in the painted surface and spatial distortions of *A Bar at the Folies-Bergère*. We may also consider this stage leading to a revised stage five, a meta-cognitive process that involves the viewer's self-recognition in, and harmony with, the temporal and material dimensions of a lived experience of dissonance.

Mirror Neurons and the Embodiment of Gestural Painting

For art historians, visual arts specialists, and philosophical aestheticians, the preceding discussion of embodied aesthetic processing may be reminiscent of a centuries-old aesthetic theory of empathy.[39] In their groundbreaking study, "Motion, Emotion and Empathy in Esthetic Experience" (2007), art historian David Freedberg and neuroscientist Vittorio Gallese extended the theory of empathy to the corporeal role of the brain's motor system in the perception of visual art. They identified the activation of the motor cortex when participants viewed figurative and abstract artworks. Mirror neurons, brain cells that 'imitate' motion observed in another sentient being,[40] are at the center of their study. Here I consider the efficacy of Freedberg and Gallese's mirror neuron model for substantiating the viewer's embodied responses to the spatial and figural distortions and the ruptured, dissonant formal qualities of Manet's painting.

Notwithstanding the absence of laboratory-based research, Freedberg and Gallese based their analysis on recent empirical studies that showed the "discovery of mirroring mechanisms and embodied simulation for empathetic responses to images in general, and to works of visual art in particular."[41] Further, Freedberg and Gallese centered their study on:

> [t]he discovery of mirror neurons [that] illuminates the neural underpinnings of the frequent but hitherto unexplained feeling of physical reaction, often in apparent imitation of the actions represented within a work of art or suggested by the implied movements involved in its making.[42]

As a result of their findings, Freedberg and Gallese "challenge the primacy of cognition in responses to art" and assert "that a crucial element of esthetic response consists of the activation of embodied mechanisms encompassing the simulation of actions, emotions and corporeal sensation[.]"[43]

In their study, Freedberg and Gallese focused on figural painting and sculpture, specifically regarding three modes of embodied responses: "(i) the feeling of bodily engagement with the gestures, movements and intentions of others; (ii) the identification of the emotions of observed others; and (iii) a feeling of empathy for bodily sensations."[44] While their observations may pertain to the barmaid's and her client's gestures and emotions in *A Bar at the Folies-Bergère*, this subject has already received substantial art historical attention.[45] Therefore, Freedberg and Gallese's study did not provide substantive new evidence for charting the viewer's embodied responses to the spatial and figural distortions. Conversely, Freedberg and Gallese's observation of a fourth embodied response, regarding the artist's gestural markings, more directly pertains to my focus on Manet's broken brushwork and facture of the mirror reflection that evoke the embodied experience of rupture and dynamism amidst the modern urban setting.

Freedberg and Gallese extended their mirror neuron model and "the theory of embodied simulation" to "the formal qualities of a work and the observation of the gestural traces of the artist," as exemplified in Jackson Pollock's *Number 14: Gray* (1948) and

Lucio Fontana's *Concetto Spaziale 'Atteza'* ([Spatial Concept 'Waiting;']1960). Accordingly, they propose "that even the artist's gestures in producing the art work induce the empathetic engagement of the observer, by activating simulation of the motor program [in the brain] that corresponds to the gesture implied by the trace."[46]

Subsequently, Freedberg and Gallese's hypothesis proved viable—a laboratory-based study lent "empirical support to the role of motor simulation in the observation of abstract art."[47] Together with lead investigator, M. Alessandra Umilta' (and others), they used high-density EEG to investigate "whether the observation of high-resolution digitized static images of abstract paintings by Lucio Fontana is associated with specific cortical motor activation in the beholder's brain."[48] Thus, they paired "3 different black and white high-resolution digitized images of abstract artworks of Lucio Fontana (Original Stimuli)" with "3 high-resolution digitized images of graphically modified versions of the original artworks (Control Stimuli) displaying the same black and white graphic pattern of the original images."[49] Using high-density EEG, their experiment "show[ed] that in spite of the similarity between the digitized versions of the Original artworks and the Control stimuli, only the former evoke cortical motor activation in the beholder's brain, as exemplified by mu rhythm suppression[.]"[50] In conclusion, Umilta' and colleagues' experiment confirmed Freedberg and Gallese's original hypothesis. They assert:

> Our interpretation is that the observation of the cuts on canvases activates the motor representation of the same gesture in the brain of beholders. In contrast, Control stimuli, because of their lack of implicit dynamicity, as shown by participants' score on the perceived amount of movement, are merely processed as images that are not the manifest consequence of someone's gesture.[51]

While skeptics of mirror neurons' presence and function in humans abound,[52] Umilta' and colleagues' empirical research provides physical evidence that viewers of abstract art embodied and identified with the painter's gestural marks—brushstrokes, dripping, and even the most minimal incisions through a canvas surface. This research supports my interpretation of Manet's painterly, sketch-like brushwork in *A Bar at the Folies-Bergère*, particularly in the mirrored reflection. Following Umilta' and colleagues' model, these gestural marks activate the motor representation of the same gesture in the brain of beholders. However, I wish to extend the viewers' *empathetic emotional responses* to figural gestures and movements (Freedberg and Gallese) to the beholder's embodied response to Manet's gestural markings. In so doing, I focus on the technical discrepancy between, on the one hand, the firmly modeled and contoured foreground figure and still-lifes and their mirrored reflections and, on the other hand, the painterly rendering of the reflected figures and lights. Such a discrepancy signals the mirror reflection as a liminal space wherein incessant fragmentation and change prevail. Extending Umilta' and colleagues' model, I contend that Manet's painterly technique communicated his (cumulative) embodied response to the modern urban environment within this liminal space. These gestural marks may have triggered the nineteenth-century beholder's cortical motor activation to embody the painter's gestural activity and *the beholder's embodied empathy*—a mnemonic response to the urban experience they shared with the painter.

Umilta' and colleagues' empirical research on abstract art and cortical motor activation also supports my hypothesis that viewers of *A Bar at the Folies-Bergère* not only focus on the figural presence of the barmaid and her customer but may comparably focus on the gestural markings in the mirror reflections. As a result, the viewer might get caught in the dynamic exchange between the recognizable foreground and the indeterminate liminal mirrored space. If the viewer's mirror neurons are activated by the ruptured, dissonant formal qualities of Manet's painting, to what degree do these qualities dynamize the act of looking? To answer this question, I collaborated with empirical psychologists at the University of Vienna to track participants' eye movements while viewing digitized images—the original and one modified—of Manet's *A Bar at the Folies-Bergère.*

Manet's *A Bar at the Folies-Bergère*: An Eye-Tracking Study

In empirical aesthetics research, cortical motor activation and eye tracking are foundational to a neurobiological model of viewing artworks. In this section, I present a synopsis of the eye-tracking model to bridge the previous section on mirror neurons with the following discussion of eye-tracking studies. I then discuss my collaborative eye-tracking study of Manet's *A Bar at the Folies-Bergère.*

On the Perception Process of Viewing Works of Art

As Umilta' and colleagues explain regarding the perception process of viewing works of art:

> There is great heterogeneity across results from [empirical] investigations trying to clarify the neural correlates associated with the perception of visual art[.] The perception of visual art works begins with the visual analysis of the stimulus, which then undergoes further processing. This progression may lead to an 'aesthetic experience' likely based on biological and embodied mechanisms that are modulated by factors such as the context, individuals' interest in the artwork, prior knowledge, and familiarity.

This account of the beholder's post-perceptual processing that leads to an "aesthetic experience" conforms with Pelowski and Akiba's five-stage model. Yet Umilta' and colleagues' study of gestural artworks by Fontana and Pollock also provides a crucial link between the physiology of art-viewing and the affective-cognitive processing that form aesthetic appraisal. As the latter team concludes, "[the] beholders' cortical motor activation constitutes an essential component of the perception of abstract art works."[53]

Following Umilta' and colleagues' and Freedberg and Gallese's earlier research, I asked this question: since *A Bar at the Folies-Bergère* combines both movement and stasis and abstract and representational techniques, can we determine the dynamical biological and embodied mechanisms activated while looking at this artwork? In 2023, I collaborated with empirical psychologists in a laboratory-based eye-tracking experiment with *A Bar at the Folies-Bergère* as a stimulus to answer this question.[54] I hypothesized that prolonged dynamic eye movement would measure the viewer's effort to resolve spatial ambiguity. I also hypothesized that these measures might simulate the nineteenth-century motor and perceptual processing of mirror reflections and painterly ruptures. These perceptual responses would also align with cognitive literary studies identifying textual rupture and paradox as aesthetic modes of delayed categorization, disorientation, and reorientation registered during the reading process.

For our study of *A Bar at the Folies-Bergère*, Dilara Beder (Ph.D. candidate, Bilkent University) incorporated this artwork into her broader eye-tracking research as a visiting scholar in the University of Vienna's psychology department (hereafter Beder and

Weingarden). We based our study on the premise that eye movements are an integral and essential part of visual perception. For this reason, eye-tracking systems are used to explore the eye movement behaviors of people with subconscious brain activity associated with aesthetic experiences.[55]

Before discussing the results of our experiment, it is imperative first to define eye-tracking technology. For our purposes, eye tracking can be defined as the process of measuring the point of gaze (where one is looking). An eye tracker is a device used to measure eye position and eye movement.[56] As explained below, we used a video-based eye-tracking camera as our measuring device. Participants viewed digitized images of Manet's *A Bar at the Folies-Bergère* on a computer screen. More importantly, it is also essential to validate eye tracking as a neuroaesthetic technique and to define the correlated eye-tracking terminology. To do so, I rely on art historian Raphael Rosenberg and psychologist Christoph Klein's eye-tracking projects designed to bridge art history and empirical psychology, as reviewed in their article "The Moving Eye of the Beholder: Eye Tracking and the Perception of Paintings" (2015).[57] Rosenberg and Klein's experiments are especially significant for their focus on the gaze movements of the beholders of paintings.

Validity of Eye Tracking as a Neuroaesthetic Technique

As empirical psychologists have shown, "the measurement of gaze movements is a prominent psychological technique [that demonstrates the gaze's] neural bases and relationships with visual attention."[58] Eye movement studies are prevalent in psychological research for two features directly related to Beder and Weingarden's study. Here, the data verify the neuroanatomical importance of eye movement and record the immediacy of ocular activation and processes. In the first instance:

> [g]aze movements are *important* to investigate because the ocular-motor system is doubtlessly the most used system among the sensory-motor systems in primates [a biological category that includes humans],[59] supported by an extensively distributed functional neural system[60] that is currently better understood than any other functional system of the primate brain.

In the second instance, ocular activity is dynamic and measurable in real time:

> Due to the anatomy of the eye, this system's effects can be investigated with an unmatched degree of *immediacy*, both psychologically and technically [original emphasis]. As there is only one small spot of highest visual acuity on the retina, the fovea centralis, both the exploration of the visual world and our responses to its changes require our gaze to move as a series of fixations and saccades which, together, provide the best available *model of the structure of human consciousness unfolding in space and time*.[61]
>
> (emphasis added)

These two features combined—the immediacy of measuring anatomical neural systems and the measurement of the structure of consciousness unfolding in space and time—make eye tracking an especially relevant technology for aligning nineteenth-century and present-day viewers' aesthetic experiences of *A Bar at the Folies-Bergère*. As we demonstrate in the eye-tracking experiment with Manet's painting, this technology supports my modeling of an embodied experiential aesthetic of Baudelairean modernity as transitory, dynamic, and indeterminant.

Terminology: Fixations and Saccades

In Beder and Weingarden's study of *A Bar at the Folies-Bergère*, eye-tracking data were diagrammed in two kinds of gaze patterns.[62]

1 Linear tracks of gaze movements across the picture's surface plane and into the picture's space are defined as "fixations" and "saccades." According to Rosenberg and Klein, these linear patterns record "[g]aze movements [that] are basically alternations between periods of the eyes remaining relatively stationary, which we call 'fixations,' and shorter periods of greater mobility, which we call 'saccades' (from the French word 'saccade' meaning 'jerk')."[63] Notably, "there is a continuum between 'fixation,' characterized by the presence of micro-saccades[64]... and '*exploration*,' characterized by the presence of larger amplitude[65] saccades [.]"[66] (original emphasis)
2 Heat maps record the viewers' fixation length—the duration of their attention span—as they focus on individual elements in the painting composition. The duration of viewers' foci is registered across a chromatic spectrum, ranging from intense red (long duration) to orange, yellow, and green (shortest duration). The shift from warm to cooler green hues signifies increasingly shorter gaze durations.

Beder and Weingarden: Eye Tracking Beholders' Viewing of A Bar at the Folies-Bergère

We conducted our study by using the video-based eye tracker Eye-Link 1000.[67] The camera was configured in a desktop mount that allows monocular recording of the eye metrics of the participants. Before the experiment, a dominant eye test was applied to each participant to decide which eye would be used to collect the data. The study involved a final sample of 84 participants (64 females, 20 males, aged 18 to 26), all of whom were 'lay' viewers of art (mostly psychology students, none of whom had any formal training or education in fine art or art history). These individuals viewed two digitized images of Manet's *A Bar at the Folies-Bergère* on the screen.[68] One of the images was the original painting (a; Original stimulus). The second image was a manipulated version of the original image designed to test the research aim (b; Manipulated image). In the manipulated image, produced with Photoshop, we eliminated the reflected male figure and female figures from the original artwork.[69] Because we were interested primarily in the spontaneous, not comparative, reactions people would make upon first seeing this artwork, the study was conducted with a between-participants design, with half of the participants viewing the original image first and the other half viewing the manipulated image first.[70] However, this was then followed by asking the participants to view the other previously unseen version to compare the results (Figure 10.2).

To start the study, the participants saw a cue (a cross) in the middle of the screen before each art image. The art-viewing study began automatically after they fixed their eyes on the cross. The artwork images were shown on the screen for thirty seconds. After each image, we asked the participants questions about their appreciation of the art and their experience when looking.[71] The recorded eye metrics were collected while the artwork images were viewed. There were no time limitations during the questionnaire;

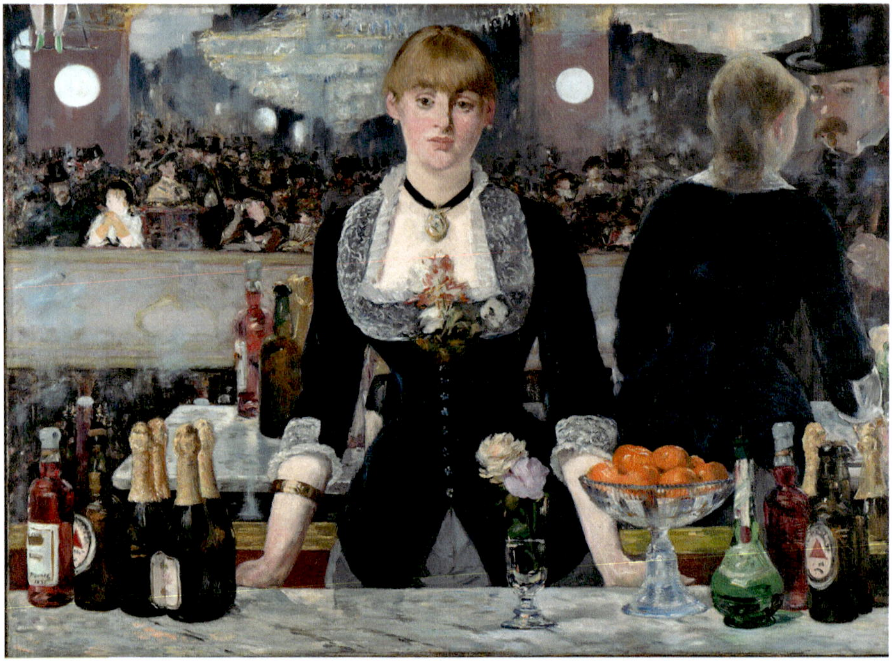

(a)

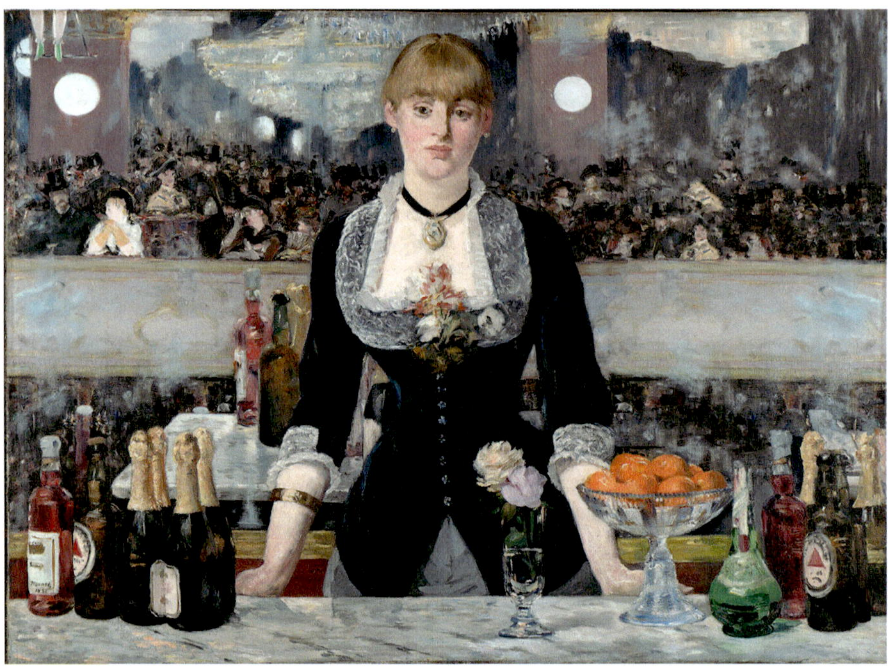

(b)

Figure 10.2 Beder and Weingarden, *Eye Tracking Beholders' Viewing of "A Bar at the Folies-Bergère"* (2023); **a.** (top) Original stimulus: *A Bar at the Folies-Bergère* and **b.** (bottom) Manipulated stimulus: photoshopped *A Bar at the Folies-Bergère*, eliminating the male and female figures' reflections. Image: Wikimedia Commons.

participants advanced through the questions by pressing the space bar key. The total experiment took about three to five minutes for each participant. The data were recorded in four composite images (Figure 10.3):

1 a heat map superimposed on the Original stimulus (a);
2 a heat map superimposed on the Manipulated stimulus (b);
3 linear patterns (fixations and saccades) combined with the heat map and superimposed on the Original stimulus (c); and
4 linear patterns combined with the heat map and superimposed on the Manipulated stimulus (d).

As seen in the Original stimulus heat map (a), participants' (hereafter viewers') gaze movements spread across the painted surface and into the reflected space. As indicated by red-orange-yellow masses, viewers' prolonged attention focused on the barmaid's face and the interaction between her reflected three-quarter turned head and the male figure's reflected frontal visage. Similarly, viewers focused on reflected figures just behind and to the left of the barmaid. One red-orange-yellow mass appends to a woman (L) and man (R). Another mass indicating viewers' attention follows an orthogonal (perspectival) line receding into the pictorial depth. To a lesser extent, viewers also focused on liquor bottles depicted in the foreground and on their mirrored reflections. These patterns of focused attention spans (or saliences) on figures and objects correlate with prior eye-tracking studies demonstrating top-down cognitive processing. In this process, the gaze is directed "to relevant areas of the visual field [e.g., looking at eyes and heads in social scenes] to extract relevant social information[.]" This correlation also pertains to "interesting objects [that] attract the allocation of attention and fixations better than the perceptual saliency of [a general] area of the visual field."[72]

Viewers' focus on the orthogonal linear projection into pictorial depth aligns with bottom-up cognitive processing. Bottom-up factors pertain to "low-level stimulus features," such as technical or formal properties, including color, contrast, and motions. In pictorial terms (as per Rosenberg and Klein), bottom-up features are no less critical than "'top-down' higher level structures," such as faces and objects.[73] Admittedly, in the mirrored reflection of *A Bar at the Folies-Bergère*, painterly brushwork marking the (spectator-) figures' diminution into pictorial space obscures the orthogonal line along which the figures align. However, our participants' visual interest in these technical features (brushwork) suggests their recognition of pictorial depth despite or because of the sketch-like paint marks. (In the post-viewing surveys participants rated space in the Original stimulus to be deeper than the Manipulated stimulus.)

By contrast, the Manipulated stimulus heat map (b) indicates an overall attention span across the picture surface. However, as can be seen by the significantly fewer masses, which are green as opposed to warmer colors, the results suggested that viewers spent much less time looking anywhere else but at the face of the woman. As expected, based on prior studies, viewers focused on the barmaid's frontal visage but relegated little interest to the background image. Participants did not see the mirror in the Manipulated stimulus. Likewise, viewers paid minimal attention to "higher level structures;" depictions of still-life objects—bottles, barmaid's corsage, flower vase, and fruit—compelled minimal gaze duration.

What can we discern from the heat maps about the viewers' embodied aesthetic experience of *A Bar at the Folies-Bergère*? The more complex composition of the Original

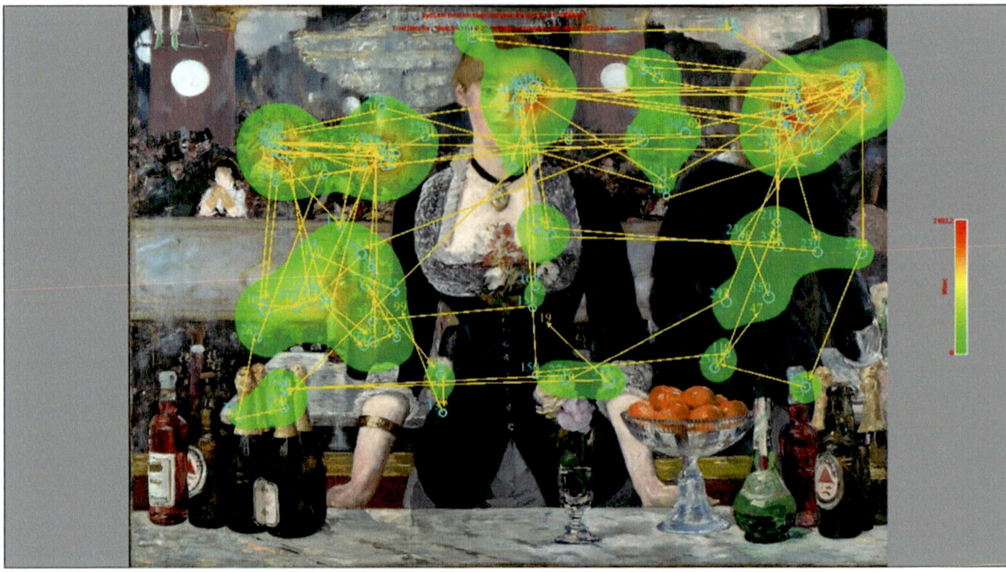

(c)

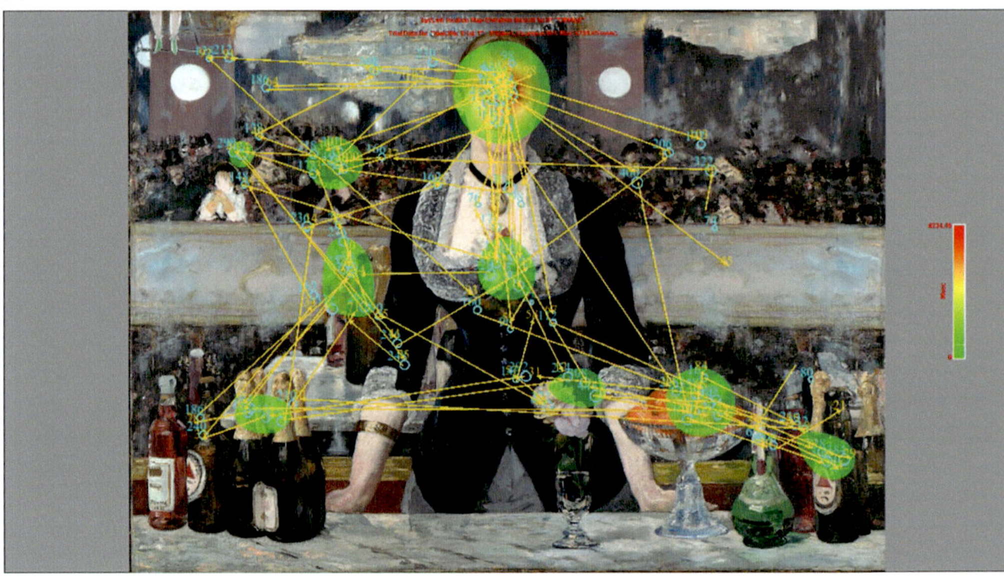

(d)

Figure 10.3 Beder and Weingarden, *Eye Tracking* c. (top) linear patterns (fixations and saccades) combined with the heat map and superimposed on the Original stimulus and d. (bottom) linear patterns combined with the heat map and superimposed on the Manipulated stimulus.

stimulus versus the Manipulated stimulus triggers much more dynamic and diffuse gaze movements. These targeted actions range across the surface and project into space. The kinetic and focalized responses recorded in the heat maps are amplified by the linear patterns of fixations and saccades also recorded in the eye-tracking data.

As seen in the combined heat maps and linear trackings superimposed on the Original stimulus (c) and the Manipulated stimulus (d), focal points of sustained interest registered in the heat maps correlate with the gaze movements registered in the network of fixations and saccades. The greater number of intersections and greater density of line clusters recorded in the eye-tracking data indicate more frequent movements between larger areas and within more minor areas of focalizations (faces, still-life objects, and, in the Original stimulus, balcony [spectator-] figures reflected in the mirror). In the latter data, gaze movements are recorded as frequent alternations between fixations (dots at the lines' termini), indicating periods when the eyes remained relatively stationary, and saccades (lines scattered across the surface), marking shorter periods of greater mobility. Different lengths of saccades indicate movements between local visual attention (shorter saccades, as seen in the barmaid and client's reflections) and global visual attention (longer saccades, as seen in the lines between the foregrounded barmaid's face and her mirror reflection).

Comparing the two composite (overlaid) images, we can determine that the linear network recorded in the Original stimulus comprises numerous abrupt movements more intricately dispersed across the picture surface. In this case (the original image), multiple gaze movements indicate more frequent alternations between fixations and saccades than recorded in the Manipulated stimulus. Also, the denser clusters of saccadic lines and fixation dots mark more frequent shorter periods of mobility and, by extension, more rapid movement between the pictorial parts (given that the thirty-second viewing period was the same for both images).

What is most salient about our study is the discrepancy between the participants' behavioral responses to the survey questions and their eye-tracking responses. Participants did not report feelings of rupture or movement in the survey. However, their actual looking does record just such physiological reactions. Since the linear trackings of gaze movements are recorded in real time, they provide evidence that *A Bar at the Folies-Bergère* stimulates rapid multi-directional optical activity. I suggest that this ocular response then undergoes further processing, leading to an aesthetic experience of fleetingness and rupture based on biological and embodied mechanisms.[74] As previously discussed (per Umilta' and colleagues), these physiological mechanisms include activation of the beholders' cortical motor system, which, in this case, mirrors (simulates) the painter's roving painterly gestures. Additionally, as suggested by the fixation/saccadic clusters, the beholders' cortical motor system is also activated by the animated, intermingling (balcony) figures reflected in the mirror surface.

What is most pronounced and noteworthy for this study is that the complex linear network of fixations and saccades recorded in the Original stimulus reveals eye movements that are spatial and multidimensional.[75] Again, we can discern this phenomenon by comparing the composite images of the Original stimulus and the Manipulated stimulus. In the Manipulated stimulus, linear configurations cluster in the foreground, aligning with the horizontal direction of the bar-top objects and the axial, vertical direction of the barmaid's corsage, pendant, and facial features. These saccades and fixations form a shallow, pyramidal volume that frames the central figure. Less frequent movement and focalization extend horizontally from the barmaid's face to the figures on the balcony behind her and the crystal chandelier on the left side of her head.

In the Original stimulus, the network of saccades and fixations cluster to the right and left of the barmaid. In contrast to the Manipulated stimulus, these gaze movements disassemble both the barmaid's centrality within the pictorial composition and the still-lifes' horizontal order. From here, multidimensionality unfolds. Accordingly, the linear patterns to the left side of the barmaid form a trapezoidal shape receding into the mirrored reflection. In this area, the network progressively becomes more densely clustered as saccadic tracks move from the foreground bottles to the mirrored reflection and focus on the balcony's mirror reflection. In the latter case, the linear network concentrates mainly on the three figures that align with the orthogonal perspective in pictorial space. This orientation to pictorial depth conflicts with what we might expect—that attention would focus on the foreground figures aligned with the horizontal balcony parapet. Conversely, the saccadic-fixation network formed to the right of the barmaid reinforces the "expected" focus on the face-to-face interaction between the barmaid and her client reflected in the mirror surface. But this gaze movement on the mirror reflection creates a spatial dynamic with the barmaid's foreground visage. The angled, linear pattern of the foci between the foreground and the mirrored background suggests a rectangular formation that tilts back into the picture space, implying, in turn, a dynamic back-and-forth movement between the two dimensions.

Restoring the (Baudelairean) Historical Gaze, Backward and Forward in Time

Given Umilta' and colleagues' assertion that factors such as the context, individuals' interest in the artwork, prior knowledge, and familiarity modulate biological and physiological mechanisms involved in visual arts perceptions, we can now consider our eye-tracking data within the historical context of nineteenth-century Paris. First, despite all but one of our participants' unfamiliarity with the painting, their gaze movements registered real-time immediacy of neuroanatomical responses, alternating between top-bottom and bottom-up processing and between fixations and saccades. Accordingly, we can determine an embodied aesthetic experience generated by *A Bar at the Folies-Bergère*. Now, imagine Manet's contemporaries' encounter with his painting. For one thing, they would be familiar with the popular Folies-Bergère or a similar cabaret venue. As such, their gazes would shift between familiar objects and figures much more rapidly than present-day viewers, recognition giving way to top-bottom processing of bottle labels and social types and relations. Given Manet's viewers' "period eye,"[76] their shifting gazes—across the foreground and from foreground to reflected background—would be even more frequent and rapidly changing than Beder and Weingarden's study participants. Equally important, Manet's viewers would have been more aware than present-day viewers of the liminal spaces of endless reflections manifested by actual mirrors mounted on walls of café and cabaret interiors[77] and by the plate glass windows that lined modern shopfronts along Haussmann's boulevards.

Within these liminal spaces of Haussmannized Paris, ever-moving crowds of people would have constantly altered one's perspective and position, giving way to more visceral bottom-top processing of omnipresent affective stimuli. Coincidently, Manet's viewers may have recognized Manet's broken brushwork as an embodied response to this clamor of urban life and the ongoing vestiges of Haussmann's urban renewal schemes during the 1880s. For all these historical and contextual conditions, Manet's contemporaries would have been more responsive to an embodied aesthetic of rupture and fleetingness, as rendered in *A Bar at the Folies-Bergère*, than present-day viewers. For better or worse, Manet's painting affected an aesthetic experience of Baudelairean modernity as an open-ended condition of liminality and irresolvable dualities. We can presume that such dissonant conditions endured in Foucault's paradigmatic postmodern heterotopic spaces and the mirrored binaries that heterotopias invoke. We should also consider a photographic series of glazed doorframe and shopfront reflections by Eugène Atget (1857–1927) as a more proximate carrier of Manet's Baudelairean legacy.

In more ways than one, Atget adopted the double nature of Baudelaire's painter of modern life.[78] As I have argued elsewhere,[79] Atget followed Baudelaire's mandate for mastering the art of ironic surprise. Having briefly trained as an artist in the early 1880s, Atget appears to have adapted Manet's parodic techniques of fragmentation and lack of finish, mirrored in *A Bar at the Folies-Bergère*, to the photographic image. In doing so, he subverted conventional expectations for an exact reproduction of reality and, by extension, challenged the limitations Baudelaire ascribed to the photographer and his medium.

As seen in *Boutique Empire, 21 Rue du Faubourg-Saint-Honoré*, plate glass shopfront windows capture rippling reflections of ambient urban spaces that mingle with and dematerialize displays behind glass (Figure 10.4). These window surfaces are paradoxical—transparent and opaque, dynamic and static. Figures are equally paradoxical when captured in window frames. At once, subjects viewed and viewing subjects stand inert or dissolve into a myriad of reflections and mingle with other reflected figures. Taken together, these personages signify the photographer's double nature of presence and absence, as seen in *Enseigne, Restaurant Au Tambour, 63 Quai de la Tournelles* (1908) (Figure 10.5). The glazed doorway frames a double portrait, one person within and the other outside. The interior figure gazes back at the figure outside, who is no other than the photographer himself, albeit almost erased. Defying the first law of portrait photography to hold still, the photographer has moved, becoming blurred and entangled in the rippled images of the trees and skyline reflected in the window that frames himself and his figural counterpart.

We can embody the Baudelairean painter of modern life thanks to Atget's photographs of window reflections and his presence within. Like the Parisian Manet (and Baudelaire's Guys), the spectator becomes immersed in the kaleidoscopic urban ambiance and is detached from it. Just as Foucault had imagined, Manet's and Atget's mirrored reflections invite us to become absorbed in the juxtaposition of the real and the enigmatic, the dynamical shifts of back-and-forth gazing, and the liminal space of internal self-reflection upon endless dissonance. Suppose we allow our imaginations to overcome our expectations for narrative closure and pictorial harmony. In that case, we can align with a nineteenth-century experiential aesthetic, a physiological and psychical embodiment triggered by the artist's gestural techniques, ironic inversions, and lived experience.

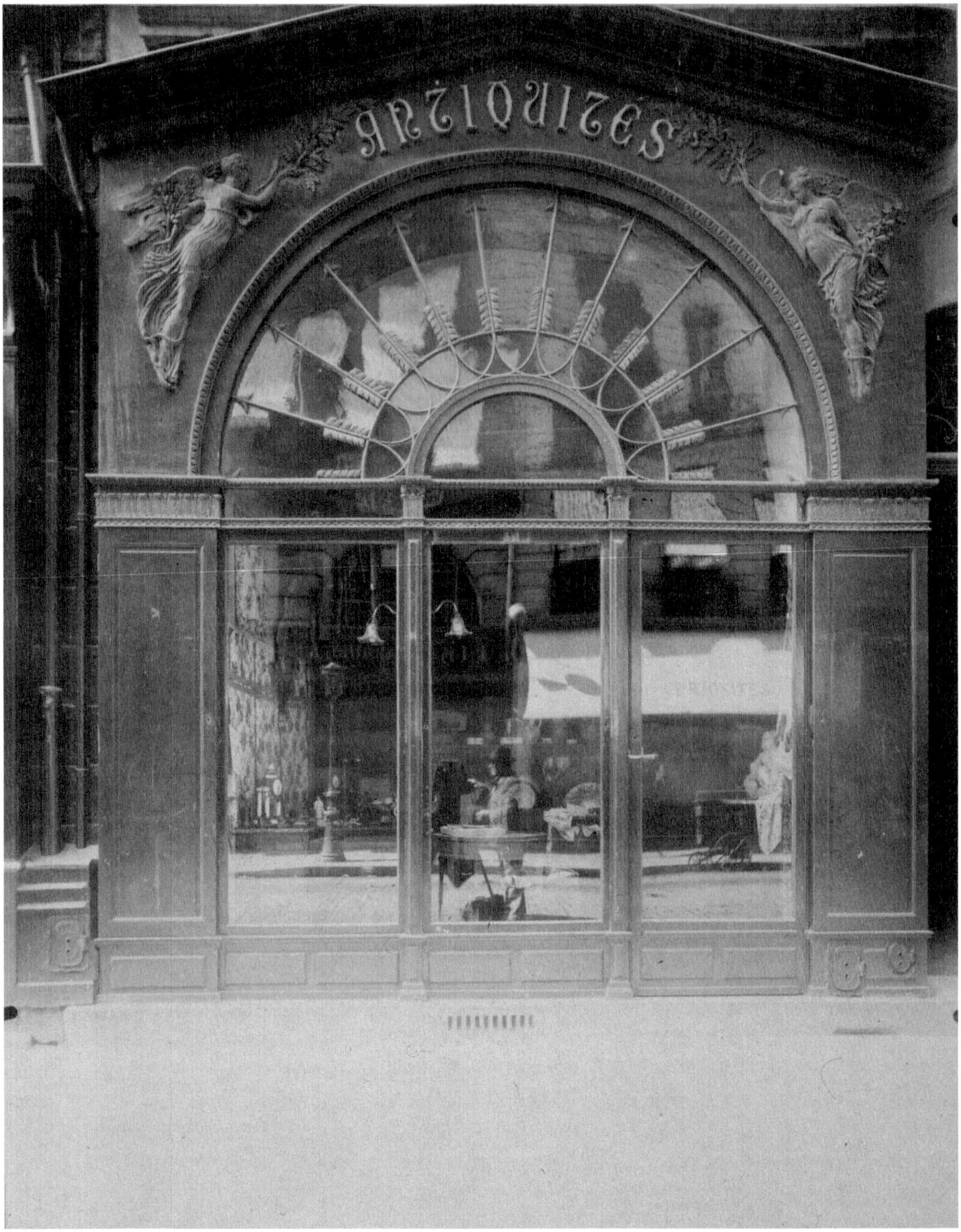

Figure 10.4 Eugène Atget, *Boutique Empire, 21 Rue du Faubourg-Saint-Honoré*, 1902, print on albumen paper, 9 x 7 in. Paris Musées/Musée Carnavalet, Histoire de Paris.

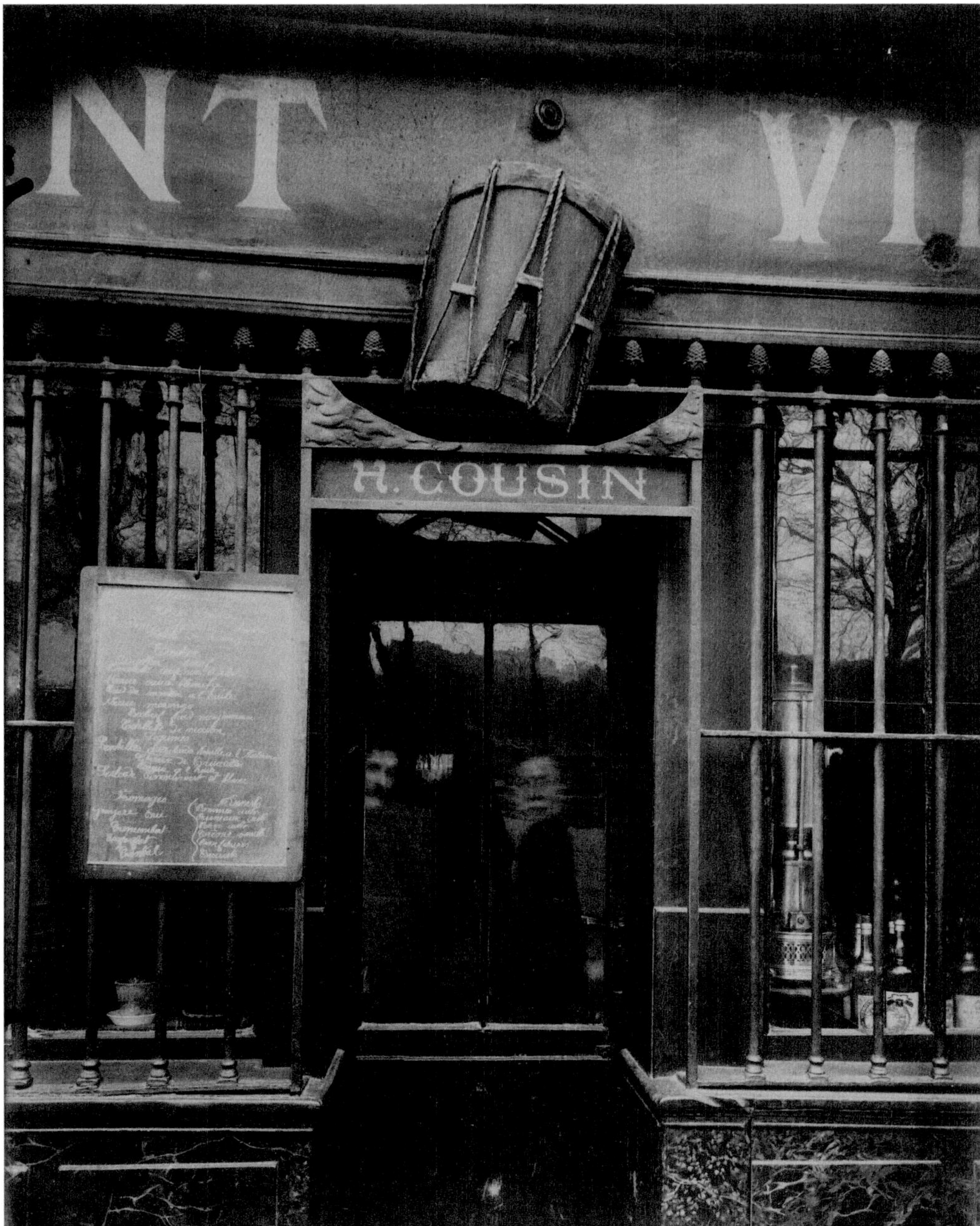

Figure 10.5 Eugène Atget, *Enseigne* [shop sign], *Restaurant "Au Tambour," 63 Quai de la Tournelle, 5ème Arrondissement, Paris*, 1908, print on albumen paper, 8 3/4 x 7 in. Paris Musées/Musée Carnavalet, Histoire de Paris.

Notes

1 These studies focus on quantifying stimulus-driven and information-processing events that take place during the viewing, psycho-physiological reception, and interpretation of artworks. For discussions of these terms, see Maurizio Corbetta and Gordon L. Shulman, "Control of Goal-directed and Stimulus-driven Attention in the Brain," *Nature Reviews Neuroscience 3*, no. 3 (2002): 201–15, https://doi.org/10.1038/nrn755; and Imed Bouchrika, "What Is Information Processing Theory? Stages, Models & Limitations," Research.com, 2024, https://research.com/education/what-is-information-processing-theory.

2 "liminal, adj." OED Online (Oxford University Press, March 2023). https://www-oed-com.proxy.lib.fsu.edu/view/Entry/108471?redirectedFrom=liminal.

3 "liminal, adj.," *APA Dictionary* (American Association of Psychology, 2023), https://dictionary.apa.org/liminal.

4 For definitions of heterotopia as the binary opposite of utopia see Michel Foucault, "Of Other Spaces," trans. Jay Miskowiec, *Diacritics* 16, no. 1 (1986): 24.

5 Foucault, 24.

6 See Enric Munar et al., "Aesthetic Appreciation: Event-Related Field and Time-Frequency Analyses," *Frontiers in Human Neuroscience* 5 (2012): 1–11. This study is especially pertinent to the mapping of dynamic liminal space since it investigates "the temporal course of the underlying neural processes [of aesthetic experience]." The research team applies "novel neuroimaging data analyses, currently used in diverse areas of the neurosciences, to explore and tentatively characterize the dynamics of the neural correlates of aesthetic preference" (1). See also Silvia and Brown, "Anger, Disgust, and the Negative Aesthetic Emotions."

7 Camilo J. Cela-Conde et al., "The Neural Foundations of Aesthetic Appreciation," *Progress in Neurobiology* 94, no. 1 (March 22, 2011): 41.

8 Miall, "Cognitive Poetics: From Interpreting to Experiencing What Is Literary," 187.

9 Reuven Tsur, "Aspects of Cognitive Poetics," in *Cognitive Stylistics: Language and Cognition in Text Analysis*, ed. Elena Semino and Jonathan Culpeper (Amsterdam: John Benjamins Publishing Company, 2002): 279.

10 Tsur, "Aspects of Cognitive Poetics," 294, 285.

11 Tsur, "Aspects of Cognitive Poetics," 295. See the related brain-activity study on irony processing: Isabel C. Bohrn et al., "Looking at the Brains behind Figurative Language—a Quantitative Meta-Analysis of Neuroimaging Studies on Metaphor, Idiom, and Irony Processing," *Neuropsychologia* 50, no. 11 (September 2012): 2669–83.

12 Tsur, "Aspects of Cognitive Poetics," 284–5.

13 Tsur, "Aspects of Cognitive Poetics," 296.

14 Spot was the pseudonym of Louis Pierre Gabriel Bernard Morel-Retz; see Clayson, *Painted Love*, 60. Given the caption heading, we might ask: what kind of consolation does the server offer? Physical, sexual comfort? In that case, the barmaid would embody the commodification of sex as prostitution. Or does she-qua-illustrator offer the consolation prize to the "game loser" (the man left out of the scene) in the form of her gaze or by restoring him to the spectacle?

15 David S. Miall and Don Kuiken, "A Feeling for Fiction: Becoming What We Behold," *Poetics* 30 (2002): 238–9.

16 "foregrounding." *Oxford Reference*. https://www.oxfordreference.com/view/10.1093/oi/authority.20110803095828296.

17 Geoffrey N. Leech, *Linguistic Guide to English Poetry* (Harlow: Longmans, 1969): 69.

18 "Event-related potentials (ERPs) are derived from electroencephalographic (EEG) measurement of neural activity. An ERP measures voltage changes in the brain that follow (or precede) the onset of specific visual, auditory, or other sensory stimuli and cues signaling motor preparation, motor execution, or covert mental operations (e.g., imagery). The ERP is obtained by extracting and then averaging multiple temporal segments from the ongoing EEG that define the event of interest," in M. M. Bradley and A. Keil, "Event-Related Potentials (ERPs)," in *Encyclopedia of Human Behavior*, 2nd ed., ed. V. S. Ramachandran (San Diego: Academic Press, January 1, 2012): 79, https://doi.org/10.1016/B978-0-12-375000-6.00154-3.

"Functional magnetic resonance imaging (fMRI): A brain imaging technology, based on conventional MRI, that gathers information relating to short-term changes in oxygen consumption by cells in the brain. It typically uses this information to depict the brain areas that become more or less active—and presumably more or less involved—while a subject in the fMRI scanner performs a cognitive task," Jordan H. Grafman and Ishan Dasgupta, eds., "Functional Magnetic Resonance Imaging (FMRI)," in *Key Brain Terms Glossary* (Dana Foundation, 2022), https://dana.org/explore-neuroscience/brain-basics/key-brain-terms-glossary/#F.

19 David S. Miall, "Neuroaesthetics of Literary Reading," in *Neuroaesthetics*, eds. Martin Skov and Oshin Vartanian (Amityville, NY: Baywood Publishing, 2009): 238. For their full study, see Miall and Kuiken, "Foregrounding, Defamiliarization, and Affect," 389–407.

20 Miall, "Neuroaesthetics of Literary Reading," 240.

21 Meta-cognition: "higher-order thinking that enables understanding, analysis, and control of one's cognitive processes, especially when engaged in learning." "meta-cognitive," *Dictionary.com Unabridged*. Random House, Inc. http://www.dictionary.com/browse/meta-cognitive.

22 Pelowski and Akiba, "A Model of Art Perception," 92.

23 Pelowski and Akiba, "A Model of Art Perception," 92.

24 Miall and Kuiken, "A Feeling for Fiction," 222–6.

25 Miall, "Neuroaesthetics of Literary Reading," 235.

26 James L. Keidel et al., "How Shakespeare Tempests the Brain: Neuroimaging Insights," *Cortex* 49, no. 4 (April 2013): 913.

27 Keidel et al., "How Shakespeare Tempests the Brain," 913.

28 Philip Davis's project is exemplary for transitioning from humanities-based theory to scientific inquiry. (Davis and the author personally met in July 2013 at Liverpool Moore University to discuss his project.) His initiative speaks to my own modeling of experiential aesthetics. Davis partnered with neuroscientists to gain "a greater sense of how and why Shakespeare really does something to our inner reality." During his collaboration, he garnered "a sense of an underlying shape to experience, as though the syntax in front of my eyes were keying into mental pathways behind them and shifting and reconfiguring them dramatically in the theatre of the brain;" Philip Davis, "The Shakespeared Brain," *The Reader*, no. 23 (n.d): 39–43, https://thereaderonline.wordpress.com/features/the-shakespeared-brain.

29 Davis, "The Shakespeared Brain."

30 Keidel et al., "How Shakespeare Tempests the Brain," 914. As Keidel and colleagues conclude, the brain areas activated during the reading included the *left caudate* nucleus, responsible for the brain's learning and memory system, and possibly the comprehension of translingual word switches; *the right inferior frontal gyrus*, typically implicated in go/no go tasks; and *the inferior temporal gyrus*, one of the higher levels of visual processing, associated with the representation of complex object features, and in the recognition of numbers and words, and the processing of speech and language. (Italicized brain regions are from Keidel et al. [913]). The analogous brain functions are sourced from Margaret E. Driscol et al., "Neuroanatomy, Nucleus Caudate," in *StatsPearls* (online), (Treasure Island [FL]: StatPearls Publishing, 2024), Last updated July 24, 2023, https://www.ncbi.nlm.nih.gov/books/NBK557407/; Rune Boen et al., "Inhibitory Control and the Structural Parcelation of the Right Inferior Frontal Gyrus," *Frontiers in Human Neuroscience* 16 (February 24, 2022), https://doi.org/10.3389/fnhum.2022.787079; and "Inferior Frontal Gyrus," Wikipedia (Wikimedia Foundation, September 19, 2019), https://en.wikipedia.org/wiki/Inferior_frontal_gyrus.

31 Keidel et al., "How Shakespeare Tempests the Brain," 913.

32 Here, "Parisians" refers to bourgeois viewers of Manet's painting *A Bar at the Folies-Bergère* in the Salon or gallery setting. However, I contend that the embodied aesthetic effects of rupture and fragmentation and the dynamized gaze would have been shared by most Parisians who navigated the boulevards, café-concerts, and demolition and construction sites, regardless of class.

33 Pelowski and Akiba, "A Model of Art Perception," 88.

34 Pelowski and Akiba, "A Model of Art Perception," 89.

35 The autonomic nervous system (ANS) is "the portion of the nervous system innervating smooth muscle and glands, including the circulatory, digestive, respiratory, and reproductive organs. It is divided into the sympathetic nervous system and parasympathetic nervous system. Autonomic responses typically involve changes in involuntary bodily functions, such as heart rate, salivation, digestion, perspiration, pupil size, hormone secretion, bladder contraction, and engorgement of the penis and clitoris;" "Autonomic Nervous System (ANS)," in *APA Dictionary of*

Psychology, 2023, https://dictionary.apa.org/autonomic-nervous-system. The "sympathetic nervous system is a network of nerves that helps your body activate its 'fight-or-flight' response. This system's activity increases when you're stressed, in danger or physically active;" Cleveland Clinic, "Sympathetic Nervous System (SNS)." The parasympathetic nervous system "is a network of nerves that relaxes your body after periods of stress or danger. It also helps run life-sustaining processes, like digestion, during times when you feel safe and relaxed;" Cleveland Clinic, "Parasympathetic Nervous System (PSNS)."

36 Pelowski and Akiba, "A Model of Art Perception," 89.

37 Pelowski and Akiba, "A Model of Art Perception," 90.

38 Pelowski and Akiba, "A Model of Art Perception," 94.

39 Immanuel Kant set the foundation for the non-cognitive, affective theory of aesthetic empathy in the *Critique of the Power of Judgment* (1790). His German Romantic disciples, including Friedrich Schlegel and Novalis, developed the theory of empathy as foundational to the aesthetic experience. See Keren Gorodeisky, "19th Century Romantic Aesthetics," *The Stanford Encyclopedia of Philosophy* (Fall 2016 Edition), ed. Edward N. Zalta, Section 1, 3.2, https://plato.stanford.edu/archives/fall2019/entries/aesthetics-19th-romantic/. By the turn of the twentieth, Theodore Lipps developed a psychological theory of aesthetic empathy, as projecting the self into an object, that converges with present day-models ("Einfühlung, Innere Nachahmung und Organempfindung," ["Empathy, Inner Imitation and Sense-Feelings], 1903). Lipps also distinguished between "positive" and "negative" empathy, equating the positive with beauty, which is life affirming and "objectified self-enjoyment," in contrast with negative empathy when an object is perceived as ugly. See Karsten Stueber, "Empathy," *The Stanford Encyclopedia of Philosophy*, (Fall 2016 Edition), ed. Edward N. Zalta, Section 1. https://plato.stanford.edu/archives/fall2019/entries/empathy/.

40 V. Rajmohan and E. Mohandas, "Mirror Neuron System," *Indian Journal of Psychiatry* 49, no. 1 (2007): 66, https://doi.org/10.4103/0019-5545.31522. Collectively, the mirror neuron system (MSN) refers to the group of specialized neurons that "mirrors" the actions and behaviour of others. Mirror neurons were first discovered in research on the grasp response of macaques. "Subsequent research has elucidated the diverse regions involved in the MNS of monkeys. Recently different cortical structures have been described as part of the MNS in humans," (66).

41 Freedberg and Gallese, "Motion, Emotion and Empathy in Esthetic Experience," 197.

42 Freedberg and Gallese, "Motion, Emotion and Empathy in Esthetic Experience," 199.

43 Freedberg and Gallese, "Motion, Emotion and Empathy in Esthetic Experience," 197.

44 Freedberg and Gallese, "Motion, Emotion and Empathy in Esthetic Experience," 201.

45 See, for example, Ruth E. Iskin, "Selling, Seduction, and Soliciting the Eye: Manet's *Bar at the Folies-Bergère*," *Art Bulletin* 77, no. 1 (March 1995): 25–44; and J. Paul Getty Museum, "Manet's Bar at the Folies-Bergere (Getty Center Exhibitions)," Getty.edu, 2019, https://www.getty.edu/art/exhibitions/manet_bar/.

46 Freedberg and Gallese, "Motion, Emotion and Empathy in Esthetic Experience," 202.

47 Umilta' et al., "Abstract Art and Cortical Motor Activation: An EEG Study," 1.

48 Umilta' et al., "Abstract Art and Cortical Motor Activation: An EEG Study," 6.

49 Umilta' et al., "Abstract Art and Cortical Motor Activation: An EEG Study," 2.

50 Umilta' et. al., "Abstract Art and Cortical Motor Activation: An EEG Study," 6. The significance of "mu rhythm suppression" is illustrated in a diagram appended to Hannah M., Hobson and Dorothy V. M. Bishop's article: "[t]wo conditions—a baseline condition (a) and an active condition (b)—are represented here. Bold cells represent neurons firing. *In the baseline condition, the participant sits motionless. When at rest, the cells in the sensorimotor cortex fire together, leading to higher power in the mu frequency band. In the active condition, the participant is asked to move, generating motor-cortex activity. This leads the sensorimotor cells to fire out of synchrony, leading to reduced mu power*" (emphasis added); Hobson and Bishop, "The Interpretation of Mu Suppression," Royal Society Open Science vol. 4, 3 160662 (Mar 1, 2017), n.p., https://doi.org/10.1098/rsos.160662.

51 Umilta' et. al., "Abstract Art and Cortical Motor Activation," 7.

52 Skepticism abounded particularly in the early stages of research. See for example A. Lingnau, B. Gesierich, and A. Caramazza, "Asymmetric FMRI Adaptation Reveals No Evidence for Mirror Neurons in Humans," *Proceedings of the National Academy of Sciences* 106, no. 24 (June 2, 2009): 9925–30, https://doi.org/10.1073/pnas.0902262106; and Gregory Hickok,

"Eight Problems for the Mirror Neuron Theory of Action Understanding in Monkeys and Humans," *Journal of Cognitive Neuroscience* 21, no. 7 (July 2009): 1229–43, https://doi.org/10.1162/jocn.2009.21189.

53 Umilta' et. al., "Abstract Art and Cortical Motor Activation," 8.

54 For a related eye-tracking study with *A Bar at the Folies-Bergère*, see Beth Harland et al., "Modes of Address in Pictorial Art: An Eye Movement Study of Manet's *Bar at the Folies-Bergère*," *Leonardo* 47, no. 3 (June 2014): 241–8. Harland and colleagues focus on the participants' (visual art experts and non-experts) perceptual and psychological engagement with "the 'double relation' set up through (1) the barmaid facing the spectator while her gaze is averted and (2) the configuration of the spectator in relation to the central barmaid and the figures reflected in the mirror" (241). My investigation sets aside the viewers' psychological responses to the frontal barmaid and the mirrored figures as they affect interpersonal relations. Rather, I am interested in the ruptured spatial effects of the mirror and painted surfaces in between those figures. For Harland's subsequent eye-tracking study, see Nick Donnelly et al., "The Influence of Pupil Alignment on Spectator Address in Manet's Portraiture," *Psychology of Aesthetics, Creativity, and the Arts* 11, no. 2 (May 2017): 167–78.

55 Dilara Beder, email message to Lauren Weingarden, April 21, 2023. Beder's research took place in November 2022. Professor Matthew Pelowski, Beder's University of Vienna supervisor, facilitated my collaboration with Beder. Pelowski also assisted me in interpreting the data. In her email correspondence, Beder cites Efi Nisiforou and Andrew Laghos, "Field Dependence–Independence and Eye Movement Patterns: Investigating Users' Differences through an Eye Tracking Study," *Interacting with Computers* 28, no. 4 (June 16, 2015): 407–20, https://doi.org/10.1093/iwc/iwv015; and Bighna Kalyan Nayak and Sougata Karmakar, "Eye Tracking Based Objective Evaluation of Visual Aesthetics: A Review," *Advances in Ergonomics in Design* 777 (June 24, 2018): 370–81.

56 "Eye Tracking." Wikipedia. Wikimedia Foundation, August 21, 2019. https://en.wikipedia.org/wiki/Eye_tracking.

57 Rosenberg and Klein, "The Moving Eye of the Beholder," in *Art, Aesthetics, and the Brain*, ed. Joseph P. Hutton, et al., (Oxford: Oxford University Press, 2015):79–108, https://doi.org/10.1093/acprof:oso/9780199670000.001.0001.

58 Rosenberg and Klein, "Moving Eye," 82. For a historical and literature review of eye-tracking experiments, in general, and, specifically, in visual arts, see Rosenberg and Klein, 79–86, 88–90. In the latter category, see especially François Molnar's pioneering studies in aesthetically based eye tracking. Notably, Molnar's studies included Manet's *Olympia* as a control stimulus to measure ocular responses to luminance (contrast) as indicators of compositional harmony (or "quality"). See "A Science of Vision for Visual Art [1992]," *Leonardo* 30, no. 3 (1997): 225–32, https://doi.org/10.2307/1576454; "About the Role of Visual Exploration in Aesthetics," in Hy I. Day, ed., *Advances in Intrinsic Motivation and Aesthetics* (Boston: Springer US 1981), 385–414; and with Dominique Ratsikas, "Some Aesthetical Aspects of Visual Exploration," in J. K. O'Regan and A. Levy-Schoen, eds., *Eye Movements: From Psychology to Cognition* (Amsterdam: Elsevier, 1987): 363–74.

59 "Primate," in *APA Dictionary of Psychology*, 2023, https://dictionary.apa.org/primate.

60 "functional" neural systems: "In neuroscience, functional specialization is a theory which suggests that different areas in the brain are specialized for different functions," in "Functional Specialization (Brain)," Wikipedia, April 7, 2020, https://en.wikipedia.org/wiki/Functional_specialization_(brain). "neural systems:" "Circuits that subserve similar functions are grouped in neural systems that serve broader behavioral purposes. The most general functional definition divides neural systems into sensory systems like vision or hearing that acquire and process information from the environment, and motor systems that allow the organism to respond to such information by generating movements," Dale Purves et al., eds. "Neural Systems," *Neuroscience*, 2nd ed. (Sunderland, MA; Sinauer Associates, 2011), https://www.ncbi.nlm.nih.gov/books/NBK11061/.

61 Rosenberg and Klein, "Moving Eye," 84.

62 Graphic illustrations of eye-tracking data represent composites of all individual participants in the study.

63 Rosenberg and Klein, "Moving Eye," 82.

64 "Microsaccades are saccades that range from about 0.01°–0.3° in amplitude and are unconsciously generated during fixation. They appear to correct for fixation errors from slow eye drifts and may prevent fading of the visual scene, which can occur when the image is stabilized on the retina," qtd. from Andrew Go Lee et al., "Saccade," in *EyeWiki* (American Academy of Ophthalmology, April 7, 2023), https://eyewiki.aao.org/Saccade.

65 "amplitude saccade." "Amplitude: The size of the saccade, usually measured in degrees or minutes of arc. A higher amplitude means the eye has traveled a greater distance during the saccade.," qt. from Lee et al., "Saccade."

66 Rosenberg and Klein, "Moving Eye," 84. While the data derived from heat maps and fixation-saccade maps are compiled from multiple viewers of a single painting, the data suggests viewers' shared viewing activity. Based on 57 experiments conducted at the Laboratory for Cognitive Research in Art History, University of Vienna, Rosenberg and Klein assert that "[f]or a significant number of paintings and despite major differences between subjects, not only fixations but also saccades build patterns that are specific to each painting: Beholders tend to reiterate particular paths with their eyes," (92). Rosenberg and Klein also discuss the differences in gaze movements between artists and non-artists and between art experts and non-experts (98–99).

67 For technical information and images of the device, see SR Research, "EyeLink 1000 plus - the Most Flexible Eye Tracker," https://www.sr-research.com/eyelink-1000-plus/.

68 For this study, we used the Wikimedia image file of *A Bar at the Folies-Bergère*. We manipulated the image, which the supplier of the Courtauld image file (Bridgeman) does not allow.

69 We acknowledge Chase Van Tilburg's authorship of the Manipulated stimulus. A visual analysis of the Manipulated stimulus, compared to the Original stimulus, reveals a predominantly horizontal composition, owing to the continuous prolongation of the foreground bar and background balcony parapet. In contrast to the original painting, this horizontality is complemented by a more pronounced triangular enframement and vertical axis of the barmaid's figure.

70 This description of the methodology used in Beder and Weingarden's experiment is adapted from Beder's email to Weingarden on April 21, 2023.

71 The questions were based on a comparative method designed to determine whether the participants noticed the differences between the two images. We asked 13 questions regarding the painting's formal and spatial features, with participants ranking on a scale of 1–7 (1 as the lowest negative rating and 7 as the highest positive rating). The following issues were targeted: liking or disliking; understanding or not; meaningful or meaningless; ambiguity or clarity; good or bad quality; shallow or deep space; separate or cohesive elements; understanding of space; location (in meters) of the back wall; proximity of the woman to the viewer; feeling inside or outside the painting; whether techniques helped or hindered understanding the painting; how much time spent looking at the surface; how much time spent looking at the scene. Two yes-or-no questions asked to the participants: 1) if they noticed a mirror in the painting and 2) if they had seen the painting before. Regarding the last two questions, most participants noticed the mirror in the Original image, and not in the Manipulated image; only one participant had prior knowledge of the painting, We were surprised by the rated responses; except for a single rating, there was insignificant variance between the participant's responses to the technical and scenic valuations of the two images. The only significant variance was that the Original image was rated as more "deep" than the Manipulated image. This response corresponds to the eye-tracking data. Overall, we concluded that the behavioral responses did not match the perceptual responses, the latter of which supported my original hypothesis. Therefore, I have not included the survey data in the present review of our study.

As Molnar explains, this discrepancy between physiological responses and post-viewing survey responses results from the participants' pre-conscious ocular-motor processes and introspective cognitive faculties activated for survey responses; "A Science of Vision for Visual Art," 230.

72 Rosenberg and Klein, "Moving Eye," 87.

73 Rosenberg and Klein, 87. For the primary salience of facial recognition, see I. Mertens, et al., "Gaze Motor Asymmetries in the Perception of Faces during a Memory Task," *Neuropsychologia* 31, no. 9 (September 1993): 989–98, https://doi.org/10.1016/0028-3932(93)90154-r; and Gregor Rainer, "Visual Neuroscience: Computational Brain Dynamics of Face Processing," *Current Biology* 17, no. 21 (2007): R933–34, https://www.sciencedirect.com/science/article/pii/S096098220701857X.

74 As per note 71 above, participants did not record feelings of rupture or fragmentation in the survey questions. However, the embodied experience of rupture does register in their actual looking experience of the painting.

75 Beder and Weingarden's outcomes conflict with the results of Harland et al. in the "Modes of Address" study. The latter study concluded that only experts with formal training were able to make second-order inferences about relative spatial/temporal positioning to understand the implications of the painting's shifting modes of address, the "double relation" of acknowledgment and absorption, and the implications of this for their own role as spectator more fully" (247). Our study shows that non-experts recognize the mirror reflection and the figures therein. I suggest these conflicting findings result from the former study's use of a Manipulated image.

76 Of course, as noted earlier, historical and cultural changes also impact the beholder's perception of artworks, an issue that Rosenberg and Klein address in "Moving Eye," 99–100. Foundational to this historical context approach is Michael Baxandall's study of "the period eye," in *Painting and Experience in Fifteenth Century Italy: A Primer in the Social History of Pictorial Style* (Oxford: Clarendon Press, 1972).

77 Mirrors and mirrored reflections are integral to French architecture and material culture, ranging from the Hall of Mirrors at Versailles and its replication in Charles Garnier's new Opera building, to commonplace mirror-lined café and cabaret walls, to domestic interiors dating from the Rococo period and replicated in modern Haussmannian bourgeois abodes.

78 More often than not, scholars have interpreted Atget's photographic subjects, especially the *petits métiers*, in terms of Baudelaire's prose poems and the characters therein; see Molly Nesbit, *Atget's Seven Albums* (New Haven: Yale University Press, 1992): 169–70; Guillaume Le Gall, "The Eye of the Archaeologist Eugène Atget and the Forms of the Old City," in *Eugène Atget: Old Paris* (Madrid: Fundación MAPFRE, 2011): 21. In addition to this interpretation, Alain Buisine associates Atget with the *flâneur* of Baudelaire's "Painter of Modern Life," *Eugène Atget ou la Mélancolie en Photographie* (Nîmes: J. Chambon, 1994): 29–30, 72–3. None, however, state directly that Atget read Baudelaire's texts. I assume that Atget was well-versed in Baudelairean theories of modernity, I thus treat him as a (competent) reader of Baudelaire's texts.

79 Weingarden, "The Mirror as a Metaphor of Baudelairean Modernity," 16–36.

Epilogue

A Neuroarthistory of The Painters of Modern Life: Embodying Baudelairean Modernity advances a transdisciplinary approach—combining art history, literary studies, and neuroaesthetics—to examine the modern urban experience of nineteenth-century Paris through language and images of fragmentation and transformation. While these particular methodologies help weave together the various methods and topics that define Baudelairean modernity, one goal of this study, in its exploratory and experimental dimensions, is that it might serve as a useful model for future projects, not only in neuroarthistory, that seek to join the humanities and the sciences more broadly.

This volume is organized along two axes—the historical and the neuroaesthetic, each of which are imbued with inter- and transdisciplinary perspectives.[1] Foremost among them, word-and-image analysis bridges literary texts and visual artworks within the shared lived experience of Haussmannization, and reveals ensuing discourses on Baudelairean modernity. Analogously, reader reception theory, derived from literary criticism, serves to connect the maker's technical means of representation with the reader/viewer's embodied aesthetic reception. This model extends from German Romantic irony to Baudelaire's tenet of *se doubler/dédoublement* (self-doubling, doubling), a modern condition that the artist inscribes in the artwork and is mediated to the viewer. Reader reception theory underscores my analysis of the Baudelairean artists' dialogic relation with Baudelaire's texts manifesting in ironic (visual) parody. Irony and ironic parody are at the heart of Baudelairean discourse on modernity. These modes are agents of reader reception wherein the creator performs a subversion or inversion of a prior text/image, in a new text/image, and in turn, irony and ironic parody activate the reader's/viewer's self-reflection and transformation. Whether motivated by the creator's ironic acts or the reader/viewer's ironic reception, the artworks that mediate these transactions propel a Baudelairean trajectory toward cultural renewal.

Reader reception theory also allows us to examine the creative, rather than subversive, interpretation of a text/image wherein the reader-creator translates it according to their own means of artistic production. Such is the case when Baudelairean artists adapted Baudelaire's critical and literary texts to their pictorial means of fragmentation and paradox, when Baudelaire adapted Méryon's old Paris etchings to his *Tableaux Parisiens* poems, or when Marville's photographic techniques (framing, lighting, and viewpoint) developed for his Old Paris series mirrored the nostalgic turn of Baudelaire's *Tableaux Parisiens*. Similarly, Manet and Zola channeled Baudelaire's ironic parody, formulated in his prose poems, into their respective art forms. Both the artist and the novelist deployed the mirror (both present and absent) to reflect upon the aesthetic and formal conventions

DOI: 10.4324/9780429442315-12

they subverted and the transitoriness of the modern condition. Yet, in Manet's *A Bar at the Folies Bergère*, the artist rejected Zola's appropriation of his art to the writer's positivist theory of Naturalist realism (and entropy) by instead mirroring modernity's incessant, kaleidoscopic effects that were encoded in Baudelaire's "The Painter of Modern Life" and his ironic poetic texts.

The reader reception model, moreover, perfectly aligns with neuroaesthetics, which measures the spectator's responses to art forms. My neuroaesthetic approach to the lived experience of Haussmannization serves to bridge the maker/creators' and the viewer/readers' visceral responses to the physical, psychic, and sensory effects of rupture and fragmentation. Baudelaire's references to the nervous system in "The Painter of Modern Life" and elsewhere provide evidence of his basic knowledge of nineteenth-century brain science, the injurious effects of urban clamor, and remedial effects of ironic self-reflection, which validate my investigations of nineteenth-century neuroscience and adaptations of contemporary research. I also adapt theoretical and experimental models from empirical psychology and neuroscience to reify the embodied and affective responses to material conditions of Haussmannization and the art forms that translate those conditions. In doing so, my work aligns with a branch of neuroaesthetics that goes beyond traditional investigations of aesthetics as beauty in art to include rupture, fragmentation, and even disgust.[2]

An important outcome of bridging maker and receiver through the lens of neuroaesthetics is that *A Neuroarthistory of The Painters of Modern Life* establishes how Baudelairean painters assimilated Baudelaire's adaptations of science. These artists represented their sensory responses to the urban environment with visual devices—such as broken brushwork, flattened surfaces, and spatial distortion—that embodied physical and aesthetic rupture. These visual devices, I argue, affected the viewer's own experience of urban rupture and transformation. Cognitive poetics also facilitated my joining a word-and-image methodology with neuroaesthetic findings. Within this branch of literary theory that analyzes the mental processes at work during literary reading, I focus on researchers' modeling of delayed cognitive processing of textual ruptures and novel forms, compelling the readers' adaptation and self-transformation. Two empirical chapters, 3 and 10, adapt these cognitive patterns of rupture and delay to the viewers' aesthetic processing of visual parody and its ironic effects. In doing so, this project concretizes the viewers' self-reflection upon ruptured artistic norms and the societal and physical ruptures they bear.

Cognitive poetics, as practiced by George Lakoff, also serves as a guide to analyze how linguistic metaphors are directly produced by embodied sensations and affective associations. In empirical Chapter 7, Lakoff's explanations of metaphors and the targets they mask were adapted to analyze the arousal and anger shrouded by the critical rhetoric of dirt and obscenity that Manet's paintings *Le Déjeuner sur L'herbe* and *Olympia* instilled. I first addressed this critical discourse in Chapter 6, asserting that Manet's scandalous paintings were parodic responses to pornographic photography and its legal censure. Furthermore, this intermedia matrix triggered a critical discourse shaped by the pejorative language of social and bodily contamination, rhetoric earlier attached to literary and artistic Realism.

This volume also demonstrates how cognitive poetics and neuroscience can be joined to further substantiate how nineteenth-century viewers of Manet's paintings embodied the sensations and affective associations of pornography, the category in which *Le Dejeuner sur L'herbe* and *Olympia* were critically framed. It does so by integrating neural research on internet pornography addiction and other neurological studies that register the physiological, neurobiological, and neuropsychological spectrums of arousal and disgust that accompany pornographic viewing. This transdisciplinary perspective sheds new light on the critical reception of Manet's portrayals of the model-cum-prostitute and, in the public's eye, the models' proxy for publicly displayed erotic photographs.

By weaving together these varied approaches, *A Neuroarthistory of The Painters of Modern Life* contributes to the interdisciplinary humanities and the field of neuroaesthetics and, thereby, can provide a model for collaborations between humanities and science. The challenges of these approaches are as important and enlightening as the successes. As my own collaborations with empirical psychologists have shown, in Chapters 5 and 10, an art historian's hypotheses do not always align with data gathered from present-day participants' responses to historical art forms. By initiating these experiments, one of my tasks became resolving the discrepancies between historical evidence and real-time data. I believe my resolutions and my collaborators' suggestions for further research offer lessons for future transdisciplinary collaborations. That is, rather than understanding these disciplinary gaps as moments of impossibility, they can and did yield further collaboration and creativity.

The experiment featured in empirical Chapter 5, for example, relied on a contextual analysis of Marville's Old Paris and New Paris photographic series, presented in Chapter 4, which recovered a predominantly nostalgic favoring of the Old Paris series. In Chapter 5, I posed the question, "Can present-day photographic viewing and, by extension, empirical aesthetics be used to identify and understand the mobilized spectator's affective and cognitive responses to the urban and cultural disruptions incurred by Haussmannization?" For this experiment, participants viewed Marville's Old Paris and New Paris photographs and answered survey questions regarding their cognitive and affective assessment of ease/difficulty and pleasure/displeasure in occupying and/or moving through the urban spaces. To our surprise, present-day viewers favored the New Paris photographs, while the art historical evidence and historiographic record favored Old Paris photographs. To resolve this misalignment, I adapted the reader reception tenet of the "ideal" reader to model the "curatorial gaze." This exemplar accommodated museum curators' "informed" historical reading of Marville's photographs. In their curatorial assessments, they exposed affective, mobilized, and embodied responses to the Old Paris series and disregarded (or shunned) his New Paris series. Given this historical bias, my collaborators and I suggested that future experiments might include "informed" viewers as participants and provide information for "uninformed" viewers prior to their participation.[3]

The second case, in empirical Chapter 10, focused on Manet's *A Bar at the Folies-Bergère* to inquire whether the painting's ruptured, dissonant formal qualities dynamize the act of looking. If so, does such looking mirror the lived experience of nineteenth-century Paris during Haussmannization? To answer this question, I collaborated with empirical psychologists in an eye-tracking study that recorded participants' eye movements while viewing digitized images of *A Bar at the Folies-Bergère*. Participants viewed the original image and a modified image in which the mirrored reflection of the man in the painting had been removed. Our data confirmed that eye movements were more rapid, complex, and prolonged for the original image than the modified image, providing a measure of the viewer's effort to resolve the spatial ambiguity produced by the painting. As in the Marville photographs experiment, this study revealed a discrepancy between the participants' subjective responses to post-viewing survey questions and their eye-tracking responses. Participants did not report feelings of rupture or movement in the survey. However, their actual looking does record such physiological reactions. The eye-tracking technology supports my modeling of an embodied experiential aesthetic of Baudelairean modernity as transitory, dynamic, and indeterminate. With this data I could hypothesize that the dynamized act of looking mirrors Parisians' embodied responses to the clamor and cacophony of transport, commerce, building demolitions, and constructions in the urban center.

Throughout this study, *A Bar at the Folies-Bergère* emerges as a reference point for constructing an urbanized, sensory body experience and a template for considering future transdisciplinary research in embodied, experiential aesthetics. First, as discussed in Chapter 2 regarding Méryon's old Paris depictions of river and street life, sound studies offers opportunities to explore the musical scores performed in the Folies-Bergère auditorium (comprising a horseshoe layout that faces the stage). In this light, might Manet's broken brushwork and spatial distortions echo the lively orchestral rhythms and melodies that accompanied the spectacles on stage and overhead trapeze acts? These inquiries into sound environments, and analogous pictorial devices, also pertain to Haussmannized urban settings as depicted in Manet's *Music in the Tuileries* and Degas' *Place de la Concorde*.[4]

To what extent can our eye-tracking study of *A Bar at the Folies-Bergère* be expanded to include other technologies for recording embodied responses to pictorial rupture and fragmentation? Brain mapping technologies, such as an EEG (electroencephalogram) or MRI (magnetic resonance imaging) could be used to monitor and measure brain functions while participants view the painting. These devices have yielded fruitful data not only in neuroaesthetics, but also in cognitive poetics studies, discussed in Chapter 10, measuring real-time brain activity in reading semantically and grammatically ruptured texts and bearing witness to the reader's delayed categorization. Empirical psychology studies show similar cognitive delays occur in viewing unconventional artworks (as per Pelowski and Akiba in Chapters 3 and 10); these delayed responses include the viewer's self-reflection and transformation. What might an EEG or MRI record during a viewer's observation of *A Bar at the Folies-Bergère*? Each of these devices record real-time brain activity and locate different parts of the brain activated during the viewing process. An EEG measures brain waves and the electrical activity of brain cells. (During an EEG, electrodes are pasted onto the scalp or embedded in a netted cap.)[5] An MRI brain scan produces very clear images of the brain anatomy, including blood vessels and nerves.[6] With this data researchers can also identify different brain functions associated with the most activated cerebral regions during the viewing process. In empirical Chapters 3 and 7, I introduced brain mapping technologies research as a promising strategy for re-embodying nineteenth-century viewers' physiological and affective responses to painterly fragmentation and rupture and to images deemed obscene. While my research on brain mapping technologies applied to neuroaesthetics is second-hand, Umilta' and colleagues' investigations engage directly with art historical concerns. Their team used high-density EEG to demonstrate how observation of high-resolution digitized static images of abstract paintings by Lucio Fontana triggered specific cortical motor activation in the beholder's brain, mirroring the artist's gestural canvas cutting.

Admittedly, many neuroaesthetic studies that implement these technologies are limited to viewers responding to digital images on computer screens. Recent studies have been conducted in museum settings, providing promising avenues for collaborations between art historians, neuroscientists, and museum professionals. We can speculate on such a collaboration for assessing viewers' physiological response to *A Bar at the Folies-Bergère* when viewed in its Courtauld Institute of Art setting. For these experiments researchers used heart rate monitors (HRM) involving personal monitoring devices that measure/display heart rate in real time or record the heart rate for later study.[7] (Such devices are found in smart watches, but more reliable devices include pulse oximeter fingertip clips.[8]) The devices are used to define the aesthetic experience through autonomic nervous system (ANS)[9] responses. A recent study had participants wear electronic gloves to record physiological responses to artworks by monitoring heart rate, heart rate variability, skin conductivity, and skin conductivity variability in museum visitors. These results were assessed together with a customized questionnaire surveying visitors' emotional and aesthetic responses to selected works in an exhibition. These two methods—mobile sensory monitoring devices and questionnaires—combined to show that physiological responses during perception of an artwork were significantly related to aesthetic-emotional experiencing.

Returning to the overarching theme of *A Neuroarthistory of The Painters of Modern Life*, I want to locate my research on an embodied, experiential aesthetic of rupture and fragmentation within a wider field of embodiment studies. My approach aligns with a recently developed philosophical modeling of embodiment—soma[10] studies, in general, and somaesthetics, in particular. Taken together, these endeavors provide an epistemological system that situates the body at its center. As philosopher Richard Shusterman, who coined the term somaesthetics,[11] explains: "Embracing both the mental and the physical, the soma is both subject and object. It is the bodily, sensory subjectivity through which we perceive things, including the soma itself as a bodily object in the world."[12] Accordingly, "[s]omaesthetics… redirects aesthetics back to the core issues of perception, consciousness, and feeling, which are embodied in the root meaning of 'aesthetic' and its familiar contrast—'anaesthetic'." Somaesthetics thus supports a dynamically embodied aesthetic experience, which is "[best] understood through… its underlying ground in more basic forms of perceptual experience that are also essentially and actively embodied."[13] In view of the body's multisensory, affective, and cognitive processes activated during the aesthetic experience, somaesthetics provides a platform for transdisciplinary research, especially neuroscience and its physiological focus, as well as "research [that] can be pursued through other disciplines in the humanities, arts, and social and natural sciences."[14] Notably, Shusterman's anthology, *Bodies in the Streets: The Somaesthetics of City Life*, offers a rich source for reifying historical urban sensoriums, expanding upon the multifaceted matters of somatic city life from global and multi-disciplinary perspectives.[15] As a neuroarthistory of nineteenth-century Paris during Haussmannization, *A Neuroarthistory of The Painters of Modern Life* provides an exemplary platform for joining art history research with somaesthetics in the transdisciplinary intersection of humanities and neuroscience.

As an exploratory and experimental study, *A Neuroarthistory of The Painters of Modern Life* models a neuroarthistory that promotes fluid relations between art history, literary history and theory, neuroscience, and somaesthetics. From this transdisciplinary perspective, the matrix of embodied responses I establish relates both to past art and to our own "real" world experiences and reveals nineteenth-century antecedents of our postmodern condition.

Notes

1 For my working definitions of "interdisciplinary" and "transdisciplinary" see the Introduction, note 1, of this volume.
2 Recent neuroaesthetic research, beyond the scope of my investigations, has aimed to broaden the aesthetic spectrum. See Ines Schindler et al., "Measuring Aesthetic Emotions: A Review of the Literature and a New Assessment Tool," ed. Mariska E. Kret, *PLOS ONE* 12, no. 6 (June 5, 2017): e0178899, https://doi.org/10.1371/journal.pone.0178899; and Eftychia Stamkou et al., "Emotional Palette: A Computational Mapping of Aesthetic Experiences Evoked by Visual Art," *Scientific Reports* 14, no. 1 (August 27, 2024), https://doi.org/10.1038/s41598-024-69686-9.
3 A recent study has investigated how contextual information influences the aesthetic appreciation of artworks; see Kohinoor M. Darda and Anjan Chatterjee, "The Impact of Contextual Information on Aesthetic Engagement of Artworks," *Scientific Reports* 13, no. 4273 (2023): n.p. https://doi.org/10.1038/s41598-023-30768-9.
4 Aimée Boutin's *City of Noise: Sound and Nineteenth-Century Paris* (Urbana: University of Illinois Press, 2015) is a stellar example of sensory studies in urban environments. Boutin reconstructs street vendors' bellowed songs known as the *Cris de Paris* that had been associated with their trades since the Middle Ages and defined Paris's street ambience. Mining accounts from guidebooks to verse, Boutin weaves together literary, cultural, and social history to reconstruct a lost auditory environment. See also Boutin, "Historical Sounds: A Case Study," in Michael Bull and Marcel Cobussen, *The Bloomsbury Handbook of Sonic Methodologies* (New York: Bloomsbury Academic, 2021): 647–57.
5 John Hopkins Medicine, "Electroencephalogram (EEG)," John Hopkins Medicine, 2019, https://www.hopkinsmedicine.org/health/treatment-tests-and-therapies/electroencephalogram-eeg.
6 Cleveland Clinic, "Brain MRI: What It Is, Purpose, Procedure & Results," Cleveland Clinic, 2022, https://my.clevelandclinic.org/health/diagnostics/22966-brain-mri.
7 Robert Gajda et al., "Sports Heart Monitors as Reliable Diagnostic Tools for Training Control and Detecting Arrhythmias in Professional and Leisure-Time Endurance Athletes: An Expert Consensus Statement," *Sports Medicine* (Auckland, N.Z.) 54, no. 1 (2024): 1–21. doi:10.1007/s40279-023-01948-4.
8 A pulse oximeter measures blood oxygen levels and pulse. See Ana Gotter, "Pulse Oximetry," Healthline (Healthline Media, August 2, 2017), https://www.healthline.com/health/pulse-oximetry.
9 "The autonomic nervous system is a component of the peripheral nervous system that regulates involuntary physiologic processes including heart rate, blood pressure, respiration, digestion, and sexual arousal;" Joshua A. Waxenbaum, et al., "Anatomy, Autonomic Nervous System," Nih.gov (StatPearls Publishing, July 24, 2023), https://www.ncbi.nlm.nih.gov/books/NBK539845/.
10 Soma is derived from the Greek word, *sōma*, for body; 'Somatics [also] refers to practices that center the body; "Soma." Merriam-Webster.com Dictionary, Merriam-Webster, https://www.merriam-webster.com/dictionary/soma.
11 See Richard Shusterman, "Somaesthetics: A Disciplinary Proposal," *The Journal of Aesthetics and Art Criticism* 57, no. 3 (1999): 299.
12 Richard Shusterman, "Introduction," in Richard Shusterman, ed., *Bodies in the Streets: The Somaesthetics of City Life* (Leiden and Boston: Brill, 2019): 14.
13 Richard Shusterman, *Thinking Through the Body: Essays in Somaesthetics* (New York: Cambridge University Press, 2012): 3. Although somaesthetics and sensory studies are related approaches to sensorium, I have established the following distinction for this study: somaesthetics is grounded in the phenomenological branch of philosophy while sensory studies is grounded in material culture and archival evidence.
14 Shusterman, *Thinking Through the Body*, 7.
15 As Max Ryynänen discusses, *Bodies in the Streets* "explores the complex network of architecture, weather, social conditioning, politics, gender, and ritual that make up our bodily presence and life in urban contexts;" "Educating Bodies, Educating Streets," *The Journal of Aesthetic Education* 55, no. 2 (2021): 115–21. https://muse.jhu.edu/article/788122.

Bibliography

Abraham, Pierre (ed). "Arts et Littératures Dans la Société Contemporaine, II," in *Encyclopédia Française*, vol. 17. Paris: Larousse, 1936: 293.

Ackerman, James. "The Photographic Picturesque." *Artibus et Historiae* 24, no. 48 (January 1, 2003): 73–94. https://doi.org/10.2307/1483732

Adhémar, Jean. "Baudelaire et les Photographes: Paris et Bruxelles." *Bulletin: Musées Royaux des Beaux-Arts de Belgique, Bruxelles*, 1985–8/1–3: 249–59.

American Psychological Association. n.d. *APA Dictionary of Psychology*. https://dictionary.apa.org/

Arago, François. "Chamber of Deputies–Bill Presented to the Minister of Interior [France], June 15, 1839." In *An Historical and Descriptive Account of the Various Processes of the Daguerréotype and the Diorama*, 1–7. London/New York: McLean Kraus, 1839. https://babel.hathitrust.org/cgi/pt?id=osu.32435011769130&seq=9&q1=Aikin

Arnheim, Rudolf. *Visual Thinking*. Berkeley: University of California Press, 1969.

ARTFL Project, The University of Chicago. n.d. "SCABREUX, EUSE." *Dictionnaire de L'Académie Française*, 6th Edition: 1832–5. http://portail.atilf.fr/cgi-bin/dico1look.pl?strippedhw=scabreux&headword=&docyear=1800-1899&dicoid=ALL&articletype=1

Aubenas, Sylvie, and Philippe Comar. *Obscénités: Photographies interdites d'Auguste Belloc*. Paris: Bibliothèque nationale de France, 2001.

Auvray, Louis. *Exposition des Beaux-Arts. Salon de 1859*. Paris: A. Taride, 1859.

Baguley, David. "Parody and the Realist Novel." *University of Toronto Quarterly* 55 (Fall 1985): 94–108.

Baguley, David. "An Essay on Naturalist Poetics." *The Journal of the British Neo-Formalist Circle* 12, no. 1 (April 1987): 41–56.

Baguley, David. *Naturalist Fiction: The Entropic Vision*. Cambridge: Cambridge University Press, 1990.

Baguley, David. *Napoleon III and His Regime: An Extravaganza*. Baton Rouge: Louisiana State University Press, 2000.

Baker, L. "Metacognition," *ScienceDirect Topics*, 2010. https://www.sciencedirect.com/topics/neuroscience/metacognition

Balducci, Temma. *Gender, Space, and the Gaze in Post-Haussmann Visual Culture: Beyond the Flâneur*. London/New York: Routledge, 2017.

Bamford, Ian J., and Nigel S. Bamford. "The Striatum's Role in Executing Rational and Irrational Economic Behaviors." *The Neuroscientist* 25, no. 5 (January 24, 2019): 475. https://doi.org/10.1177/1073858418824256

Barrett, Lisa Feldman. "The Theory of Constructed Emotion: An Active Inference Account of Interoception and Categorization." *Social Cognitive and Affective Neuroscience* 12, no. 1 (January 2017): 1–23. https://doi.org/10.1093/scan/nsw154

Baudelaire, Charles. "Poemes Nocturne." *Le Present: Revue Hebdomadaire de la Littérature et des Beaux-arts* 1, no. 8 (August 24, 1857): 284–90.

Baudelaire, Charles. "Painters and Etchers" [orig. 1862]. In *Art in Paris, 1845–1862: Salons and Other Exhibitions*, edited and translated by Jonathan Mayne, 217–22. London: Phaidon, 1965.

Baudelaire, Charles. "De L'essence du Rire et Généralement du Comique dans les Arts Plastiques" [orig. 1855]. *Oeuvres Complètes de Charles Baudelaire. II. Curiosités Esthétiques*, 359–87. Paris: Michel Lévy Frères, 1868a. https://fr.wikisource.org/wiki/De_l%E2%80%99essence_du_rire

Baudelaire, Charles. "The Loss of a Halo." In *Oeuvres Complètes. IV*. Paris: Michel Levy Frères, 1868b–1870.

Baudelaire, Charles. *Oeuvres Complètes de Charles Baudelaire. II. Curiosités Esthétiques*. Paris: Michel Lévy Frères, 1868c.

Baudelaire, Charles. Salon de 1859. In *Oeuvres Complètes de Charles Baudelaire. II. Curiosités Esthétiques*. Paris: Michel Lévy Frères, 1868d.

Baudelaire, Charles. "The Generous Gambler" [orig. 1869]. In *The Prose Poems and La Fanfarlo*, translated with an introduction and notes by Rosemary Lloyd, 74–7. Oxford/New York: Oxford University Press, 1991.

Baudelaire, Charles. *Baudelaire: His Prose and Poetry*, edited by T. R. Smith. New York: Boni and Liveright, Pub., 1910.

Baudelaire, Charles. *Baudelaire: His Prose and Poetry*, edited by Joseph T. Shipley, translated by Arthur Symons. New York: Modern Library, 1919a.

Baudelaire, Charles. "The Confiteor of the Artist" [orig. 1862]. In Charles Baudelaire, *The Poems and Prose Poems of Charles Baudelaire*, edited by James Huneker. New York: Brentano's Publishers, 1919b. https://www.gutenberg.org/files/36287/36287-h/36287-h.htm#THE_CONFITEOR_OF_THE_ARTIST

Baudelaire, Charles. "The Loss of a Halo" [orig. 1869]. In *Baudelaire: His Prose and Poetry*, edited by T. R. Smith. New York: Boni and Liveright, Pub., 1910, 99. New York: Modern Library, 1919c. https://www.gutenberg.org/files/47032/47032-h/47032-h.htm

Baudelaire, Charles. "Dédicace/Dedication" [orig. 1857]. In *Les Fleurs du mal/Flowers of Evil* [2nd ed., orig.1861], translated by William Aggeler. Fresno, CA: Academy Library Guild, 1954a. https://fleursdumal.org/poem/098

Baudelaire, Charles. *Les Fleurs du Mal / Flowers of Evil* [2nd ed., orig. 1861], translated by William Aggeler. Fresno, CA: Academy Library Guild, 1954b. https://fleursdumal.org/

Baudelaire, Charles. "Little Old Women" [orig. 1859]. In *Les Fleurs du Mal/Flowers of Evil* [2nd ed., orig.1861], translated William Aggeler. Fresno, CA: Academy Library Guild, 1954c; https://fleursdumal.org/poem/222

Baudelaire, Charles. "The Seven Old Men" [orig. 1859]. In *Les Fleurs du Mal/The Flowers of Evil* [2nd ed., orig. 1861], translated by William Aggeler. Fresno, CA: Academy Library Guild, 1954d. https://fleursdumal.org/poem/221

Baudelaire, Charles. "The Swan" [orig. 1860]. In *Les Fleurs du Mal/The Flowers of Evil* [2nd ed., 1861], translated by William Aggeler. Fresno, CA: Academy Library Guild, 1954e. https://fleursdumal.org/poem/220

Baudelaire, Charles. "To a Passerby" [orig. 1857]. In *Les Fleurs du Mal/The Flowers of Evil* [2nd ed., 1861], translated by William Aggeler. Fresno, CA: Academy Library Guild, 1954f. https://fleursdumal.org/poem/224

Baudelaire, Charles. "The Painter of Modern Life" [orig. 1863]. In *The Painter of Modern Life, And Other Essays*, edited and translated by Jonathan Mayne, 1–40. London: Phaidon, 1964a.

Baudelaire, Charles. *The Painter of Modern Life, And Other Essays*, edited and translated by Jonathan Mayne. London: Phaidon, 1964b.

Baudelaire, Charles. *The Exposition Universelle 1855 [Universal Exposition 1855]*. In *Art in Paris, 1845–1862*, edited and translated by Jonathan Mayne, 121–43. Ithaca: Cornell University Press/Phaidon, 1965.

Baudelaire, Charles. *Correspondance*, 2 vols., edited by Claude Pichois and Jean Ziegler. Paris: Gallimard, 1973.

Baudelaire, Charles. *Oeuvres Complètes de Charles Baudelaire. II*, edited by Claude Pichois, with Jean Ziegler. Paris: Gallimard, 1975–6.

Baudelaire, Charles. *Art in Paris, 1845–1862: Salons and Other Exhibitions*, edited and translated by Jonathan Mayne. Ithaca: Cornell University Press/Phaidon, 1981a.

Baudelaire, Charles. *Salon of 1859*. In *Art in Paris 1845–1862*, edited and translated by Jonathan Mayne, 144–216. Ithaca: Cornell University Press/Phaidon, 1981b.

Baudelaire, Charles. *Baudelaire: The Poems in Prose with "La Fanfarlo*," vol. II, introduction and translations by Francis Scarfe. London: Anvil Press Poetry, 1989.

Baudelaire, Charles. *The Prose Poems and La Fanfarlo*, translated with an introduction and notes by Rosemary Lloyd. Oxford/New York: Oxford University Press, 1991.

Baudelaire, Charles. "*Madame Bovary* by Gustave Flaubert" [orig. 1857]. In *Selected Writings on Art and Literature*, edited and translated by P. E. Charvet, 244–55. London/New York: Penguin Classics, 1992a.

Baudelaire, Charles. "Of the Essence of Laughter, and the Comic in the Plastic Arts," [orig. 1855]. In *Selected Writings on Art and Literature*, edited and translated by P. E. Charvet, 140–161. London/New York: Penguin Classics, 1992b.

Baudelaire, Charles. "Of the Heroism of Modern Life" [orig. 1846]. In *Selected Writings on Art and Literature*, edited and translated by P. E. Charvet, 104–7. London/New York: Penguin Classics, 1992c.

Baudelaire, Charles. "The Painter of Modern Life" [orig. 1863]. In *Selected Writings on Art and Literature*, edited and translated by P. E. Charvet, 390–435. London/New York: Penguin Classics, 1992d.

Baudelaire, Charles. "The Queen of the Faculties" [orig. 1859]. In *Selected Writings on Art and Literature*, edited and translated by P. E. Charvet, 298–302. London/New York: Penguin Classics, 1992e.

Baudelaire, Charles. "The Salon of 1859: The Modern Public and Photography." In *Selected Writings on Art and Literature*, edited and translated by P. E. Charvet, 291–8. London/New York: Penguin Classics, 1992f.

Baudelaire, Charles. *Selected Writings on Art and Literature*, edited and translated by P. E. Charvet. London/New York: Penguin Classics, 1992g.

Baudelaire, Charles. "Some French Caricaturists" [orig. 1857]. In *Selected Writings on Art and Literature*, edited and translated by P. E. Charvet, 209–43. London/New York: Penguin Classics, 1992h.

Baudelaire, Charles. "The Bad Glazier" [orig. 1862]. In *The Parisian Prowler; Le Spleen de Paris; Petits Poèmes en Prose*, 2nd ed., edited and translated by Edward K. Kaplan, 13–5. Athens: University of Georgia Press, 1997a.

Baudelaire, Charles. "The Mirror" [orig. 1864]. In *The Parisian Prowler; Le Spleen de Paris; Petits Poèmes en Prose*, 2nd ed., edited and translated by Edward K. Kaplan, 101. Athens: University of Georgia Press, 1997b.

Baudelaire, Charles. *The Parisian Prowler; Le Spleen de Paris; Petits Poèmes en Prose*, 2nd ed., translated by Edward K. Kaplan. Athens: University of Georgia Press, 1997c.

Baudelaire, Charles. *Oeuvres Complètes de Charles Baudelaire*, edited by Michel Jamet, preface by Claude Roy. Paris: Bouquins/Robert Laffont, 1999a.

Baudelaire, Charles. "Le Public Moderne et la Photographie" [orig. 1859], with notes by Paul-Louis Roubert. *Études Photographiques* 6 (May 1999b) [online]. https://journals.openedition.org/etudesphotographiques/185#ftn25

Baudelaire, Charles. "The Jewels" [orig. 1857]. In *Les Fleurs du Mal/The Flowers of Evil*, translated by James McGowan, 47–8. Oxford: Oxford University Press, 2008a. First published in 1993.

Baudelaire, Charles. *Paris Spleen and La Fanfarlo*, translated by Raymond N. MacKenzie. Indianapolis: Hackett Pub. Co, 2008b.

Baudelaire, Charles. *The Poem of Hashish* [orig. 1858], translated by Aleister Crowley, 1895. Reprint, n.p.: Erowid.org, 2014. http://www.erowid.org/culture/characters/baudelaire_charles/baudelaire_charles_poem1.shtml

Baxandall, Michael. *Painting and Experience in Fifteenth Century Italy: A Primer in the Social History of Pictorial Style*. Oxford: Clarendon Press, 1972.

Beasley-Murray, Jon. "Lakoff and the Body." *Berfrois*. Published online April 5, 2011. https://www.berfrois.com/2011/04/vented/

Benjamin, Walter. *Charles Baudelaire: A Lyric Poet in the Era of High Capitalism*, translated by Harry Zohn. London: NLB, 1973a.

Benjamin, Walter. "The Paris of the Second Empire in Baudelaire" [orig. 1938]. In *Charles Baudelaire: A Lyric Poet in the Era of High Capitalism*, translated by Harry Zohn, 155–76. London: NLB, 1973b.

Benjamin, Walter. "Short History of Photography" [orig. 1931]. Translated by Phil Patton. *Artforum* 15, no. 6 (February 1977): 46–51. https://www.artforum.com/print/197702/walter-benjamin-s-short-history-of-photography-36010

Benjamin, Walter. "The Work of Art in the Age of Its Technological Reproducibility" [orig. 1936]. In *Walter Benjamin: Selected Writings. Volume 3: 1935–1938*, edited by Howard Eiland and Michael W. Jennings, translated by Edmund Jephcott and Harry Zorn. Cambridge, MA: Belknap, 2002: 101–33.

Benjamin, Walter. "Some Motifs in Baudelaire" [orig. 1939]. In *Walter Benjamin: Selected Writings. Volume 4: 1938–1940*, edited by Howard Eiland and Michael W. Jennings, translated by Edmund Jephcott. Cambridge, MA: Belknap, 2006: 313–59.

Berg, William J. *The Visual Novel: Emile Zola and the Art of His Times*. University Park, PA: Pennsylvania State University Press, 1992.

Bernheimer, Charles. "Manet's *Olympia*: The Figuration of Scandal." *Poetics Today* 10, no. 2 (1989): 255–77. https://doi.org/10.2307/1773024

Bertall, "Le Salon Depeint et Dessine par Bertall"/"La Terrible Savoyarde, par Courbet." *Le Journal pour Rire* 91 (June 25, 1853): 1.

Blanchère, Henri de la. "Études Photographiques." *La Lumière* 6 (January 31, 1857): 18–9. Quoted and translated in Paul-Louis Roubert, "A Bourgeois Ambition: Photography and the Salon des Beaux-Arts during the Second Empire." In *Real/Ideal: Photography in Mid-Nineteenth-Century France*, edited by Karen Hellman. Los Angeles: J. Paul Getty Museum, 2016: 23–35.

Blanquart-Evrard, Louis Désiré. "On the Intervention of Art in the Practice of Photography," *British Journal of Photography*, August 1, 1863; Part II–Practical Means–Manipulations." *British Journal of Photography*, August 15, 1863, 326. http://albumen.conservation-us.org/library/c19/blanq2.html

Blanquart-Evrard, Louis Désiré. *La Photographie, ses Origines, ses Progrès, ses Transformations*. Lille, France, 1869. https://gallica.bnf.fr/ark:/12148/bpt6k1231104

Blood, Susan. "The Sonnet as Snapshot: Seizing the Instant in Baudelaire's 'A une passante.'" *Nineteenth-Century French Studies* 36, no. 3–4 (2008): 255–69. https://doi.org/10.1353/ncf.0.0033

Blumenfeld, Hal, and Abhijeet Gummadavelli. "Thalamus: Definition, Anatomy, Function, & Disorders: Britannica." In *Encyclopædia Britannica*. 2020. https://www.britannica.com/science/thalamus

Boen, Rune, Lisa Raud, and Rene J. Huster. "Inhibitory Control and the Structural Parcelation of the Right Inferior Frontal Gyrus." *Frontiers in Human Neuroscience* 16 (February 24, 2022). https://doi.org/10.3389/fnhum.2022.787079

Boggs, Jean Sutherland, Douglas W. Druick, Henri Loyrette, Michael Pantazzi, and Gary Tinterow. *Degas*. New York/Ottawa: The Metropolitan Museum of Art/National Gallery of Canada, 1988.

Bohrn, Isabel C., Ulrike Altmann, and Arthur M. Jacobs. "Looking at the Brains Behind Figurative Language—a Quantitative Meta-Analysis of Neuroimaging Studies on Metaphor, Idiom, and Irony Processing." *Neuropsychologia* 50, no. 11 (September 2012): 2669–83.

Bouchrika, Imed. "What Is Information Processing Theory? Stages, Models & Limitations." Research.com. Accessed online 2024. https://research.com/education/what-is-information-processing-theory

Bourdin, G. [Introduction to Baudelaire's Prose Poems]. *Le Figaro* (February 7, 1864): 3.

Boutin, Aimée. "Aural Flânerie." *Dix-Neuf* 16, no. 2 (July 2012a): 149–61.

Boutin, Aimée. "Rethinking the Flâneur: Flânerie and the Senses." *Dix-Neuf* 16, no. 2 (July 2012b): 124–32.

Boutin, Aimée. *City of Noise: Sound and Nineteenth-Century Paris*. Urbana: University of Illinois Press, 2015.

Boutin, Aimée. "Historical Sounds: A Case Study." In *The Bloomsbury Handbook of Sonic Methodologies*, edited by Michael Bull and Marcel Cobussen. New York: Bloomsbury Academic, 2021: 647–57.

Bradley, M. M., and A. Keil. "Event-Related Potentials (ERPs)." In *Encyclopedia of Human Behavior*, 2nd ed., edited by V. S. Ramachandran. San Diego: Academic Press, January 1, 2012, 79. https://doi.org/10.1016/B978-0-12-375000-6.00154-3

Brewster, David. *The Stereoscope: Its History, Theory and Construction*. London: 1856.

Brielmann, Aenne. "Aesthetics, Empirical," *The Internet Encyclopedia of Philosophy*, n.d. https://iep.utm.edu/empirical-aesthetics/

Brombert, Beth Archer. *Edouard Manet: Rebel in a Frock Coat*. Chicago: University of Chicago Press, 1997.

Brown, Frederick. *Zola: A Life*. London: Papermac, 1997.

Brown, Steven, Xiaoqing Gao, Loren Tisdelle, Simon B. Eickhoff, and Mario Liotti. "Naturalizing Aesthetics: Brain Areas for Aesthetic Appraisal across Sensory Modalities." *NeuroImage* 58, no. 1 (September 1, 2011): 250–8.

Browne, Susan, Xiufang Chen, Faten Baroudi, and Esra Sevinc. "Reader Response Theory." In *Oxford Bibliographies Online: Literary and Critical Theory*, edited by Eugene O'Brien. n.p. Oxford University Press, 2021.

Buisine, Alain. *Eugène Atget ou la Mélancolie en Photographie*. Nîmes: J. Chambon, 1994.

Bullot, Nicolas J., and Rolf Reber. "The Artful Mind Meets Art History: Toward a Psycho-Historical Framework for the Science of Art Appreciation." *Behavioral and Brain Sciences* 36, no. 2 (March 18, 2013): 123–37. https://doi.org/10.1017/s0140525x12000489

Burke, Edmund. "Section XIII 'The Effects of SUCCESSION in Visual Objects Explained'." In *A Philosophical Enquiry into the Origin of Our Ideas of the Sublime and Beautiful* [2nd ed., orig. 1759], edited with an introduction and notes by Adam Phillips, 128. Oxford/New York: Oxford University Press, 1990.

Burnham, Douglas. "Kant, Immanuel: Aesthetics." *The Internet Encyclopedia of Philosophy (IEP): A Peer-Reviewed Academic Resource*. Published online 2019. https://www.iep.utm.edu/kantaest/

Burty, Philippe. *Charles Méryon, Sailor, Engraver, and Etcher: A Memoir and Complete Descriptive Catalogue of His Works*, translated by Marcus B. Huish. London: Fine Arts Society, 1879.

Cachin, Françoise, Charles S. Moffett, and Juliet Wilson-Bareau. *Manet, 1832–1883*. New York: Metropolitan Museum of Art/Abrams, 1983.

Campbell Fine Art. "Charles Méryon: Biography." n.d. https://www.campbell-fine-art.com/artists.php?id=80

Carnavalet. "The Musée Carnavalet." n.d. https://www.carnavalet.paris.fr/musee-carnavalet

Cela-Conde, Camilo J., Luigi Agnati, Joseph P. Huston, Francisco Mora, and Marcos Nadal. "The Neural Foundations of Aesthetic Appreciation." *Progress in Neurobiology* 94, no. 1 (March 22, 2011): 39–48.

Cham, "Prémiere Promenade à L'exposition, par Cham"/"Une Baigneuse, par M. Courbet." *Le Charivari* 22, no. 149 (May 29, 1853): [3].

Chambord, Jacqueline. *Charles Marville: Photographs of Paris at the Time of the Second Empire on Loan from the Musée Carnavalet, Paris*. New York: French Institute/Alliance Française, 1981.

Chatterjee, Anjan. "Neuroaesthetics: Researchers Unravel the Biology of Beauty and Art." *The Scientist Magazine* (May 1, 2014): [6/7]. https://www.the-scientist.com/neuroaesthetics-37572

Chatterjee, Anjan, and Oshin Vartanian. "Neuroaesthetics," *Trends in Cognitive Sciences*, 18, no. 7 (July 1, 2014): 370–5. https://doi.org/10.1016/j.tics.2014.03.003

Chesneau, Ernest. *L'art et les Artistes Modernes en France et en Angleterre*. Paris: Didier, 1864.

Choi, Bernard C. K., and Anita W. P. Pak. "Multidisciplinarity, Interdisciplinarity and Transdisciplinarity in Health Research, Services, Education and Policy: 1. Definitions, Objectives, and Evidence of Effectiveness," *Clinical and Investigative Medicine. Medecine Clinique et Experimentale* 29, no. 6 (2006): 351.

Cinzia, Di Dio, and Gallese Vittorio. "Neuroaesthetics: A Review." *Current Opinion in Neurobiology* 19, no. 6 (2009): 682–7.

Clarétie, Jules. "Deux Heures au Salon." *L'Artiste* 1 (January 1, 1865): 224–9.

Clark, Timothy J. *The Painting of Modern Life: Paris in the Art of Manet and His Followers*. Princeton: Princeton University Press, 1984.

Clayson, Hollis. *Painted Love: Prostitution in French Art of the Impressionist Era* [orig. 1991]. Los Angeles: Getty Research Institute, 2003.

Cleveland Clinic. "Brain MRI: What It Is, Purpose, Procedure & Results." Published online 2022a. https://my.clevelandclinic.org/health/diagnostics/22966-brain-mri

Cleveland Clinic. "Cerebral Cortex: What It Is, Function & Location." Published online 2022b. https://my.clevelandclinic.org/health/articles/23073-cerebral-cortex

Cleveland Clinic. "Parasympathetic Nervous System (PSNS): What It Is & Function." Published online 2022c. https://my.clevelandclinic.org/health/body/23266-parasympathetic-nervous-system-psns#:~:text=Your%20parasympathetic%20nervous%20system%20is

Cleveland Clinic. "Sympathetic Nervous System (SNS): What It Is & Function" Published online 2022d. https://my.clevelandclinic.org/health/body/23262-sympathetic-nervous-system-sns-fight-or-flight#:~:text=Your%20sympathetic%20nervous%20system%20is

CNRTL. n.d. www.cnrtl.fr

Collier, Peter, and Robert Lethbridge. *Artistic Relations: Literature and the Visual Arts in Nineteenth-Century France*. New Haven: Yale University Press, 1994.

Corbetta, Maurizo, and Gordon L. Shulman, "Control of Goal-directed and Stimulus-driven Attention in the Brain." *Nature Reviews Neuroscience 3*, no. 3 (2002): 201–15. https://doi.org/10.1038/nrn755

Corbin, Alain. "Charting the Cultural History of the Senses." In *Empire of the Senses: The Sensual Culture Reader*, edited by David Howes. Oxford: Berg, 2005: 128–37.

Courbet, Gustave. *Exhibition et Vente de 40 Tableaux et 4 Dessins de L'œuvre de M. Gustave Courbet, Avenue Montaigne*. 7 Champs-Elysées, Paris: n.p., 1855.

Courbet, Gustave. "Realist Manifesto (1855)." In *Realism and Tradition in Art, 1848–1900: Sources and Documents*, edited and translated by Linda Nochlin, 33–4. Englewood Cliffs, NJ: Prentice-Hall, 1966.

Couture, Thomas. *Méthode et Entretiens D'atelier*. [Paris]: n.p., 1867.

Crary, Jonathan. *Techniques of the Observer: On Vision and Modernity in the Nineteenth Century*. Cambridge: MIT Press, 1999.

Crawford, Lawrence. "Viktor Shklovskij: Différance in Defamiliarization." *Comparative Literature* 36, no. 3 (1984): 209–19.

Curtiss, Mina. "Manet Caricatures: Olympia." *The Massachusetts Review* 7, No. 4 (1966): 725–52.

Dällenbach, Lucien. *The Mirror in the Text*, translated by Jeremy Whiteley with Emma Hughes. Cambridge: Polity Press, 1989.

Damade, Jacques (ed), *Paris, 1860: Eaux-fortes sur Paris & "Les Tableaux Parisiens,"* texts by Charles Baudelaire, illustrations by Charles Meryon, preface and notes by Jacques Damade. Paris: Editions la Bibliothèque, 2001.

Darda, Kohinoor M., and Anjan Chatterjee, "The Impact of Contextual Information on Aesthetic Engagement of Artworks." *Scientific Reports* 13, no. 4273 (2023): n.p. https://doi.org/10.1038/s41598-023-30768-9

Davis, Philip. "The Shakespeared Brain," *The Reader*, no. 23 (n.d.): 39–43 [online]. https://thereaderonline.wordpress.com/features/the-shakespeared-brain/

de Man, Paul. *Blindness and Insight: Essays in the Rhetoric of Contemporary Criticism*, 2nd ed., revised [orig. 1971]. Minneapolis: University of Minnesota Press, 1983.

Delacroix, Eugène. *Journal*, vol. 2, edited by Paul Flat and René Piot. Paris: Plon, 1893.

Desbuissons, Frédérique. "A Ruin: Jacques-Louis David's *Sabine Women*." *Art History* 20, no. 3 (September 1997): 432–48.

Disdéri, [Eugène]. *L'art de la Photographie* (Paris: Chez L'auteur, 1862): 304. Quoted and translated in Marie Robert, "The Photographer, the Model and Body Trafficking." In *Splendours & Miseries: Pictures of Prostitution in France, 1850–1910*, 70. Paris: Flammarion, 2015.

Donnelly, Nick, Beth Harland, Natalie Mestry, Natalie Thompson, Tobiasz Trawiński, and Simon P. Liversedge. "The Influence of Pupil Alignment on Spectator Address in Manet's Portraiture." *Psychology of Aesthetics, Creativity, and the Arts* 11, no. 2 (May 2017): 167–78. https://doi.org/10.1037/aca0000096

Driscol, Margaret E., Pradeep C. Bollu, and Prasanna Tadi. "Neuroanatomy, Nucleus Caudate." *StatsPearls*. Treasure Island [FL]: StatPearls Publishing, 2024. Last updated July 24, 2023. https://www.ncbi.nlm.nih.gov/books/NBK557407/

Du Camp, Maxime. *Les Beaux-arts à L'exposition Universelle et aux Salons de 1863, 1864, 1865, 1866 et 1867*. Paris: Vve J. Renouard, 1867.

Dumas, Ann, and Metropolitan Museum of Art. *The Private Collection of Edgar Degas*, translated by Mark Polizzotti. New York: Metropolitan Museum of Art, 1997, 188, fig. 265.

Durandard. "Petit Manuel de Philosophie Photographique." *Le Monde Illustré* 6, no. 272 (June 28, 1862): 408–10.

Durieu, M. E. "Rapport Présenté par M. E. Durieu, au Nom de la Commission Chargée de L'examen de l'Exposition Ouverte dans les Salons de la Société Française de Photographie, du 1 Août au 15 Novembre 1855." *Bulletin de la Société Française de Photographie* 2 (January 1856): 37–72.

Eco, Umberto. *The Open Work* [orig. 1962]. Cambridge: Harvard University Press, 1989.

Ego. "Courrier de Paris." *Le Monde Illustré* 9, no. 422 (May 13, 1865): 290–1.

Encyclopedia.com. "Muller, Johannes Peter." *Complete Dictionary of Scientific Biography*. Updated June 16, 2025. https://www.encyclopedia.com/people/medicine/medicine-biographies/johannes-peter-muller

Etienne, Louis. *Le Jury et les Exposants: Salon des Refusés*. Paris: E. Dentu, 1863.

Felluga, Dino Franco. *Critical Theory: The Key Concepts*. London/New York: Routledge, Taylor & Francis Group, 2015.

Fish, Stanley. *Is There a Text in This Class?: The Authority of Interpretive Communities*. Cambridge: Harvard University Press, 1980.

Flaubert, Gustave. *The Complete Works of Gustave Flaubert*, vol. 2. New York/London: M. W. Dunne, 1904.

Flaubert, Gustave. *Oeuvres Complètes de Gustave Flaubert: Madame Bovary*, t. 2. Paris: L. Conard, 1910.

Flescher, Sharon. "More on a Name: Manet's 'Olympia' and the Defiant Heroine in Mid-Nineteenth-Century France." *Art Journal* 45, no. 1 (1985): 27–35.

Floyd, Phylis A. "The Puzzle of Olympia." *Nineteenth-Century Art Worldwide* 3, no. 1 (2004). https://www.19thc-artworldwide.org/spring04/285-the-puzzle-of-olympia).

Font-Réaulx, Dominique de, Laurence des Cars Michel Hilaire, Bruno Mottin, and Bertrand Tillier. *Gustave Courbet*. New York/Ostfildern: Metropolitan Museum of Art/Hatje Cantz, 2008.

Foucault, Michel. "Of Other Spaces" [orig. 1967], translated by Jay Miskowiec. *Diacritics* 16, no. 1 (1986): 22–7.

Fournier, Edouard. *Paris Démoli. Mosaïque de Ruines* [orig. 1854], 2nd ed., revised and expanded, preface by Théophile Gautier. Paris: A. Aubry, 1855.

France, Peter (ed). "Mise en Abyme" In *The New Oxford Companion to Literature in French*. Clarendon: Oxford University Press, 1995. https://www.oxfordreference.com/view/10.1093/acref/9780198661252.001.0001/acref-9780198661252-e-3152

Freedberg, David, and Vittorio Gallese. "Motion, Emotion and Empathy in Esthetic Experience." *Trends in Cognitive Sciences* 11, no. 5 (May 2007): 197–203.

Gajda, Robert, Jacek Gajda, Miłosz Czuba, Beat Knechtle, and Wojciech Drygas. "Sports Heart Monitors as Reliable Diagnostic Tools for Training Control and Detecting Arrhythmias in Professional and Leisure-Time Endurance Athletes: An Expert Consensus Statement." *Sports Medicine* (Auckland, N.Z.) 54, no. 1 (2024): 1–21. doi:10.1007/s40279-023-01948-4

Gallese, Vittorio, and Alessandro Gattara. "Embodied Simulation, Aesthetics, and Architecture: An Experimental Aesthetic Approach." In *Mind in Architecture: Neuroscience, Embodiment, and the Future of Design*, 161–79. Cambridge: MIT Press, 2017.

Gautier, Théophile. "Salon de 1853: Onzième Article." *La Presse* 17 (July 21, 1853): [1–3].

Gautier, Théophile. "Préface." In Edouard Fournier, *Paris Démoli. Mosaïque de Ruines*, i–xi. Paris: A. Aubry, 1855.

Gautier, Théophile. "Introduction." In *Paris et les Parisiens au XIXe siècle: Moeurs Arts et Monuments*. Paris: Morizot, 1856.

Gautier, Théophile. "Le Salon de 1865." *Le Moniteur Universel*, no. 175 (June 24, 1865): [1–2].

Gautier, Théophile. *Charles Baudelaire: His Life* [orig. 1868], translated by Guy Thorne. London: Greening & Co., 1915.

Gernsheim, Helmut. *The Rise of Photography 1850–1880: The Age of Collodion*. London/New York: Thames and Hudson, 1988.

Gloaguen, Laurent. *Vergue.com* [Marville photographs]. First published September 25, 2013.

Godfrey, Sima. "From Memory Lane to Memory Boulevard: Paris Change!" In Mary Ann Caws, ed., *City Images*, edited by Mary Ann Caws. New York: Gordon and Breach, 1991, 168–9, n. 1.

Gombrich, E. H. *Art and Illusion: A Study in Psychology of Pictorial Representation*. New York: Pantheon Books [Bollingen series], 1960.

Gorodeisky, Keren. "19th Century Romantic Aesthetics." *The Stanford Encyclopedia of Philosophy* (Fall 2016 Edition), edited by Edward N. Zalta: Section 1, 3.2. https://plato.stanford.edu/archives/fall2019/entries/aesthetics-19th-romantic/

Gotter, Ana. "Pulse Oximetry." *Healthline*. Published online August 2, 2017. https://www.healthline.com/health/pulse-oximetry

Grafman, Jordan H., and Ishan Dasgupta (eds). "Functional Magnetic Resonance Imaging (FMRI)." In *Key Brain Terms Glossary*. Dana Foundation, 2022. https://dana.org/explore-neuroscience/brain-basics/key-brain-terms-glossary/#F

Grigsby, Darcy Grimaldo. "Still Thinking about Olympia's Maid." *The Art Bulletin* 97, no. 4 (2015): 430–51.

Grøtta, Marit. "Reading/Developing Images: Baudelaire, Benjamin, and the Advent of Photography." *Nineteenth-Century French Studies* 41, no. 1–2 (2012): 80–90. https://doi.org/10.1353/ncf.2012.0053

Hambourg, Maria Morris. "Charles Marville's Old Paris." In *Charles Marville: Photographs of Paris at the Time of the Second Empire on Loan from the Musée Carnavalet, Paris*, 7–11. New York: French Institute/Alliance Française, 1981.

Hamilton, George Heard. *Manet and His Critics.* New Haven/London: Yale University Press, 1987.

Hampe, Beate. "Image Schemas in Cognitive Linguistics: Introduction." In *From Perception to Meaning: Image Schemas in Cognitive Linguistics*, edited by Beate Hempe. Berlin/New York: De Gruyter Mouton, 2005: 1–14.

Hannoosh, Michèle. "Etching and Modern Art: Baudelaire's 'Peintres et Aquafortistes.'" *French Studies* 53, no. 1 (1989a): 47–60.

Hannoosh, Michèle. "The Reflexive Function of Parody." *Comparative Literature* 41, no. 2 (Spring 1989b): 113–27.

Hannoosh, Michèle. *Baudelaire and Caricature: From the Comic to an Art of Modernity.* University Park, PA: Pennsylvania State University Press, 1992.

Hannoosh, Michèle. "Reading the Trial of the *Fleurs du Mal.*" *Modern Language Review* 106, no. 2 (2011): 374–87.

Hansen, Dorothee. "Manet and Astruc: Friendship and Inspiration." In *Manet and Astruc: Friendship and Inspiration*, edited by Dorothee Hansen. Madrid: Kunsthalle Bremen with Centro de Estudios Europa Hispánica, 2021a: 12.

Hansen, Dorothee. "Manet and Astruc: Friendship and Inspiration." In *Manet and Astruc: Friendship and Inspiration*, edited by Dorothee Hansen. Madrid: Kunsthalle Bremen with Centro de Estudios Europa Hispánica, 2021b: 12–3.

Harland, Beth, John Gillett, Carl M. Mann, Jason Kass, Hayward J. Godwin, Simon P. Liversedge, and Nick Donnelly. "Modes of Address in Pictorial Art: An Eye Movement Study of Manet's *Bar at the Folies-Bergère.*" *Leonardo* 47, no. 3 (June 2014): 241–8. https://doi.org/10.1162/leon_a_00676

Harrison, Nicholas. *Circles of Censorship: Censorship and its Metaphors in French History, Literature and Theory.* Oxford: Clarendon Press, 1995.

Harvey, David. *Paris, Capital of Modernity.* New York: Routledge, 2006.

Haxthausen, Charles W. "Reproduction/Repetition: Walter Benjamin/Carl Einstein." *October* 107 (Winter 2004): 47–74. https://doi.org/10.1162/016228704322790890

Heck, Francis S. "Baudelaire's La Fanfarlo: An Example of Romantic Irony." *The French Review* 49, no. 3 (February 1976): 328–36.

Heilbrunn Timeline of Art History (ed). "Gustave Le Gray (1820–1884)." Metmuseum.org, 2020. https://www.metmuseum.org/toah/hd/gray/hd_gray.htm

Hellman, Karen. "Real/Ideal: Photography in Mid-Nineteenth-Century France." In *Real/Ideal: Photography in Mid-Nineteenth-Century France*, edited by Karen Hellman. Los Angeles: J. Paul Getty Museum, 2016a: 3–12.

Hellman, Karen (ed). *Real/Ideal: Photography in Mid-Nineteenth-Century France.* Los Angeles: J. Paul Getty Museum, 2016b.

Helsinger, Elizabeth K. *Ruskin and the Art of the Beholder.* Cambridge: Harvard University Press, 1982.

Hendricks, LaVelle, Sam Bore, Dean Aslinia, and Guy Morriss. "The Effects of Anger on the Brain and Body." *National Forum Journal of Counseling and Addiction* 2, no. 1 (2013): 1–12. http://www.nationalforum.com/Electronic%20Journal%20Volumes/Hendricks,%20LaVelle%20The%20Effects%20of%20Anger%20on%20the%20Brain%20and%20Body%20NFJCA%20V2%20N1%202013.pdf

Herbert, Robert L. *Impressionism: Art, Leisure, and Parisian Society.* New Haven: Yale University Press, 1991.

Hickok, Gregory. "Eight Problems for the Mirror Neuron Theory of Action Understanding in Monkeys and Humans." *Journal of Cognitive Neuroscience* 21, no. 7 (July 2009): 1229–43. https://doi.org/10.1162/jocn.2009.21189

Higonnet, Anne. "Manet and the Multiple." *Grey Room* 48 (July 2012): 102–16. https://doi.org/10.1162/grey_a_00082
Hildebrand, David. "John Dewey," in *Stanford Encyclopedia of Philosophy*, edited by Edward N. Zalta and Uri Nodelman (Summer 2024 Edition). https://plato.stanford.edu/archives/sum2024/entries/dewey
Hobson, Hannah M., and Dorothy V. M. Bishop. "The Interpretation of Mu Suppression as an Index of Mirror Neuron Activity: Past, Present and Future." *Royal Society Open Science* vol. 4, 3 160662 (Mar 1, 2017). doi:10.1098/rsos.160662
Hoffmann, E. T. A. *The Golden Pot, and Other Tales*, translated by Ritchie Robertson. Oxford: Oxford University Press, 1992.
Hogan, Patrick Colm, Bradley J. Irish, and Lalita Pandit (eds). *The Routledge Companion to Literature and Emotion*. London: Routledge, 2022.
Howes, David. "Olfaction and Transition: An Essay on the Ritual Uses of Smell." *The Canadian Review of Sociology and Anthropology* 24, no. 3 (August 1987): 398–416.
Howes, David. "Charting the Sensorial Revolution." *The Senses and Society* 1, no. 1 (March 2006): 113–28.
Hutcheon, Linda. *A Theory of Parody: The Teachings of Twentieth-Century Art Forms*. New York: Methuen, 1985.
Hutto, Daniel, and Ian Ravenscroft. "Folk Psychology as a Theory." In *The Stanford Encyclopedia of Philosophy*, edited by Edward N. Zalta. Published online 2021. https://plato.stanford.edu/archives/fall2021/entries/folkpsych-theory/
Huysmans, J[oris].-K[arl]. "La *Nana* de Manet." *L'Artiste de Bruxelles* (May 13, 1877): 148–9. http://www.huysmans.org/artcriticism/nana.htm
Hyslop, Lois Boe, and Francis E. Hyslop. "Baudelaire and Meryon: Painters of the Urban Landscape." *Symposium: A Quarterly Journal in Modern Literatures* 38, no. 3 (Fall 1984): 196–220.
Iser, Wolfgang. "The Reading Process: A Phenomenological Approach." *New Literary History* 3, no. 2 (1972): 279–99.
Ishizu, T., and S. Zeki. "Toward a Brain-based Theory of Beauty." *Plos ONE* 6, no. 7 (2011): e21852.
Iskin, Ruth E. "Selling, Seduction, and Soliciting the Eye: Manet's *Bar at the Folies-Bergère*." *The Art Bulletin* 77, no. 1 (March 1995): 25–44. https://doi.org/10.2307/3046078
Jahyer, Félix. *Salon de 1865: Étude sur les Beaux-arts*. Paris: Dentu, 1865.
John Hopkins Medicine. "Electroencephalogram (EEG)." Published online 2019. https://www.hopkinsmedicine.org/health/treatment-tests-and-therapies/electroencephalogram-eeg
Kaplan, Edward K. *Baudelaire's Prose Poems: The Esthetic, the Ethical, and the Religious in "The Parisian Prowler."* Athens/London: University of Georgia Press, 2009.
Keidel, James L., Philip M. Davis, Victorina Gonzalez-Diaz, Clara D. Martin, and Guillaume Thierry. "How Shakespeare Tempests the Brain: Neuroimaging Insights." *Cortex* 49, no. 4 (2013): 913–9. https://doi.org/10.1016/j.cortex.2012.03.011
Keilty, Patrick. "Embodied Engagements with Online Pornography." *The Information Society* 32, no. 1 (December 22, 2015): 64–73. https://doi.org/10.1080/01972243.2015.1107162
Kemp, Wolfgang, and Joyce Rheuban. "Images of Decay: Photography in the Picturesque Tradition." *October* 54 (Autumn 1990): 102–33. https://doi.org/10.2307/778671
Kennel, Sarah. "Charles Marville: Hidden in Plain Sight." In *Charles Marville: Photographer of Paris*, 2–41. Washington, DC: National Gallery of Art, 2013a.
Kennel, Sarah. *Charles Marville: Photographer of Paris*. Washington, DC: National Gallery of Art, 2013b.
Kessler, Marni. "Dusting the Surface, or the Bourgeoisie, the Veil, and Haussmann's Paris." In *The Invisible Flâneuse?: Gender, Public Space, and Visual Culture in Nineteenth-Century Paris*, edited by Aruna D'Souza and Tom McDonough. New York: Manchester University Press, 2006: 49–64.
Kieran, Matthew. "On Obscenity: The Thrill and Repulsion of the Morally Prohibited." *Philosophy and Phenomenological Research* 64, no. 1 (January 2002): 31–55. https://doi.org/10.1111/j.1933-1592.2002.tb00141.x
Kirkland, Stephane. *Paris Reborn: Napoléon III, Baron Haussmann, and the Quest to Build a Modern City*. New York: Picador, 2014.
Koelsch, Stefan, Arthur M. Jacobs, Winfried Menninghaus, Katja Liebal, Gisela Klann-Delius, Christian von Scheve, and Gunter Gebauer. "The Quartet Theory of Human Emotions: An

Integrative and Neurofunctional Model." *Physics of Life Reviews* 13 (June 2015): 1–27. https://doi.org/10.1016/j.plrev.2015.03.001
Kolb, Jocelyn. "Romantic Irony." In *A Companion to European Romanticism*, edited by Michael Ferber. Malden, MA: Blackwell, 2005: 376–92.
Krell, Alan. "Manet, Zola and the 'Motifs d'une exposition particulière' 1867." *Gazette des Beaux-Arts* 99 (1982): 109–15.
Krell, Alan. "Manet's *Déjeuner sur L'herbe* in the *Salon des Refusés*: A Re-appraisal." *The Art Bulletin* 65, no. 2 (June 1983): 316–20.
Krell, Alan. *Manet and the Painters of Contemporary Life*. New York: Thames and Hudson, 1996.
LaCapra, Dominick. *Madame Bovary on Trial*. Ithaca: Cornell University Press/Phaidon, 1982.
LaCapra, Dominick. "Memory, Law, and Literature: The Cases of Flaubert and Baudelaire." In *History, Memory, and the Law*, edited by Austin Sarat and Thomas R. Kearns. Ann Arbor: University of Michigan Press, 1999: 95–130.
Lacoste-Veysseyre, Claudein. "Introduction." In *Paris et les Parisiens*, Gautier, Théophile. Paris: Boite à documents, 1996.
Laincel, Louis de. *Promenade aux Champs-Élysées: [...] Le Salon de 1865*. Paris: E. Dentu, 1865.
Lajer-Burcharth, Ewa. "David's *Sabine Women*: Body, Gender and Republican Culture under the Directory." *Art History* 14, no. 3 (September 1991): 397–430.
Lakoff, George. *Women, Fire, and Dangerous Things*. Chicago: University of Chicago Press, 2008.
Landon, C.-P. "Peinture." *Journal des Arts, de Littérature et de Commerce*, no. 32, 10 nivôse, an 8 (December 31, 1799).
Lazare, Félix, and Louis Lazare. *Dictionnaire Administratif et Historique des Rues et Monuments de Paris*. Paris: Bureau de la Revue Municipale, 1855. http://vergue.com/post/760/Rue-du-Bon-Puits
Le Gall, Guillaume. "Paris, Objet d'Histoire: Sauvegarde, Representation." *Le Vieux Paris* [online exhibition]. Paris: Bibliothèque nationale de France, 2007. http://expositions.bnf.fr/atget/arret/17.htm
Le Gall, Guillaume. "The Eye of the Archaeologist Eugène Atget and the Forms of the Old City." In *Eugène Atget: Old Paris*,16–29. Madrid: Fundación MAPFRE, 2011.
Lecour, C. J. *La Prostitution à Paris et à Londres 1789–1870*. Paris: P. Asselin, 1870.
Leder, Helmut, and Marcos Nadal. "Ten Years of a Model of Aesthetic Appreciation and Aesthetic Judgments: The Aesthetic Episode – Developments and Challenges in Empirical Aesthetics," *British Journal of Psychology*, 105, no. 4 (October 3, 2014): 443–64. https://doi.org/10.1111/bjop.12084
Lee, Andrew Go, Ashwini Kini, and Bayan Al Othman. "Saccade." In *EyeWiki*. American Academy of Ophthalmology, April 7, 2023. https://eyewiki.aao.org/Saccade
Lee, Daryl. "The Ambivalent Picturesque of the Paris Commune Ruins." *Nineteenth Century Prose* 29, no. 2 (September 22, 2002): 138–61.
Lee, Daryl. "Baudelaire's 'Paysage' and the Commune Ruins Picturesque." *Romance Notes* 45, no. 1 (2004): 107–15.
Leech, Geoffrey N. *Linguistic Guide to English Poetry*. Harlow: Longmans, 1969, 69.
Lethbridge, Robert. "Zola, Manet and *Thérèse Raquin*." *French Studies* 34, no. 3 (July 1, 1980): 278–99.
Lethbridge, Robert. "Manet and Textual Frames." In *Artistic Relations: Literature and the Visual Arts in Nineteenth-Century France*, edited by Peter Collins and Robert Lethbridge. New Haven: Yale University Press, 1994: 144–58.
Lethbridge, Robert. "Zola and Contemporary Painting." In *The Cambridge Companion to Emile Zola*, edited by Brian Nelson. Cambridge: Cambridge University Press, 2007: 67–85.
Lilley, E. D. "Two Notes on Manet." *The Burlington Magazine* 132, no. 1045 (1990): 266–9.
Lingnau, A., B. Gesierich, and A. Caramazza. "Asymmetric FMRI Adaptation Reveals No Evidence for Mirror Neurons in Humans." *Proceedings of the National Academy of Sciences* 106, no. 24 (June 2, 2009): 9925–30. https://doi.org/10.1073/pnas.0902262106
Lipton, Eunice. *Alias Olympia: A Woman's Search for Manet's Notorious Model & Her Own Desire*. New York: Charles Scribner's & Sons, 1992.
Lloyd, Rosemary. *Baudelaire et Hoffmann: Affinités et Influences*. Cambridge: Cambridge University Press, 1979.
Lonner, Walter J. "Universals." *The Encyclopedia of Cross-Cultural Psychology*. Published online September 3, 2013: 1308–11. https://doi.org/10.1002/9781118339893.wbeccp553

Love, Todd, Christian Laier, Matthias Brand, Linda Hatch, and Raju Hajela. "Neuroscience of Internet Pornography Addiction: A Review and Update." *Behavioral Sciences* 5, no. 3 (September 18, 2015): 388–433. https://doi.org/10.3390/bs5030388

Luchet, Auguste, and Claude-François Denecourt. *Hommage à C.-F. Denecourt. "Fontainebleau", Paysages, Légendes, Souvenirs, Fantaisies....* Paris: Hachette, 1855.

Lyons, Andrew P. "The Two Lives of Sara Baartman: Gender, 'Race,' Politics and the Historiography of Mis/Representation." *Anthropologica* 60, no. 1 (May 2018): 337–46. https://doi.org/10.3138/anth.60.1.a05

MacFarlane, Dana. "'Waiting Still:' Baudelaire and the Temporality of the Photographic Portrait." *History of Photography* 36, no. 1 (February 2012): 3–14. https://doi.org/10.1080/03087298.2012.633439

Mambrol, Nasrullah. "Transactional Reader Response Theory," *Literary Theory and Criticism* (October 28, 2016).

Maneglier, Hervé. *Paris Impérial: La Vie Quotidienne sous le Second Empire.* Paris: Armand Colin, 1990.

Marder, Elissa. "A Thousand Hungry Eyes: Baudelaire's Stereoscope." *Yale French Studies*, no. 139, Photography and the Body in Nineteenth-Century France (2021): 85–99. https://www.jstor.org/stable/45455814

Marrinan, Michael. *Caillebotte: Painting the Paris of Naturalism, 1872–1887.* Los Angeles: Getty Research Institute, 2016.

Marshall, David. "The Problem of the Picturesque." *Eighteenth-Century Studies* 35, no. 3 (2002): 413–37. https://doi.org/10.1353/ecs.2002.0029

Martin, Pauline. "'Le Flou du Peintre ne Peut être le Flou du Photographe'. Une Notion Ambivalente dans la Critique Photographique Française au Milieu du XIXe Siècle." *Études Photographiques*, no. 25 (May 2010): document 2512, 180–209 [online]. http://journals.openedition.org/etudesphotographiques/3060

Massumi, Brian. *Parables for the Virtual: Movement, Affect, Sensation.* Durham, NC: Duke University Press, 2002.

McCauley, Elizabeth Anne. "François Arago and the Politics of the French Invention of Photography." In *Multiple Views: Logan Grant Essays on Photography, 1983–89* edited by Daniel P. Younger. Albuquerque: University of New Mexico Press, 1991: 43–69.

McCauley, Elizabeth Anne. *Industrial Madness: Commercial Photography in Paris, 1848–1871.* New Haven/London: Yale University Press, 1994.

McCauley, Elizabeth Anne. "Sex and the Salon: Defining Art and Immorality." In *Manet's "Le Déjeuner sur L'herbe,"* edited by Paul Tucker. New York: Cambridge University Press, 1998: 38–74.

McKee, George D. "Jurisprudence: Crimes of the Image." Chronology to Accompany Online Resource *Image of France 1795–1880.* Published online September 21, 2014a, web.archive.org. https://web.archive.org/web/20140921044520/https://sites.google.com/site/imoffrdocs/home/jurisprudence

McKee, George D. "Legislation." Chronology to Accompany Online Resource *Image of France 1795–1880.* Published online September 21, 2014b, web.archive.org. https://web.archive.org/web/20201007192519/https://sites.google.com/site/imoffrdocs/home/legislation

McLees, Ainslie Armstrong. *Baudelaire's "Argot Plastique": Poetic Caricature and Modernism.* Athens: University of Georgia Press, 1989.

Medicine LibreTexts. "12.1C: Sensory Modalities." Published online July 20, 2018. https://med.libretexts.org/Bookshelves/Anatomy_and_Physiology/Anatomy_and_Physiology_(Boundless)/12%3A_Peripheral_Nervous_System/12.1%3A_Sensation/12.1C%3A_Sensory_Modalities

Mertens, I., H. Siegmund, and O.-J. Grüsser. "Gaze Motor Asymmetries in the Perception of Faces during a Memory Task." *Neuropsychologia* 31, no. 9 (September 1993): 989–98. https://doi.org/10.1016/0028-3932(93)90154-r

Miall, David S. "Cognitive Poetics: From Interpreting to Experiencing What Is Literary." In *Anglistentag 2007 Münster: Proceedings*, edited by Klaus Stierstorfer. Trier: Wissenschaftlicher Verlag, 2008: 187–98.

Miall, David S. "Neuroaesthetics of Literary Reading." In *Neuroaesthetics*, edited by Martin Skov and Oshin Vartanian. Amityville, NY: Baywood Publishing, 2009: 233–47.

Miall, David S., and Don Kuiken. "Foregrounding, Defamiliarization, and Affect: Response to Literary Stories." *Poetics* 22, no. 5 (August 1994): 389–407.

Miall, David S., and Don Kuiken. "A Feeling for Fiction: Becoming What We Behold." *Poetics* 30, no. 4 (August 2002): 221–41.

Molnar, François. "A Science of Vision for Visual Art [1992]." *Leonardo* 30, no. 3 (1997): 225–32. https://doi.org/10.2307/1576454

Molnar, François. "About the Role of Visual Exploration in Aesthetics." *Advances in Intrinsic Motivation and Aesthetics*, edited by Hy I. Day. Boston: Springer US, 1981: 385–414.

Molnar, François, and Dominique Ratsikas, "Some Aesthetical Aspects of Visual Exploration." In *Eye Movements: From Psychology to Cognition*, edited by J. K. O'Regan and A. Levy-Schoen. Amsterdam: Elsevier, 1987: 363–74.

Moncan, Patrice de, and Claude Heurteux. *Le Paris d'Haussmann.* [Paris]: Editions du Mécène, 2002.

Monchaux, Didier de. "Salon de 1863–Les Refusés." *La Patrie* 23 (May 21, 1863): 2.

Moreau, Jacques-Joseph. *La Psychologie Morbide Dans Ses Rapports Avec la Philosophie de L'histoire.* Paris: V. Masson, 1859.

Moreau, Jacques-Joseph. *Hashish and Mental Illness* [orig. 1845], edited by Hélène Peters and Gabriel G. Nahas, translated by Gordon J. Barnett. New York: Raven Press, 1973.

Moulène, Monique. "L'Ancien Paris de Martial Potémont." Le Blog Gallica, Bibliothèque Nationale de France. Published online January 1, 2013. https://gallica.bnf.fr/blog/01012013/lancien-paris-de-martial-potemont?mode=desktop

Muecke, D[ouglas]. C[olin]. *Irony and the Ironic.* London: Methuen & Co., 1970.

Müller, Johannes. *Elements of Physiology* [orig. 1838], 2 vols., translated by William Baly. London: Taylor & Walton, 1840–2.

Munar, Enric, Marcos Nadal, Nazareth P. Castellanos, Albert Flexas, Fernando Maestú, Claudio Mirasso, and Camilo J. Cela-Conde. "Aesthetic Appreciation: Event-related Field and Time-frequency Analyses." *Frontiers in Human Neuroscience* 5 (2012): 1–11. https://doi.org/10.3389/fnhum.2011.00185

Muriand, Raoul. *Les Folies Bergère.* Sèvres: Editions La Sirène, 1994.

Murphy, Libby, and Greggor Mattson. "Types of Prostitutes in 19th Century France." *A la Récherche des Femmes Perdues* [online exhibition]. Oberlin, OH: Oberlin College Library, 2016. https://onprostitution.oberlincollegelibrary.org/exhibits/show/prostitutes-in-19th-century-fr

Murrell, Denise. *Posing Modernity: The Black Model from Manet and Matisse to Today.* New Haven/London: Yale University Press, 2018.

Nayak, Bighna Kalyan, and Sougata Karmakar. "Eye Tracking Based Objective Evaluation of Visual Aesthetics: A Review." In *Advances in Ergonomics in Design. AHFE 2018*, edited by F. Rebelo and M. Soares, vol. 777. Cham, CH: Springer, 2019: 370–81. https://doi.org/10.1007/978-3-319-94706-8_41

Nazarieff, Serge. *Early Erotic Photography.* Köln: Taschen, 1993.

Nègre, Charles. "Héliographie sur papier ciré et à sec." *La Lumière*, no. 14 (May 11, 1851): 56. Quoted and translated in Karen Hellman, "Real/Ideal: Photography in Mid-Nineteenth-Century France," *Real/Ideal: Photography in Mid-Nineteenth-Century France*, edited by Karen Hellman. Los Angeles: J. Paul Getty Museum, 2016: 7.

Nelson, Brian. *The Cambridge Companion to Zola.* Cambridge/New York: Cambridge University Press, 2007.

Nemoto, Kiyotaka. "[Understanding Voxel-Based Morphometry]." *Brain and Nerve = Shinkei Kenkyu No Shinpo* 69, no. 5 (May 1, 2017): 505–11. https://doi.org/10.11477/mf.1416200776

Nesbit, Molly. *Atget's Seven Albums.* New Haven: Yale University Press, 1992.

Nisiforou, Efi, and Andrew Laghos. "Field Dependence–Independence and Eye Movement Patterns: Investigating Users' Differences through an Eye Tracking Study." *Interacting with Computers* 28, no. 4 (June 16, 2015): 407–20. https://doi.org/10.1093/iwc/iwv015

Nochlin, Linda (ed). *Realism and Tradition in Art, 1848–1900: Sources and Documents.* Englewood Cliffs, NJ: Prentice-Hall, 1966.

Nora, Pierre. "Introduction." In *Realms of Memory: The Construction of the French Past, vol. 3: Symbols* [orig. 1962], under the direction of Pierre Nora, ix–xii, English language edition edited by Lawrence D. Kritzman, translated by Arthur Goldhammer. New York: Columbia University Press, 1998.

Novalis [Friedrich von Hardenberg]. "Miscellaneous Writings" ("Vermischte Bemerkungen" [orig. 1797]). In *German Aesthetic and Literary Criticism. The Romantic Ironists and Goethe*, edited by Kathleen M. Wheeler. Cambridge: Cambridge University Press, 1984: 83–93.

Novella, Steven. "The Global Workspace – Consciousness Explained?," *NeuroLogica Blog*. Published online March 23, 2010. https://theness.com/neurologicablog/index.php/the-global-workspace-consciousness-explained/

Ohr Somayach. "Six Six Six." Accessed online July 10, 2020. https://ohr.edu/ask_db/ask_main.php/277/Q1/

Olin, Margaret. "Gaze." In *Critical Terms for Art History*, edited by Robert S. Nelson and Richard Shiff. Chicago: University of Chicago Press, 1996: 208–19.

Onians, John. *Neuroarthistory: From Aristotle and Pliny to Baxandall and Zeki*. New Haven: Yale University Press, 2007.

Onians, John, and Eric Fernie. "Neuro Ways of Seeing: Neuroarthistory, May 1, 2008 [interview]," *Tate Etc.*, 13: Summer 2008. https://www.tate.org.uk/tate-etc/issue-13-summer-2008/neuro-ways-seeing

Paul, Adrien. "Salon de 1863: Les Refusés." *Le Siècle* 28, no. 103009 (July 19, 1863): [1–2].

Pearce, Marcus T., Dahlia W. Zaidel, Oshin Vartanian, Martin Skov, Helmut Leder, Anjan Chatterjee, and Marcos Nadal. "Neuroaesthetics: The Cognitive Neuroscience of Aesthetic Experience." *Perspectives on Psychological Science* 11, no. 2 (March 2016): 265–79.

Pélin, Gabriel. *Les Laideurs du Beau Paris*. Paris: L'écrivain et Toubon, 1861.

Pellerin, Denis. *La Photographie Stéréoscopique sous le Second Empire*. Paris: Bibliothèque nationale de France, 1995.

Pelowski, Matthew, and Fuminori Akiba. "A Model of Art Perception, Evaluation and Emotion in Transformative Aesthetic Experience." *New Ideas in Psychology* 29, no. 2 (August 2011): 80–97. https://doi.org/10.1016/j.newideapsych.2010.04.001

Pelowski, Matthew, Eva Specker, Gernot Gerger, Helmut Leder, and Lauren S. Weingarden. "Do You Feel Like I Do? A Study of Spontaneous and Deliberate Emotion Sharing and Understanding between Artists and Perceivers of Installation Art," *Psychology of Aesthetics, Creativity, and the Arts* 14, no. 3 (August 2020): 276–93.

Petit, Pierre. "Localisation Structurale d'un Roman de Zola: *Nana*." *New Zealand Journal of French Studies* 5, no. 1 (May 1984): 19–24.

Pia, Pascal. *Baudelaire*, translated by Patrick Gregory. New York: Grove Press, 1961.

Pichois, Claude and Jean-Paul Avice. *Baudelaire-Paris*. Paris: Editions Paris-Musées/Quai Voltaire, 1993.

Pichois, Claude, with Jean Ziegler. *Baudelaire*, translated by Graham Robb. London: H. Hamilton, 1989.

Pinard, Ernest. "The Public vs. M. Gustave Flaubert: Speech of the Prosecuting Attorney, M. Ernest Pinard" [orig. 1857]. In *The Complete Works of Gustave Flaubert: Embracing Romances, Travels, Comedies, Sketches and Correspondence*, vol. 2, Appendix: 3–36. New York/London: M. W. Dunne, 1904.

Pinard, Ernest. "Procès Intenté à L'auteur: Réquisitoire" [orig. 1857]. In *Oeuvres Complètes de Gustave Flaubert: Madame Bovary*, t. 2, 558–78. Paris: L. Conard, 1910.

Pinard, Ernest. "Le Procès des 'Fleurs du Mal': Réquisitoire d'Ernest Pinard" [orig. 1857]. In Baudelaire, *Oeuvres Complètes*, edited by Claude Pichois, with Jean Ziegler. Paris: Gallimard, 1975: 1206–9.

Planche, Gustave. "Le Paysage et les Paysagistes." *Revue des Deux Mondes* 9 (May 1, 1857): 756–87.

Poirier, Richard. "The Politics of Self-Parody." *Partisan Review* 35 (1968): 339–53.

Poisson, Charles. *Mémoire sur L'oeuvre Historique de la ville de Paris*. Paris: Impr. Impériale, 1867.

Pollit, Ben. "Manet, Music in the Tuileries Gardens." Khan Academy. https://www.khanacademy.org/humanities/becoming-modern/avant-garde-france/realism/a/manet-music-in-the-tuileries-gardens

Prendergast, Christopher. *Paris and the Nineteenth Century*. Oxford: Blackwell, 1992.

Pride, Jennifer, and Kingsborough Art Gallery. *Daumier's Paris: Caricature and Cultural Trauma in the Age of Haussmann*. Brooklyn: Kingsborough Community College, 2016.

Privat, Gonzague. *Place aux Jeunes, Causeries Critiques sur Le Salon de 1865*. Paris: F. Cournol, 1865.

Proudhon, Pierre-Joseph. *Du Principe de L'art et de sa Destination Sociale*. Paris: Garnier Frères, 1865.

Purves, Dale, George J. Augustine, and David Fitzpatrick (eds). "Neural Systems," *Neuroscience*, 2nd edition. Sunderland, MA: Sinauer Associates, 2011. https://www.ncbi.nlm.nih.gov/books/NBK11061/

Rainer, Gregor. "Visual Neuroscience: Computational Brain Dynamics of Face Processing." *Current Biology* 17, no. 21 (2007): R933–34. https://www.sciencedirect.com/science/article/pii/S096098220701857X

Rajmohan, V., and E. Mohandas. "Mirror Neuron System." *Indian Journal of Psychiatry* 49, no. 1 (2007): 66. https://doi.org/10.4103/0019-5545.31522

Ravenel, Jean [Alfred Sensier]. [Review of *Olympia*], *L'Epoque* (June 7, 1865). Quoted and translated in Timothy J. Clark, *The Painting of Modern Life: Paris in the Art of Manet and his Followers*, 139–43. Princeton: Princeton University Press, 1984.

Reed, Edward S. *From Soul to Mind: The Emergence of Psychology, from Erasmus Darwin to William James*. New Haven: Yale University Press, 1997.

Reff, Theodore. "Manet's *Portrait of Zola*." *The Burlington Magazine* 117, no. 862 (January 1975): 35–44.

Reynaud, Françoise. "Marville and Old Paris." In Sarah Kennel, *Charles Marville: Photographer of Paris*, 188–205. Washington, DC: National Gallery of Art, 2013.

Riat, Georges. *Gustave Courbet, Peintre*. Paris: H. Floury, 1906.

Rice, Shelley. *Parisian Views*. Cambridge: MIT Press, 1999.

Robert, Marie. "The Photographer, the Model and Body Trafficking." In *Splendours & Miseries: Pictures of Prostitution in France, 1850–1910*, 68–79. Paris: Flammarion, 2015.

Rodemacq, Maxence. "L'industrie de L'obscénité à Paris (1855–1930)." *Romantisme* 167, no. 1 (2015a): 13–6.

Rodemacq, Maxence. "L'industrie de L'obscénité à Paris (1855–1930)." *Romantisme* 167, no. 1 (2015b): 13–20.

Röder, Birgit. *A Study of the Major Novellas of E.T.A. Hoffmann*. Rochester: Camden House, 2003.

Roos, Anne Marie. "*Le Virgile Travesti*, Paul Scarron, Paris, 1648." http://lister.history.ox.ac.uk/index-page_id=709.html

Rosenberg, Raphael, and Christoph Klein. "The Moving Eye of the Beholder: Eye-Tracking and the Perception of Paintings." In *Art, Aesthetics and the Brain*, edited by Joseph P. Hutton, Marcos Nadal, Francisco Mora, Luigi F. Agnati, and Camilo José Cela Conde. Oxford: Oxford University Press, 2015: 79–108. https://doi.org/10.1093/acprof:oso/9780199670000.001.0001

Rosenblum, Naomi. *A World History of Photography*, 3rd edition. New York: Abbeville, 1997.

Ross, Novelene. *Manet's "A Bar at the Folies Bergère" and the Myths of Popular Illustration*. Ann Arbor: UMI Research Press, 1982.

Roubert, Paul-Louis. "A Bourgeois Ambition: Photography and the Salon des Beaux-Arts during the Second Empire." In *Real/Ideal: Photography in Mid-Nineteenth-Century France*, edited by Karen Hellman. Los Angeles: J. Paul Getty Museum, 2016: 23–35.

Rouillé, André, and Bernard Marbot. *Le Corps et Son Image: Photographies du Dix-neuvième Siècle*. Paris: Contrejour, 1986.

Ryynänen, Max. "Educating Bodies, Educating Streets." *The Journal of Aesthetic Education* 55, no. 2 (2021): 115–21. https://muse.jhu.edu/article/788122

Saint-Victor, Paul de. "Salon de 1865." *La Presse* (May 28, 1865): n.p.

Sammons, Jeffrey L. *Heinrich Heine: A Modern Biography*. Princeton: Princeton University Press, 1979.

Sardou, Victorien. *Maison Neuve: Comédie en 5 Actes*. Paris: Théâtre du Vaudeville, 1866.

Scamaroni, Claire, and Maison de Balzac, *Elle Coud, elle Court, la Grisette: 14 octobre 2011–15 janvier 2012*. Paris: Paris-Musées, 2011.

Scarfe, Francis. "Introduction." In *Baudelaire: The Poems in Prose with "La Fanfarlo,"* vol. II, introduction and translations by Francis Scarfe, 11–21. London: Anvil Press Poetry, 1989.

Schindler, Ines, Georg Hosoya, Winfried Menninghaus, Ursula Beermann, Valentin Wagner, Michael Eid, and Klaus R. Scherer. "Measuring Aesthetic Emotions: A Review of the Literature and a New Assessment Tool." Edited by Mariska E. Kret, *PLOS ONE* 12, no. 6 (June 5, 2017): e0178899. https://doi.org/10.1371/journal.pone.0178899

Schlegel, Friedrich von. *Athenaeum Fragments* (no. 238) (*Athenäums-Fragmente* [orig. 1798]), in *Friedrich Schlegel's "Lucinde" and the Fragments*, introduction and translation by Peter Firchow, 195. Minneapolis: University of Minnesota Press, 1971a.

Schlegel, Friedrich von. *Friedrich Schlegel's "Lucinde" and the Fragments*, introduction and translation by Peter Firchow. Minneapolis: University of Minnesota Press, 1971b.

Schlegel, Friedrich von. "Critical Fragment" ("Kritische Fragmente" [orig. 1797]). In *German Aesthetic and Literary Criticism. The Romantic Ironists and Goethe*, edited by Kathleen M. Wheeler. Cambridge: Cambridge University Press, 1984: 40–4.

Schlutz, Alexander. "The Mirror of Laughter: Mediation, Self-Reflection, and Healing in E. T. A. Hoffmann's *Princess Brambilla*." *European Romantic Review* 22, no. 3 (2011): 415–22.

Scholl, Aurélien. "Les Coulisses." *Le Figaro* 9, no. 726 (January 30, 1862): 5–6.

Scott, Maria C. *Baudelaire's Le Spleen de Paris: Shifting Perspectives*. Aldershot, Hampshire: Ashgate, 2005.

Scullion, Val. "'Kinaesthetic, Spastic and Spatial Motifs as Expressions of Romantic Irony in E. T. A. Hoffmann's 'The Sandman' and Other Writings.'" *The Journal of Literature and Science* 2, no. 1 (2009): 1–22.

Senard, Jules. "The Public vs. M. Gustave Flaubert: Plea for the Defense, by M. Senard" [orig. 1857]. In *The Complete Works of Gustave Flaubert: Embracing Romances, Travels, Comedies, Sketches and Correspondence*, vol. 2, Appendix: 37–119. New York/London: M. W. Dunne, 1904.

Senard, Jules. "Procès Intenté à L'auteur: Plaidoirie" [orig. 1857]. In *Oeuvres Complètes de Gustave Flaubert: Madame Bovary*, t. 2, 579–627. Paris: L. Conard, 1910.

Sennwald, Vanessa, Eva Pool, Tobias Brosch, Eva Pool, Tobias Brosch, Sylvain Delplanque, Francesco Bianchi-Demicheli, and David Sander. "Emotional Attention for Erotic Stimuli: Cognitive and Brain Mechanisms." *Journal of Comparative Neurology* 524, no. 8 (2016): 1668–75. doi:10.1002/cne.23859

Shusterman, Richard. "Somaesthetics: A Disciplinary Proposal." *The Journal of Aesthetics and Art Criticism* 57, no. 3 (1999): 299–313. https://doi.org/10.2307/432196

Shusterman, Richard. *Thinking Through the Body: Essays in Somaesthetics*. New York: Cambridge University Press, 2012.

Shusterman, Richard (ed). *Bodies in the Streets: The Somaesthetics of City Life*. Leiden and Boston: Brill, 2019.

Silvia, Paul J., and Elizabeth M. Brown. "Anger, Disgust, and the Negative Aesthetic Emotions: Expanding an Appraisal Model of Aesthetic Experience." *Psychology of Aesthetics, Creativity, and the Arts* 1, no. 2 (May 2007): 100–6. https://doi.org/10.1037/1931-3896.1.2.100

Sloane, Joseph C. *French Painting between the Past and the Present*. Princeton: Princeton University Press, 1973.

Sobchack, Vivian. *Carnal Thoughts: Embodiment and Moving Image Culture*. Berkeley: University of California Press, 2004.

Solger, Karl. *Erwin, or Four Dialogues on Beauty and Art (Erwin. Vier Gespräche über das Schöne und die Kunst*, [orig. 1816]). In *German Aesthetic and Literary Criticism. The Romantic Ironists and Goethe*, edited by Kathleen M. Wheeler. Cambridge: Cambridge University Press, 1984: 128–56.

SR Research. n.d. "EyeLink 1000 plus - the Most Flexible Eye Tracker." https://www.sr-research.com/eyelink-1000-plus/

Stamkou, Eftychia, Dacher Keltner, Rebecca Corona, Eda Aksoy, and Alan S. Cowen. "Emotional Palette: A Computational Mapping of Aesthetic Experiences Evoked by Visual Art." *Scientific Reports* 14, no. 1 (August 27, 2024). https://doi.org/10.1038/s41598-024-69686-9

Stark, Rudolf, Anne Schienle, Cornelia Girod, Bertram Walter, Peter Kirsch, Carlo Blecker, Ulrich Ott, Axel Schäfer, Gebhard Sammer, Mark Zimmermann, Dieter Vaitl. "Erotic and Disgust-Inducing Pictures—Differences in the Hemodynamic Responses of the Brain." *Biological Psychology* 70, no. 1 (September 2005): 19–29. https://doi.org/10.1016/j.biopsycho.2004.11.014

Stephens, Sonya. "Argot Littéraire, Argot Plastique: Caricature in Baudelaire's Prose Poetry." *Australian Journal of French Studies* 30, no. 2 (1993): 197–206.

Stephens, Sonya. "Boundaries, Limits and Limitations: Baudelaire's Poèmes-Boutades." *French Studies* 52, no. 1 (January 1, 1998): 28–41.

Stephens, Sonya. *Baudelaire's Prose Poems: The Practice and Politics of Irony*. Oxford/New York: Oxford University Press, 1999.

Stern, L. "Pornography and Disgust." *Contemporary Aesthetics* 11 (January 2013): 7 ff.

Steward, Jill, and Alexander Cowan. "Introduction." In *The City and the Senses: Urban Culture since 1500*, edited by Alexander Cowan and Jill Steward. Aldershot, Hampshire: Ashgate, 2007: 1–22.

Stueber, Karsten. "Empathy." *The Stanford Encyclopedia of Philosophy* (Fall 2016 Edition), edited by Edward N. Zalta: Section 1, 3.2. Section 1. https://plato.stanford.edu/archives/fall2019/entries/empathy/

Sutcliffe, Anthony. *Paris: An Architectural History*. New Haven: Yale University Press, 1993.

Tatar, Maria M. "E.T.A. Hoffmann's 'Der Sandmann': Reflection and Romantic Irony." *MLN* 95, no. 3 (1980): 585–608.

Thélot, Jérôme. "'Le Rêve d'un Curieux' ou la Photographie Comme Fleurs du Mal." *Études Photographiques* 6 (May 1999). http://journals.openedition.org/etudesphotographiques/186

Thézy, Marie de. "Marville et la Naissance du Paris d'Haussmann." *Photographies*, Numéro Spéciale (March 1986): 46–51.

Thézy, Marie de. *Marville: Paris*. Paris: Hazan, 1994.

Thézy, Marie de. *Charles Marville, Paris Disparu*, introduction by Jean-Paul Avice. Paris: Paris Musées, 1995.

Tieck, Ludwig. *Nachgelassene Schriften*, edited by R. Köpke. Leipzig: F. A. Brockhaus, 1855.

Tinterow, Gary, and Geneviève Lacambre. *Manet/Velázquez: The French Taste for Spanish Painting*. New York/New Haven: Metropolitan Museum of Art/Yale University Press, 2003.

Tomlinson, Janis A. "Goya and the Censors." In *Suspended License: Censorship and the Visual Arts*, edited by Elizabeth C. Childs. Seattle: University of Washington Press, 1997: 125–47.

Torkladani, Safoora, and Pyeaam Abbasi. "Dream: A Brief Comparative Study of Nerval and Keats." *International Letters of Social and Humanistic Sciences* 55 (July 2015): 140–46. https://doi.org/10.18052/www.scipress.com/ilshs.55.140

Tsur, Reuven. "Aspects of Cognitive Poetics." In *Cognitive Stylistics: Language and Cognition in Text Analysis*, edited by Elena Semino and Jonathan Culpeper. Amsterdam: John Benjamins Publishing Company, 2002: 279–318.

Umilta', M. Alessandra, Cristina Berchio, Mariateresa Sestito, David Freedberg, and Vittorio Gallese. "Abstract Art and Cortical Motor Activation: An EEG Study." *Frontiers in Human Neuroscience* 6 (2012): 1–9. https://www.frontiersin.org/journals/human-neuroscience/articles/10.3389/fnhum.2012.00311

Vaillant, Alain. "Pornographie ou Obscénité?" *Romantisme* 167, no.1 (2015): 8–12.

Vanta, Brindusa. "Physiology of Anger." Mentalhelp.net, last updated September 25, 2024. https://www.mentalhelp.net/anger/physiology/

Varnedoe, Kirk. *Gustave Caillebotte*. New Haven/London: Yale University Press, 1987.

Vartanian, Oshin, and Martin Skov. "Neural Correlates of Viewing Paintings: Evidence from a Quantitative Meta-Analysis of Functional Magnetic Resonance Imaging Data." *Brain and Cognition* 87 (June 2014): 52–6. https://doi.org/10.1016/j.bandc.2014.03.004

Villines, Z. "What Does the Frontal Lobe Do?" *Medical News Today*. Updated online November 30, 2023. https://www.medicalnewstoday.com/articles/318139.php#Functions-of-the-frontal-lobe

Vitu, Auguste. *Paris: 450 Dessins D'après Nature*. Paris [n.p.], 1890, 23–6. http://catalogue.bnf.fr/ark:/12148/cb365853927

Waxenbaum, Joshua A., Matthew Varacallo, and Vamsi Reddy. "Anatomy, Autonomic Nervous System." Nih.gov (StatPearls Publishing). Published online July 24, 2023. https://www.ncbi.nlm.nih.gov/books/NBK539845/

Weingarden, Lauren S. "William Gilpin: The Poetics of Picturesque (Painterly) Perception and Architectural Representation." In *Interart Poetics: Essays on the Interrelations of the Arts and Media*, edited by Ulla-Britta Lagerroth, Hans Lund, and Erik Hedlink. Amsterdam/Atlanta: Editions Rodopi B.V., 1997: 179–93.

Weingarden, Lauren S. "The Mirror as a Metaphor of Baudelairean Modernity." In *Orientations: Space/Time/Image/Word, Word & Image Interactions 5*, edited by Claus Clüver, Véronique Plesch, and Leo H. Hoek. Amsterdam/New York: Editions Rodopi, 2005a: 16–36.

Weingarden, Lauren S. "Re-Viewing Baudelaire's Paris: Photography, Modernity and Parody." In *Writing and Seeing: Essays on Word and Image*, edited by Rui Carvalho Homem and Maria de Fátima Lambert. Leiden: Brill, 2005b: 145–56.

Weingarden, Lauren S. "The Photographic Subversion: Benjamin, Manet and Art(istic) Reproduction." *Aletria: Revista de Estudos de Literatura* 14, no. 2 (December 31, 2006): 224–45 [online]. https://doi.org/10.17851/2317-2096.14.2.224-245

Weingarden, Lauren S. "John Ruskin: The Picturesque Discourse and the Language of Architectural Naturalism." In *Louis H. Sullivan and a 19th-Century Poetics of Naturalized Architecture*. Farnham, England: Ashgate, 2009, 39–70.

Weingarden, Lauren S. "The Performative Turn at Inhotim: Installation Art and Baudelairean Modernity." *Aletria: Revista de Estudos de Literatura* 23, no. 3 (December 31, 2013): 13–30 [online]. https://doi.org/10.17851/2317-2096.23.3.13-30

Weingarden, Lauren S. "Framing an Experiential Aesthetics: The Affective Turn in Contemporary Installation Art." In *Notions Esthétiques (II): La Perception Sensible Organisée*, edited by Véronique Alexandre Journeau and Christine Vial Kayser. Paris: L'Harmattan, 2015: 197–210.

Weingarden, Lauren S. "Cross-contaminated bodies: Constructing Obscenity in the Public Imaginary." In *Pornography: Interdisciplinary Perspectives*, edited by Frank Jacobs. Bern, CH: Peter Lang, 2019a: 85–108.

Weingarden, Lauren S. "Mirroring Naturalism in Word and Image: A Critical Exchange Between Émile Zola and Edouard Manet." In *Ekphrastic Encounters: New Interdisciplinary Essays on Literature and the Visual Arts*, edited by David Kennedy and Richard Meek. Manchester: Manchester University Press, 2019b: 144–64.

Weingarden, Lauren S., Matthew Pelowski, and Corrina Kühnapfel. "Embodying the Photographic Image: Charles Marville's Old Paris and New Paris Series." In *Art as Experience of the Living Body/L'art Comme Experience du Corps Vivant*, edited by Christine Vial Kayser. Wilmington, DE: Vernon Press, 2024: 261–92.

Wey, Francis. "De L'influence de L'héliographie sur les Beaux-arts." *La Lumière*, no. 1 (February 8, 1851): 2; quoted and translated in Karen Hellman, "Real/Ideal: Photography in Mid-Nineteenth-Century France," in *Real/Ideal: Photography in Mid-Nineteenth-Century France*, edited by Karen Hellman. Los Angeles: J. Paul Getty Museum, 2016: 8.

Wheeler, Kathleen M. (ed). *German Aesthetic and Literary Criticism. The Romantic Ironists and Goethe*. Cambridge: Cambridge University Press, 1984.

Wilson-Bareau, Juliet. "Manet and Spain." In Gary Tinterow and Geneviève Lacambre, *Manet/Velázquez: The French Taste for Spanish Painting*, 203–57. New York/New Haven: Metropolitan Museum of Art/Yale University Press, 2003.

Wilson-Mendenhall, Christine D., Alexa Henriques, Lawrence W. Barsalou, and Lisa Feldman Barrett. "Primary Interoceptive Cortex Activity During Simulated Experiences of the Body," *Journal of Cognitive Neuroscience* 31, no. 2 (February 2019): 221–35. https://doi.org/10.1162/jocn_a_01346

Wolf, Joke de. "Paris on Display: Marville's Photographs at the Universal Exhibitions." In Kennel, *Charles Marville: Photographer of Paris*, 206–21. Washington, DC: National Gallery of Art, 2013.

Wrigley, Richard. "Unreliable Witness: The Flâneur as Artist and Spectator of Art in Nineteenth-Century Paris." *Oxford Art Journal* 39, no. 2 (July 28, 2016): 267–84.

Young, Robert M. *Mind, Brain, and Adaptation in the Nineteenth Century: Cerebral Localization and Its Biological Context from Gall to Ferrier*. New York: Oxford University Press, 1990.

Yriarte, Charles. *Goya*. Paris: Plon, 1867.

Zhang, Minming, Shaohua Hu, Lijuan Xu, Qidong Wangd, Xiaojun Xua, Erqing Weie, Leqin Yanf, Jianbo Hub, Ning Weib, Weihua Zhoub, Manli Huangb, and Yi Xu. "Neural Circuits of Disgust Induced by Sexual Stimuli in Homosexual and Heterosexual Men: An FMRI Study." *European Journal of Radiology* 80, no. 2 (November 2011): 418–25. https://doi.org/10.1016/j.ejrad.2010.05.021

Zola, Émile. "Mon Salon (1866): Les Realistes du Salon." *Les Cahiers Naturalistes. Société Littéraire des Amis d'Émile Zola*. https://www.cahiers-naturalistes.com/documents/zola-et-arts/ecrits-sur-lart-de-zola/mon-salon-1866-les-realistes-du-salon/

Zola, Émile. *Éd. Manet: Étude Biographique et Critique*. 1867. Paris: E. Dentu, 1867a.

Zola, Émile. "Edouard Manet: Etude Biographique et Critique." *La Revue du XIXe Siècle*, (January 1, 1867b).

Zola, Émile. "Mon Salon (1868): Edouard Manet." *Les Cahiers Naturalistes. Société Littéraire des Amis d'Émile Zola*. https://www.cahiers-naturalistes.com/documents/zola-et-arts/ecrits-sur-lart-de-zola/mon-salon-1868-edouard-manet/

Zola, Émile. *Madeleine Férat*, 3rd ed. [orig. 1868]. Paris: C. Marpon et E. Flammarion, 1878.

Zola, Émile. "Le Roman Experimental." In *Le Roman Experimental*, 1–54. Paris: G. Charpentier, 1880a.

Zola, Émile. "Le Salon de 1880b: Le Naturalisme au Salon." *Les Cahiers Naturalistes. Société Littéraire des Amis d'Émile Zola.* https://www.cahiers-naturalistes.com/documents/zola-et-arts/ecrits-sur-lart-de-zola/index-des-ecrits-sur-lart-de-zola/le-salon-de-1880/

Zola, Émile. Thérèse Raquin, 2nd ed. [orig. 1868]. *Edition Illustrée par Castelli.* Paris: C. Marpon et E. Flammarion, 1883.

Zola, Émile. *Nana: A Realistic Novel (1880)*, translated by Ernest Alfred Vizetelly. London: Vizetelly, 1884. https://www.online-literature.com/emile-zola/nana/

Zola, Émile. "Mon Salon (1867): Edouard Manet." *Oeuvres Complètes Illustrées de Émile Zola. 32, Oeuvres Critiques*, t. 1, 93–106. Paris: E. Fasquelle, 1906a.

Zola, Émile. "Mon Salon (1866): M. Manet." In *Oeuvres Complètes Illustrées de Émile Zola. 32, Oeuvres Critiques*, t. 1, 84–6. Paris: E. Fasquelle, 1906b.

Zola, Émile. *Oeuvres Complètes Illustrées de Émile Zola. 32, Oeuvres Critiques*, t. 1. Paris: E. Fasquelle, 1906c.

Zola, Émile. "Preface to the Second Edition (1868)." In *Thérèse Raquin*, translated by Leonard Tancock, 21–7. London: Penguin Classics, 1962a.

Zola, Émile. *Thérèse Raquin*, 2nd ed. [orig. 1868], translated by Leonard Tancock. London: Penguin Classic, 1962b.

Zola, Émile. "The Experimental Novel" [orig. 1880]. In *Documents of Modern Literary Realism*, edited by George Joseph Becker. Princeton: Princeton University Press, 1963: 162–96.

Zola, Émile. "A New Style of Painting" [orig. 1867]. In *Realism and Tradition in Art, 1848–1900; Sources and Documents*, edited and translated by Linda Nochlin, 71–7. Englewood Cliffs, NJ: Prentice-Hall, 1966.

Zola, Émile. Pour Manet, edited by Jean-Pierre Leduc-Adine. *Le Regard Littéraire*, 32. Bruxelles: Editions Complexe, 1989.

Zola, Émile. *L'Assommoir*, translated by Margaret Mauldon, introduction by Robert Lethbridge. Oxford: Oxford University Press, 1998.

Zola, Émile. *Les Rourgon-Macquart. L'Arbre Généalogique; Copie de la Main d'Alexandrine Zola avec Ajouts de Zola*; n.d. manuscript. In "Émile Zola. Oeuvres. Manuscrits et Dossiers Préparatoires. Les Rougon-Macquart. Le Docteur Pascal. Dossier Préparatoire." Bibliothèque Nationale de France, Départment des Manuscrits, Paris. NAF 10290. Published online June 3, 2013. https://gallica.bnf.fr/ark:/12148/btv1b90797703/f173.image

Zola, Émile. *Madeleine Férat: A Realistic Novel [1868]*. English translation, London: Vizetelly & Co., 1888. Reprint, Portland: Mint Editions, 2021.

Index

Pages in *italics* refer to figures and pages followed by "n" refer to notes.

For Product Safety Concerns and Information please contact our EU representative GPSR@taylorandfrancis.com Taylor & Francis Verlag GmbH, Kaufingerstraße 24, 80331 München, Germany

Batch number: 10756062

Printed by Printforce, the Netherlands